recent advances in phytochemistry

volume 26

Phenolic Metabolism in Plants

RECENT ADVANCES IN PHYTOCHEMISTRY

Proceedings of the Phytochemical Society of North America
General Editor: Helen A. Stafford, *Reed College, Portland, Oregon*

Recent Volumes in the Series

A Continuation Order Plan is available for this series. A continuation order will bring delivery of each new volume immediately upon publication. Volumes are billed only upon actual shipment. For further information please contact the publisher.

recent advances in phytochemistry

volume 26

Phenolic Metabolism in Plants

Edited by

Helen A. Stafford

Reed College
Portland, Oregon

and

Ragai K. Ibrahim

Concordia University
Montreal, Quebec, Canada

SPRINGER SCIENCE+BUSINESS MEDIA, LLC

Library of Congress Cataloging-in-Publication Data

Phenolic metabolism in plants / edited by Helen A. Stafford and Ragai
K. Ibrahim.
        p. cm. -- (Recent advances in phytochemistry ; v. 26)
    "Proceedings of the Thirty-first Annual Meeting of the
Phytochemical Society of North America's Symposium on Phenolic
Metabolism in Plants, held June 22-26, 1991, in Fort Collins,
Colorado"--T.p. verso.
    Includes bibliographical references and index.
    ISBN 978-0-306-44231-5   ISBN 978-1-4615-3430-3 (eBook)
    DOI 10.1007/978-1-4615-3430-3
    1. Phenols--Metabolism--Congresses. 2. Plants--Metabolism-
-Congresses.   I. Stafford, Helen A., 1922-   . II. Ibrahim, Ragai
K.   III. Phytochemical Society of North America. Meeting (31st :
1991 : Fort Collins, Colo.) IV. Series.
QK861.R38 vol.26
[QK898.P57]
581.19'2 s--dc20
[581.19'24]   92-13371
                                                                CIP

Proceedings of the Thirty-first Annual Meeting of the
Phytochemical Society of North America's Symposium on
Phenolic Metabolism in Plants, held June 22–26, 1991,
in Fort Collins, Colorado

ISBN 978-0-306-44231-5

© 1992 Springer Science+Business Media New York
Originally published by Plenum Press, New York in 1992

This volume is dedicated to the memory of
Hans Grisebach (1926–1990)

PREFACE

This volume contains reviews presented at the 31st annual meeting of the Phytochemical Society of North America, held at Colorado State University in Fort Collins, Colorado on June 22-26, 1991. This symposium, entitled *Phenolic Metabolism in Plants,* celebrated the origin of this society as the Plant Phenolics Group of North America; the first symposium, entitled Biochemistry of Plant Phenolic Substances, was also held at Fort Collins from August 31 to September 1, 1961. A brief history of the Society is presented in Chapter 12 by Stewart Brown, one of the original founders of the Society.

We dedicate this volume to Hans Grisebach, 1926-1990, Professor of Biochemistry at the Biologisches Institut II, Freiburg, Germany, where he headed for many years a laboratory responsible for major advances in the area of phenolic metabolism; this will be self evident from the numerous bibliographical references cited in the literature for papers by his Freiburg group from about 1958 until now, and subsequently by former students and colla- borators. His impact on the data reviewed in this volume will testify to this.

Before commenting on the individual reviews in this volume and to assess how far scientists have elucidated phenolic biosynthesis since about 1961, it would be appropriate to summarize the status of knowledge about plant phenolics and their metabolism at the time of the first symposium in 1961. The status is best summarized in the reviews by A.C. Neish in 1960,[1] H. Grisebach in 1962,[2] and by T.A. Geissman in 1963.[3] All these authors were early members of our Society.

---

[1]Neish, A. C. , 1960. Biosynthetic pathways of aromatic compounds, Annu. Rev. Plant Physiol. 11:55-80.
[2]Grisebach, von H., 1962 Die Biosynthese der Flavonoide, Planta Medica 10:385-397.
[3]Geissman, T.A., 1963. The biosynthesis of phenolic plant products. In: Biogenesis of Natural Compounds, (P. Bernfield, ed.), Pergamon Press, New York , 563.

     The basic chemistry of the major groups of phenolics was known: lignin, various flavonoid sub-groups, coumarins and cyanogenic glycosides such as dhurrin. Reviews edited by Geissman in 1962 summarized the status of knowledge about flavonoids.[1] The importance of methodology in the isolation and identification of phenolic compounds was summarized by Seikel.[2] The main outlines of the shikimate and acetate pathways had been demonstrated in plants with radioactive tracer techniques, leading to the above products.[3] Based on ^{14}C tracer experiments, the acetate pathway via 3 molecules of malonyl-CoA was known to give rise to the A-ring of flavonoids, with the shikimate pathway providing an activated (via a CoA ester) form of a cinnamic acid. A chalcone was postulated as the first C_{15} intermediate in the flavonoid pathway. All the above authors noted the necessity for cell-free enzymological studies.

     Perhaps the first cell-free enzymology was the demonstration of the oxidative polymerization of coniferyl alcohol by laccase and peroxidase by Freudenberg's group in the 1950's, resulting in the dehydrierungspolymerizat (DHP), a lignin-like preparation.[4] Cell-free enzymology continued with the demonstration of phenylalanine ammonia-lyase (PAL) by Koukol & Conn in 1961.[5] The comparable tyrosine pathway via TAL, discovered by Neish's group,[6] has been subsequently ignored, probably because it is limited to just a few plants within the Gramineae. The second enzyme of the phenylpropanoid (C_6-C_3) pathway, cinnamate 4-hydroxylase,[7] and the first C_{15} enzyme,

---

[1]Geissman, T.A.. (ed). 1962. The Chemistry of Flavonoid Compounds, MacMillan Co., New York.

[2]Seikel, M. K., 1964. Isolation and identification of phenolic compounds in biological material. In: Biochemistry of Phenolic Compounds, (J.B.Harborne,ed.), Academic Press, pp.33-76.

[3]Birch, A.J. 1962. Biosynthesis of flavonoids and anthocyanins. In: The Chemistry of Flavonoid Compounds, (Geissman, T.A..,ed), MacMillan Co., New York.

[4]Freudenberg K. 1959. Biosynthesis and constitution of lignin. Nature 183:1152- 1155.

[5]Koukol, J., and Conn, E.E., The metabolism of aromatic compounds in higher plants. IV., Purification and properties of the phenylalanine deaminase of *Hordeum vulgare*, J. Biol. Chem. 236, 2692, 1961.

[6]Neish, A. C., 1961. Formation of m- and p-coumaric acids by enzymatic deamination of the corresponding isomer of tyrosine. Phytochemistry 1:1-24.

[7]Russell, D.W., Conn, E.E., The cinnamic acid 4-hydroxylase of pea seedlings, Arch. Biochem. Biophys. 122, 256, 1967.

chalcone isomerase were demonstrated in 1967,[1] whereas chalcone synthase (called flavanone synthase originally because of the presence of the above isomerase activity) was not demonstrated in cell-free extracts until 1975.[2] Summaries of the status of the biochemistry of phenolics in plants in 1964 can be found in the reviews by G.H.N.Towers, A.C. Neish, S.A. Brown, and E.E. Conn in the volume edited by J.B. Harborne.[3]

Whereas the above work was done by chemists and biochemists, geneticists were accumulating information about genes involved mainly in flavonoid biosynthesis in plants such as maize, *Zea mays.*, snapdragon, primrose, *Dahlia, Petunia, Silene*, potato and bean. Reviews by Harborne in 1962[4] summarized the chemicogenetical aspects and the genetics by Alston in 1964.[5]

How far has our knowledge of the metabolism of phenolics advanced in the past thirty years? The present volume contains reviews summarizing some of these advances and the state of our present knowledge in some areas. The enzymology of the aromatic pathway in plants, which includes the shikimate, the phenylpropanoid and the flavonoid sections, are described by Hrazdina. Emphasis is placed on their postulated ER-localization, necessitating different signals for their transport to either the plant cell wall or central vacuole. The use of immunology to determine the subcellular localization of secondary products and two of the enzymes involved in their biosynthesis is described by Ibrahim, with the polymethylated flavonols in *Chrysosplenium* as a prime example.

Two papers emphasize techniques used to elucidate the molecular biology of the genes involved in phenylpropanoid pathways. Douglas summarizes work on molecular genetics of PAL and 4-coumarate:CoA ligase in

---

[1]Moustafa, E., Wong, E. 1967 Purification and properties of chalcone-flavanone isomerase from soya bean seed. Phytochemistry, 6:625-632.

[2]Kreuzaler & Hahlbrock, 1975. Eur J Biochem 56:205-213.

[3]Harborne, J.B., (ed.), Biochemistry of Phenolic Compounds. Academic Press, New York

[4]Harborne, J.B., 1962. Chemicogenetical studies of flavonoid pigments, in The Chemistry of Flavonoid Compounds, (Geissman, T.A., Ed.), Macmillan Co., New York, 1962, 593.

[5]Alston, R.E., 1964. The genetics of phenolic compounds. In: Biochemistry of Phenolic Compounds, (J.B. Harborne, ed.), Academic Press, New York, pp. 171-204.

parsley and poplar. Dixon *et al* review their work on the enzymology and the use of homologous cDNA probes and transgenic systems to study stable and transient gene expression for the genetic manipulation of the phenylpropanoid pathway in alfalfa.

Barz describes recent and unpublished work on the isoflavonoid pathways emphasizing the constitutive aspects of isoflavones and their glycosides as potential intermediates during early stages of induction of the last steps leading to the phytoalexins known as pterocarpans. Demonstration of the basic cell-free enzymology of this isoflavonoid pathway is now essentially complete.

Gerats summarizes the genetics of anthocyanin formation, mainly in *Petunia*, with emphasis on chalcone synthase, chalcone isomerase and dihydro-flavonol reductase enzymic steps. The effects of flavonoids released into the rhizosphere on mycorrhizal fungi and on the induction of the symbiotic relationship between N-fixing bacteria and legumes to produce nodules are summarized by Phillips, with emphasis on alfalfa. Recent work on the enzymology and molecular cloning of sulfated flavonols and their potential use in modifying sulfur metabolism are described by Varin.

The complexities of the chemical synthesis and rearrangments of proanthocyanidin (condensed tannin) oligomers and polymers found mainly in tropical heartwoods, including so-called biomimetic syntheses, are discussed by Ferrerira *et al*. The intriguing aspects of lignans and neolignans, dimers of coniferyl alcohol with their stereochemistry maintained, and the non-optically active wall-localized lignin polymers, are summarized by Davin & Lewis. Finally, the recent break throughs in the enzymology of hydrolyzable tannins are discussed by Gross.

What might our hopes be for the future? The shikimate and phenyl-propanoid pathways have been well elucidated. The hydroxylation steps at the C_6-C_3 level, including cytochrome P-450 oxygenases and the varied pathways to *o*-diphenols, need further study. The basic cell-free enzymology of flavanones, 3-hydroxyflavanones (dihydroflavonols), flavones, flavonols and isoflavonoids is now known. Many details and variations are still to be discovered. The crucial terminal step(s) leading from the leucoanthocyanidins to anthocyanins is still unknown. The enzymology of the 2,3-*trans* pathway to flavan-3-ols and proanthocyanidins is known, but the enzymes of the 2,3-*cis* routes as well as the presumed condensation step have yet to be demonstrated. One wonders whether some of the proanthocyanidins found in heartwood and periderm of the bark represent non-enzymic alterations of basic enzymic products during senescence of the cells.

Regulatory control of inducible phenolic pathways by exogenous factors such as light and external signals from pathogens and symbionts is being actively studied. Knowledge about endogenous control of constitutive enzymes is still in its infancy. In order to produce a flavonoid such as naringenin chalcone via chalcone synthase, there must be a coordination of both the acetate pathway leading to 3 malonyl-CoA and the phenylpropanoid pathway to prooduce p-coumaroyl-CoA. Both of these substrates of chalcone synthase are also used in other competitive pathways such as stilbene biosynthesis. Little thought has been given in the past to this dual pathway requirement of flavonoid biosynthesis. Information from studies of the molecular biology of genes and molecular genetics of phenolic metabolism is insufficient to explain the presumed organization that must be involved, as multienzyme complexes or aggregates associated with membranes, to ensure the production of an end-product involving 3 to as many as 15 enzymatic steps just at the C_{15} level of flavonoids. Compartmentation of some of the enzymic steps in phenolic metabolism is still unclear, including the mechanisms of transport and accumulation of the end-products in the vacuole, vesicles, wall or into the rhizosphere. It will be exciting to understand how cis-factors on the promoter interact with $trans$-factors to regulate the postulated enzyme complexes or aggregates necessary to synthesize end-products of multienzyme pathways of considerable length. This information is just beginning to be published.

We wish to thank Dr. Frank Stermitz, Department of Chemistry at Colorado State University, for the meeting arrangements, and Dr. Susan S. Martin, USDA-ARS Crops Research Laboratory in Fort Collins for program arrangements for the very successful meeting, with one of the largest attendance at a yearly meeting so far. We gratefully acknowledge the generous financial support of Colorado State University, Concordia University and The Samuel Roberts Noble Foundation.

February, 1992

Helen A. Stafford
Ragai Ibrahim

CONTENTS

Chapter One

COMPARTMENTATION IN AROMATIC METABOLISM

Geza Hrazdina

Institute of Food Science
Cornell University
Geneva, NY 14456-0462

INTRODUCTION

Plant aromatic metabolism is a unique and fascinating biochemical process that has no counterpart in avian or mammalian systems. The process is long and complex, and it interfaces three metabolic processes in cell organization: the metabolism of carbohydrates, proteins and lipids. This process also represents the most complicated flow of carbon in plants that we know of today.

Aromatic metabolism in plants consists of three segments. These are the *shikimate pathway* section that produces the aromatic amino acids phenylalanine, tyrosine and tryptophan, the *phenylpropanoid* section that produces the diverse coumarins, the activated cinnamic acid derivatives and the plant structural component lignin; and the *flavonoid* section that produces all flavonoid compounds, including the pterocarpan phytoalexins (Schemes 1-3).

Phenolic Metabolism in Plants, Edited by H.A. Stafford
and R.K. Ibrahim, Plenum Press, New York, 1992

scheme 1, part 1

PEP   E-4-P   DAHP

Dehydroquinate   Dehydroshikimate

Shikimate   Shikimate-P

Enolpyruvyl-shikimate-P   Chorismate

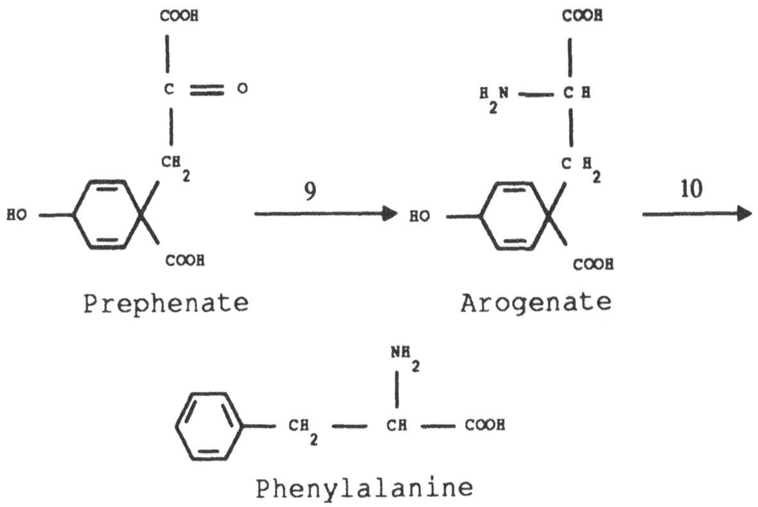

Prephenate

Arogenate

Phenylalanine

Scheme 1. (facing page and above) The shikimate section of the plant aromatic metabolism. Enzymes of the pathway section are: **1**, deoxy-*arabino*heptonate phosphate synthase; **2**, dehydroquinate synthase; **3**, dehydroquinate dehydratase; **4**, shikimate dehydrogenase; **5**, shikimate kinase; **6**, enolpyruvylshikimate phosphate synthase; **7**, chorismate synthase; **8**, chorismate mutase; **9**, prephenate aminotransferase; **10**, arogenate dehydratase.

Although in the past the three segments of aromatic metabolism were treated separately, recent developments suggest that they function as one metabolic unit. There can be no synthesis of phenylpropanoids without phenylalanine, the product of the shikimate pathway section, and the synthesis of flavonoids cannot take place without p-coumaryl-CoA that is produced by the phenylpropanoid pathway section.

The general phenylpropanoid pathway section and the specific branches of the flavonoid pathway section are induced in tissue cultures and seedlings upon exposure to mechanical, chemical or biological stress.[1] Induction of these pathway segments requires an increased production of the shikimate pathway product, phenylalanine, otherwise the downstream sections of the pathway would be starved for substrate. Indeed, recent reports show convincingly that the first enzyme of the shikimate pathway, 3–deoxy*arabino*heptulosonate phosphate synthase (DAHP synthase) is induced upon wounding in potatoes and tomatoes.[2]

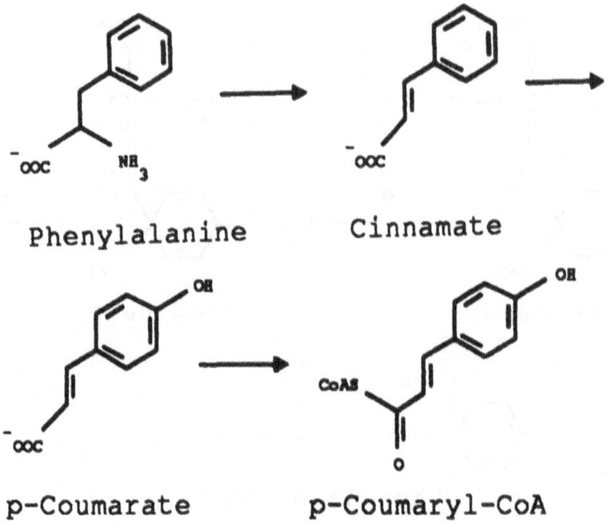

Phenylalanine   Cinnamate

p-Coumarate   p-Coumaryl-CoA

Scheme 2. (above) The general phenylpropanoid section of plant aromatic metabolism. Enzymes in the pathway section are **1**, phenylalanine ammonia-lyase, **2**, cinnamate 4-hydroxylase, **3**, p-coumarate:CoA ligase.

A large number of phenylpropanoid and flavonoid compounds are known to accumulate in plant tissues and cells under the influence of various environmental stimuli. This requires either an extremely complicated transport mechanism that shuttles metabolic intermediates between individual enzymes of the various pathways, or an organized metabolism that does not permit the mixing of the various intermediary metabolite pools has to be present. There is also a requirement for strict compartmentalization of metabolic end products to prevent interference with other cellular activities. In the following section our present understanding of the location and functioning of the aromatic pathway sections, and the accumulation of pathway end products will be discussed.

TISSUE SPECIFIC LOCATION OF THE PATHWAY AND ITS PRODUCTS

Determination of the tissue distribution of metabolic pathways and their end products is usually hampered by the difficulties encountered in obtaining single tissues types free of contamination. These difficulties were bypassed in our laboratory using a novel mutant of garden peas (*Pisum sativum* L. cv. *Argenteum*) that contains large air spaces between the epidermal and the

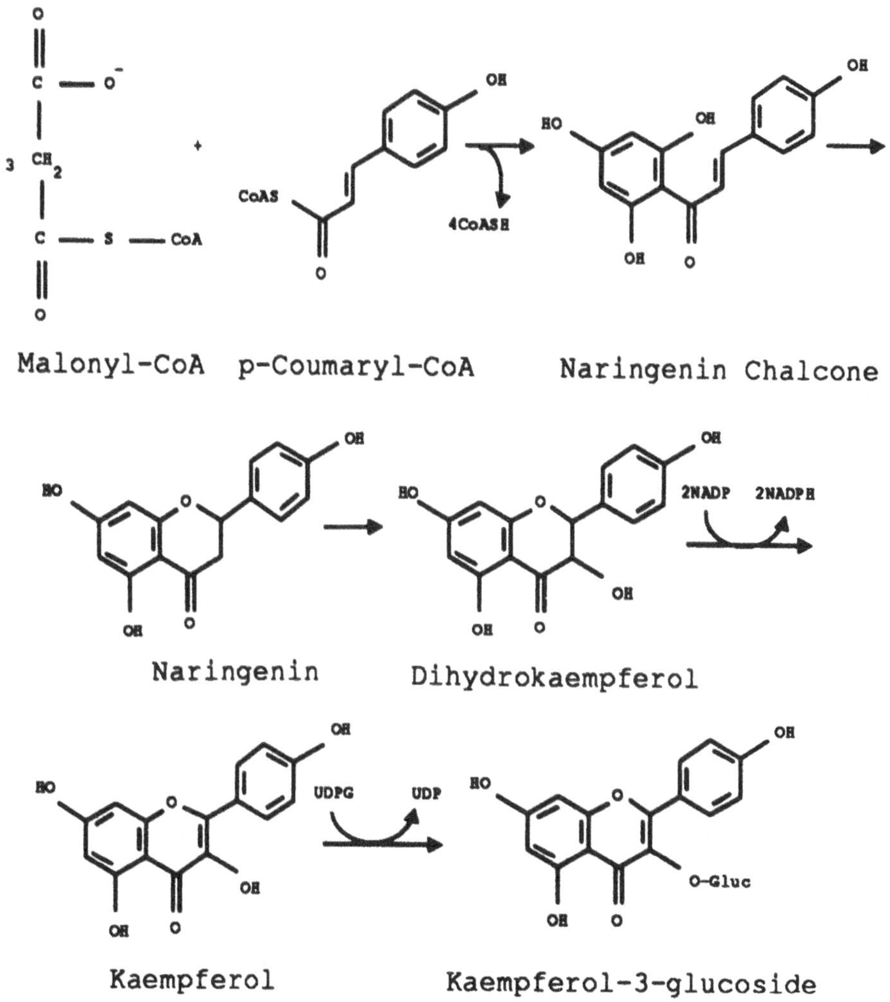

Scheme 3. Flavonoid section of plant aromatic metabolism leading to the formation of kaempferol-3-glucoside. Enzymes in the pathway section are **1**, chalcone synthase, **2**, chalcone isomerase; **3**, flavanone 3-hydroxylase; **4**, 3-hydroxyflavanonol oxidase; **5**, flavonol 3-glucosyltransferase.

Table 1. Distribution of flavonoids in the leaf fraction of the *Pisum sativum* cv. *Argenteum* mutant.

Compound	% of total activity recovered in:		
	Upper epidermis	Mesophyll-vascular layer	Lower epidermis
Anthocyanin	70.3	0	29.7
Flavonol glycoside	68.7	0	31.3

mesophyll layers.[3,4] These air spaces give the pea leaves a silvery appearance, hence its name *Argenteum*. Also, because of these air spaces, the number of attachment sites between the epidermis and mesophyll is greatly decreased. This permits the peeling of the leaves, in what amounts to a horizontal tissue fractionation. Such leaf fractions, e.g. the upper and lower epidermal peels and the mesophyll, remain largely intact, and show a minimal cross-contamination by other tissues. Hence, they are ideally suited for tissue localization studies.

Investigations of anthocyanin and flavonol glycoside content of such fractionated tissues showed that all anthocyanin and flavonol glycosides accumulated in the epidermal peels, whereas the mesophyll layer, including the vascular bundles, was essentially free from flavonoid compounds.[5] The greatest accumulation of both anthocyanin and flavonol glycosides took place in the upper epidermal layers, which were directly exposed to visible and ultraviolet light irradiation, suggesting a UV protective role for these compounds (Table 1).

Investigation of key enzyme activities in both phenylpropanoid and flavonoid metabolisms gave less defined results. The majority of phenylalanine ammonia-lyase (PAL) activity was present in the mesophyll plus vascular bundle fraction of the pea leaves. This may be taken to indicate that this middle layer is the major site for aromatic metabolism. Esters of phenolic acids such as ferulic could be present in the mesophyll-vascular layer. The vascular tissue included in this fraction would be the main site of lignin biosynthesis in leaves, and PAL is also an enzyme of the lignin biosynthetic pathway. Since *p*-coumarate-CoA ligase also participates in both lignin and flavonoid biosynthesis, its presence in all three leaf fractions has been expected (Table 2).

The key enzyme of the specific flavonoid pathways is chalcone synthase. This enzyme produces naringenin chalcone, from which all flavonoid

Table 2. Distribution of key phenylpropanoid and flavonoid enzyme activities in the leaf fraction of the *Pisum sativum* cv. *Argenteum* mutant.

Enzyme	% of total activity recovered in:		
	Upper epidermis	Mesophyll- vascular layer	Lower epidermis
Phenylalanine ammonia-lyase	16.1	79.4	4.5
p-Coumaryl-CoA ligase	28.3	43.3	28.3
Chalcone synthase	88.7	0	11.3
UDP-Glucose flavonoid 3-0-glucosyltransferase	34.5	54.7	10.5

compounds are derived, and was found to be present in epidermal peels only. This is a firm indication that the flavonoid section of the plant aromatic pathway is largely restricted to the epidermis in some plants.

Recent immunohistochemical investigations by Jahnen and Hahlbrock[6] in developing parsley seedlings confirmed the results obtained with the *Argenteum* mutant both for PAL and chalcone synthase. *In situ* hybridization of chalcone synthase mRNA with specific cDNA probes further indicated the exclusive localization of chalcone synthase in the epidermal layers,[7] and a sequential accumulation of the pathway end products. These data clearly show the synthesis and accumulation of flavonoids occurring in the epidermal layer of plants, invalidating unsupported claims of their synthesis in the mesophyll layer and intercellular transport to the epidermis.[8]

AROMATIC METABOLISM IN CELLS

Aromatic pathways in plants use products of carbohydrate meta-bolism as its precursors. Phosphoenolpyrurate is a product of the glycolytic pathway, while erythrose 4-P is supplied by the oxidative pentose phosphate cycle (Scheme 1). Both of these pathways have been reported to occur in the chloroplasts and also in the cytoplasm.[9] These chloroplastic and cytoplasmic enzymes are present in plant cells as isozymes which have similar physical and kinetic properties, and are readily separated on chromatography columns.

The first enzyme of the shikimate pathway section is 3-deoxy-arabinoheptulosonate phosphate synthase (DAHP-synthase) that uses erythrose 4-P and phosphoenol pyruvate to form the seven carbon compound 3-deoxy-*arabino*heptulosonate-P. This enzyme too has been found to exist as isozyme pairs in mung beans, and in *Nicotiana silvestris* tissue cultures.[10] The two DAHP-synthase isozymes differed in their properties and metal requirements for activity. The one isozyme was activated by Mn^{++}, had a pH-optimum of 6.9 and was inhibited by glyphosate, while the other required Co^{++}, had a pH optimum of 8.8, and was not affected by glyphosate.

Isozymes of chorismate mutase are also present in *Nicotiana silvestris*.[10] One isozyme (CM-1) is subject to allosteric control by phenyl-alanine, tyrosine and tryptophan, while the other form (CM-2) is not effected by pathway end products. Other enzymes of the pathway section which have been indicated to be present as isozymes are anthranilate synthase,[11] arogenate dehydrogenase[10] and 5-enolpyruvylshikimate-3-phosphate synthase (EPSP).[10] Pea seedlings were reported to contain a number of major and minor isozymes of the pre-chorismate pathway section.[12,13]

Chorismate mutase isozymes have been shown to exist in plastidic (CM-1) and cytoplasmic (CM-2) compartments,[14] and similar locations were reported for DAHP-synthase.[10] Although recent reports on the sub-cellular localization of EPSP synthase in tissue cultures of *Corydalis sempervirens*[15] and dehydroquinate-shikimate:NADP oxidoreductase (SH0Rase) in peas[12] indicate their being plastidic only, the idea of a dual shikimate pathway cannot yet be concluded. It has been reported that optimal conditions for assaying either isozyme did not allow detection of the other.[10] Therefore, a dual shikimate pathway, analogous to multiple subcellular localization of the glycolytic and pentose phosphate pathways, appears to be a natural extension of multiple location of primary metabolic pathways making unnecessary complicated and energy-consuming transport processes. Aromatic amino acids produced in plastids would be used for the synthesis of proteins. Those produced in the cytoplasm would be also used for protein biosynthesis, but in addition to this, they serve also as precursors for the consecutive segments of aromatic metabolism that lead to the formation of the plant structural component lignin, and the diverse phenylpropanoid and flavonoid compounds.

Although it was claimed earlier that phenylpropanoid[16] and flavonoid pathway section enzymes[17] are located in chloroplasts, this is clearly not so. Cell fractionation experiments using various plants showed that isolated chloroplasts lacked the activity of phenylpropanoid and flavonoid pathway enzymes.[18] The previously described tissue fractionation experiments using the

Argenteum mutant of *Pisum* already showed the exclusive localization of chalcone synthase in the epidermal layers. Epidermal cells contain fewer and less well developed chloroplasts than mesophyll cells. Chloroplastic localization of the phenylpropanoid and flavonoid pathway sections and pathway end products would have resulted in detection of key enzyme activities and flavonoids in the photosynthetic tissue. These tissue and cell fractionation experiments clearly show that the phenylpropanoid and flavonoid segments of the aromatic pathway are not chloroplast located.

SOLUBLE ENZYMES OF AROMATIC METABOLISM

While the enzymes of the shikimate pathway section are present in both the plastids and the cytoplasm, those of the phenylpropanoid and flavonoid pathway sections are clearly all cytoplasmic. Each of these two latter segments contains a hydroxylating enzyme that is microsomal (presumably ER localized). In the general phenylpropanoid section it is cinnamic acid 4-hydroxylase,[19] in the flavonoid section it is a flavonoid-3'-hydroxylase.[20] (See Barz *et al.*, Chapter 5 for other microsomal enzymes in isoflavonoid pathways). All other enzymes of the pathway sections in Scheme 3 were reported to be 'cytosolic'. The term 'cytosolic', however, is not a functional definition. It is strictly an operational definition, indicating that the enzyme activities are detected in the supernatant fraction recovered after homogenisation and centrifugation of the plant material (see Ureta's review[21]). 'Cytosolic' implies that the cell is a bag full of enzymes, where enzymic reactions take place by collision kinetics. In such a mechanism, a substrate molecule would wander around in the cytoplasm until it meets its proper enzyme, which also wanders around. When both substrate and enzyme molecules meet by chance, a product is formed which is the substrate of the second enzyme, and the wandering process would repeat itself again. Complex metabolic pathways that produce in relatively short time large amounts of pathway end products are not likely to be governed by chaos. Indeed, data are accumulating in the literature that show that enzymes which were previously believed to be 'cytosoluble', are part of strictly organized enzyme complexes. The best example for this is the glycolytic pathway, which was shown to function as an enzyme complex.[22]

The plant aromatic pathway, which is one of the most complex and energy requiring processes in higher plants, is not likely to function by the chance meeting of substrates with enzymes. Indeed, accumulating data indicate that it may have similar organization as the glycolytic process.

BIFUNCTIONAL ENZYMES AND CHANNELING IN THE PATHWAY

To add further complexity to aromatic metabolism, its first segment, the shikimate pathway section, seems to function differently in various organisms. In fungi, the *arom* conjugate, steps 2-6 of the pathway section (Scheme 1), are catalyzed by a pentafunctional polypeptide[23] that is under the control of a gene cluster. In contrast, investigations with bacterial systems have shown that the same enzyme activities are not aggregated, nor are the genes encoding these enzymes clustered in those bacterial species for which genetic evidence exists.[24] Subsequent investigations in the yeast *Saccharomyces cerevisieae* showed that the activities of the pentafunctional *arom* enzyme are a mosaic of monofunctional domains, residing on a single polypeptide with an M_r of ~174,555.[25] These data support the suggestion that the ARO1 gene (that codes for the pentafunctional polypeptide) in yeast and other fungi evolved by linking ancestral bacterial genes together. From purely evolutionary considerations, it would seem likely that this organization in higher plants would be preserved or improved upon. This, however, is not the case. In higher plants, as in bacteria, most of the individual enzyme activities do not seem to be linked, and occur as separate, individual enzymes when isolated by ultra-centrifugation. Both the shikimate and phenylpropanoid pathway sections (Schemes 1, 2) contain an enzyme pair that seems to function as a bienzyme complex in a variety of plants. In the shikimate pathway section, this is the dehydroquinase-shikimate:NADP oxidoreductase that has been detected in peas (*Pisum sativum* L.),[12] in mung beans (*Phaseolus mungo*[26]) and in the moss *Physcomitrella patens*.[27] Both enzyme activities were shown to reside on a single 59,000 Mr polypeptide, indicating that the association of the two enzyme activities is the result of a bifunctional enzyme and not a dissociable enzyme complex. The dehydroquinase-shikimate oxidoreductase is present in all plants investigated as multiple isozymic forms, some of which were shown to be chloroplastic.[12]

Although in bacterial systems the formation of tryptophan is catalyzed by an enzyme aggregate composed of anthranylate synthase, phosphoribosyl-anthranylate isomerase and indoleglycerolphosphate synthase,[28] this pathway section has not been investigated in higher plants. Also, the final step of tryptophan synthesis, the conversion of indoleglycerolphosphate to tryptophan, is carried out in fungi by the bienzyme complex tryptophan synthase.[29] In bacteria this reaction was reported to be channeled in a 30 Å tunnel.[30] As with the previous reaction sequence, this part of the pathway has not been investigated in higher plants.

The phenylpropanoid section of the aromatic pathway has also been reported to contain an enzyme pair. Early enzymological work in buckwheat seedlings showed that when phenylalanine was used as a substrate in the phenylalanine ammonia-lyase reaction with a tissue homogenate, p-coumaric acid was formed as the reaction product.[31] Microsomal preparations from potato,[32] cucumber cotyledons[33] and buckwheat seedlings[34] showed that phenylalanine is channeled in this reaction. Contrary to the dehydroquinase-shikimate:NADP oxidoreductase bifunctional enzyme in the shikimate section of aromatic metabolism, the phenylalanine ammonia-lyase-cinnamate 4-hydroxylase enzymes are present in a loose association that easily disintegrates during the homogenization and purification process. Obtaining evidence for channeling in a loose enzyme complex is among the most difficult undertakings in biochemistry. The channeling experiments with the phenylalanine ammonia-lyase-cinnamate 4-hydroxylase system resulted in the recognition that the phenylpropanoid section of plant aromatic metabolism, or parts thereof, may function as consecutively assembled, membrane associated enzyme complexes.[34,35,36]

MEMBRANE ASSOCIATION OF ENZYMES IN THE FLAVONOID SEGMENT

Chalcone synthase, the key enzyme of flavonoid segment of aromatic metabolism, is one of the best investigated plant enzymes. This enzyme has been widely reported to be "cytosolic" in its localization, because its activity appears in the supernatant ("cytosolic") fraction upon homogenization of plant materials. During the early cell fractionation experiments that we carried out with George Wagner at Brookhaven National Laboratory in 1976 with *Hippeastrum* protoplasts and tissues, we repeatedly obtained small but significant amounts of chalcone synthase activity associated with a microsomal fraction.[37] The activity of chalcone synthase that consistently appeared in the 100,000 x g pellet fraction in *Hippeastrum* and *Tulipa* tissues indicated that a part of the chalcone synthase activity was either associated with membranes in the microsomal fraction of the cells, or that the microsomal fraction was contaminated with chalcone synthase (Table 3).

We have devised ultracentrifugation experiments on sucrose density gradients to establish if the presence of chalcone synthase, and possibly other enzymes of the phenylpropanoid and flavonoid pathway on membrane fractions is due to protein entrapment, or *de facto* associations.[38] In these experiments the activities of PAL, cinnamate 4-hydroxylase and chalcone synthase

Table 3. Distribution of chalcone synthase activities in subcellular fractions of *Hippeastrum* and *Tulipa*.

	% of total activity recovered in:			
Tissue	Particulate Cytoplasm	Cytosol	Vacuole	100,000 x g Pellet
Hippeastrum petal	5-15	80-95	0	1-4
Tulipa petal	0	97	0	3
Tulipa leaf	0	88	0	12

consistently appeared in fractions where the activity of NADPH:Cyt c reductase accumulated.[39] Since NADPH:Cyt c reductase has been shown previously to be an endoplasmic reticulum (ER) embedded enzyme,[40] together with cinnamate 4- hydroxylase, the association of other pathway enzymes with these activities indicated ER localization.[19] Therefore, these tissue fractionation experiments gave a strong indication that the key enzymes of the phenylpropanoid and flavonoid pathways are associated with the endoplasmic reticulum membranes under *in vivo* conditions. We have published previously a working hypothesis[40] describing our current understanding of how the pathway may function. In this hypothesis, we suggested that the phenylpropanoid and flavonoid sections of aromatic metabolism may function as a consecutively assembled, membrane associated enzyme complexes, where pathway end products are most likely channeled. Based on Parham and Kaustinen's data,[41] we suggested that pathway end products may accumulate in the lumen of endoplasmic reticulum membranes, from where they would be transported to the central vacuole for storage.

We have recently obtained further supporting evidence for such a sub-cellular mechanism. This evidence was derived from biochemical, immuno-logical and immunocytological experiments that were performed to localize chalcone synthase, PAL and a flavonol specific glucosyltransferase from buckwheat and red cabbage seedlings. The information is more complete for the chalcone synthase from buckwheat and this will be discussed first. Buckwheat seedlings grown in the dark for six days and illuminated with white light show maximal chalcone synthase activity after 18 h illumination.[34] When the seedlings were homogenized in the presence of proteinase inhibitors and subjected to centrifugation on 25%/55% sucrose step gradients, a significant amount (approx. 25-30%) of chalcone synthase activity appeared in the membrane fraction that accumulated at the 25%/55% sucrose concentration interface. The remaining portion of enzyme activity was located in the sample zone, and was "soluble."

The membrane fraction accumulating at the 55% sucrose interface was removed and loaded onto 25%-55% linear sucrose gradients. The gradients were centrifuged to isopycnic equilibrium, fractionated, and the activities of chalcone synthase and the marker enzyme NADPH:Cyt c reductase measured in the fractions. In both the $MgCl_2$- and EDTA-containing gradients, chalcone synthase activity was detected in the fractions that contained ER marker enzyme activities. The results from enzyme activity measurements were confirmed by immunoblotting, using a rabbit anti-buckwheat chalcone synthase antibody preparation. The intensity of the immunostaining of the chalcone synthase band on the electroblotted gradients coincided with enzyme activities showing that chalcone synthase, or a part of its activity, was associated with the endoplasmic reticulum membranes.[42]

To ascertain that the presence of chalcone synthase activity on the ER membranes was not the result of non-specific binding or enzyme entrapment during homogenization, we included relatively large amounts of bovine serum albumin (BSA) in the homogenization media, and submitted the gradient fractions to electrophoresis on polyacrylamide gels. Staining the gels with Coomassie brilliant blue showed that, indeed, BSA penetrated into the gradient. However, the amount of BSA progressively decreased with increasing sucrose concentrations and did not show any preferential accumulation with ER-containing fractions. We concluded from these data that the association of chalcone synthase with the ER membranes is most likely not the result of an artifact, but that it occurs under natural conditions in the cells.

Further support for this conclusion was provided by immunocyto-chemical experiments using an anti-chalcone synthase antibody preparation that was purified to monospecificity by immunoaffinity. Using this antibody with membrane preparations obtained from sucrose density gradients, and a 20 nm gold-IgG conjugate showed that most of the gold label was located on the membranes, with a decreasing label concentration upon increasing distance from the membranes. In situ immunocytochemical studies using identical probes showed gold deposition directly on tubular ER membranes which are character-istic features of six day old buckwheat seedlings. The gold deposition was observed on the ER membrane only, no other cellular organelles showed

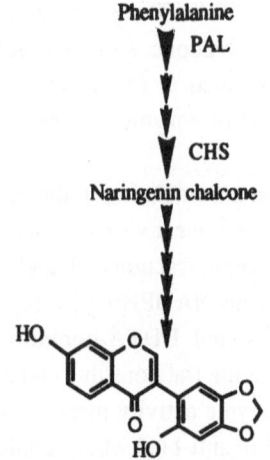

Phenylalanine
PAL

CHS
Naringenin chalcone

2',7-Dihydroxy-3',4-methylenedioxyisoflavone (DMI)

DMI Reductase

(+) Sophorol   (-) Sophorol

(+) Maackiain   (-) Maackiain

In chickpea

(+) 6a-hydroxymaackiain (HMK)

(+) Pisatin

In pea

labeling. An interesting feature of some electron micrographs is the appearance of electron dense deposits on the vacuolar face of the tonoplast in the vicinity of the gold labels. Parham and Kaustinen[41] suggested that such electron dense deposits in "vesicles" and the vacuole are an indication of "tannin" accumulation. Although another laboratory using similar immunocytochemical probes in spinach leaves reported chalcone synthase localization as ' cytosolic',[43] a closer inspection of the electromicrographs suggests ER localization here also.

We have obtained essentially identical biochemical data with a flavonoid 3-O-glucosyltransferase in illuminated red cabbage (*Brassica oleracea* cv. Red Danish) seedlings (Sun, Zobel and Hrazdina, unpublished data). The activity of the glucosyltransferase accumulated in the 1.16-1.18 g/cm^3 specific gravity area in MgCl$_2$ containing gradients. In the presence of EDTA, the activity peak shifted to a lower density, to the 1.12-1.15 g/cm^3 specific gravity area. The bulk of the protein showed a similar shift, indicating that the enzyme was associated with rough endoplasmic reticulum membranes (under *in vivo* conditions, Hrazdina, Sun and Reeves, unpublished data). We have developed polyclonal antibodies against glucosyltransferase from red cabbage for immuno-chemical and immunocytochemical investigations. However, either because of the rabbit or the protein preparation, the antibodies were weak, and did not provide clear results. However, the sucrose density gradient data support our earlier report[38] on association of this enzyme with endoplasmic reticulum membranes in *Hippeastrum* petal tissues.

Our latest data concerning enzyme localization on endoplasmic reticulum membranes comes from an unusual system and concerns an unusual enzyme. The plant system is CuCl$_2$-challenged pea (*Pisum sativum* L.), and the enzyme is a isoflavone-specific reductase that introduces chirality in isoflavonoid phytoalexins[44] (Scheme 4). We reported earlier the isolation of this enzyme from the challenged pea seedlings, and the development of polyclonal antibodies against it.[44] The pea system seems to be somewhat different from other plants investigated in our laboratory. When pea homo-genates, obtained 32 h after CuCl$_2$ treatment with isoflavone reductase activity at its maximum, were subjected to centrifugation on linear sucrose density gradients and the activity of

---

Scheme 4. The isoflavonoid section of plant aromatic metabolism that leads to the formation of pterocarpan phytoalexins. DMI reductase (7,2'-dihydroxy-4',5'-methylene-dioxyisoflavone oxido-reductase) is the enzyme responsible for the introduction of chirality in the isoflavonoid molecule.

the ER-marker enzyme cinnamate 4-hydroxylase was assayed, the results showed that the enzyme equilibrated in both the $MgCl_2$- and EDTA-containing gradients at somewhat lower densities than in buckwheat, *Hippeastrum* or red cabbage. This may be an indication that the composition of the ER in challenged peas (or legumes) may differ from that of the other plants we used previously as model systems. The $MgCl_2$- and EDTA-containing gradients showed that rough ER (or ribosome-bearing ER) accumulated at the 1.15 g/cm^3 density, while the smooth ER was at 1.11 g/cm^3. Although PAL activity, which was used as a control in the experiment, did not show such sharp shift as did the marker enzyme cinnamate 4-hydroxylase, its activity profile on the two gradients suggested its localization on the ER membranes.

Because of the rapid loss of isoflavone reductase activity in crude preparations, enzyme activity measurements of sucrose density gradients were not useful for establishing its subcellular localization. Therefore, we used immunoblotting as the tool (Sun and Hrazdina, unpublished). Although the blots did not give a sharp curve, the intensity of the isoflavone reductase bands clearly indicate an ER association on both gradients. Although ER association of general phenylpropanoid and flavonoid pathway enzymes such as PAL and chalcone synthase were documented earlier,[38,42] the data obtained with the iso-flavone reductase is the first evidence that the isoflavonoid phytoalexin section of plant aromatic metabolism may also be located on the endoplasmic reticulum (see also Barz *et al.*, Chapter 5). Localization of isoflavonoid phytoalexin biosynthesis on endoplasmic reticulum membranes raises interesting questions. Is the accumulation of the pathway end products within or outside the ER, and what is the transport mechanism that delivers them to their site of accumulation?

SITES OF ACCUMULATION OF AROMATIC COMPOUNDS IN CELLS

Early microscopic investigations suggested that the main subcellular site of anthocyanin accumulation is the vacuole.[45,46] These reports were based on visual observation of red, purple, or blue color in the center of flower petal cells. Unfortunately, visual observation can only be used in cytology when the target compounds are colored. Although protoplasts and vacuoles were prepared earlier from microorganisms such as yeast,[47] vacuoles from higher plants could be prepared only after reasonably pure fungal cell wall enzyme preparations became available. Methodology for protoplast and vacuole preparation, developed by Wagner and Siegelmen,[48] permitted the analysis of vacuolar contents

Table 4. Aromatic metabolites localized in isolated vacuoles of higher plants.

Compound	Plant genus
Anthocyanin	*Brassica*
	Hippeastrum
	Tulipa
	Daucus
	Vitis
Flavonol 3-glycoside	*Vitis*
p-Coumaryl tartrate	*Vitis*
Caffeoyl tartrate	*Vitis*
Flavone 7-0-malonylglucoside	*Petroselinum*
Cyanogenic glycoside	*Sorghum*
o-Coumaric acid glucoside	*Melilotus*
Glucosinolate	*Armoracia*
Betanine	*Beta*

for aromatic metabolites. Vacuoles were prepared from *Hippeastrum*,[48] tulips,[48] carrots,[49] *Vitis* species,[50] parsley,[51] sorghum,[52] white clover,[53] mustard seedlings,[54] and red beets;[55] the aromatic metabolites found in the isolated vacuoles of the above plants are listed in Table 4.

Although the vacuole seems to be the major site for the accumulation of flavonoid glycosides and for the glycosides of cinnamic acids, cyanogenic glycosides, glucosinolates and the betalains, it is not the exclusive site of accumulation for other aromatic metabolites. Lignin, the plant structural component, is deposited in the cell wall (see Lewis *et al.*, Chapter 11). It has recently been shown that coumarin derivatives accumulate on the surface of epidermal cells (S. Brown, personal communication). Also, preliminary investigations indicated that the isoflavonoid phytoalexin pisatin, although produced in high concentrations in the infected cells, does not seem to accumulate in the vacuoles (H. Van Etten, personal communication).

The differential accumulation of the various aromatic pathway end products at different subcellular sites poses an interesting transport problem for the cell. Cytochemical experiments indicated transport of such metabolic end products as 'tannins'[41] and flavonoid glycosides[42] from the ER to the vacuoles, and like other transport processes, these are believed to be transported by specific vesicles. Synthesis of aromatic metabolites by ER-localized complexes, and their differential transport to either the vacuole or the cell wall would most likely require specific composition of the ER regions where they are produced, or different signals for their transport. Presently, fundamental aspects of the transport of proteins and the signals involved in their transport are understood. In contrast, we have little knowledge about the transport of small molecules. This can only be obtained through rigorously designed experiments, for which there seem to be no tools available at present. To quote Davis,[56] "The sub-cellular localization of small molecules is easy to believe, but difficult to prove."

ACKNOWLEDGEMENT

Special thanks to Elizabeth A. Baroody for her help with the manuscript.

REFERENCES

1.   DIXON, R.A., BOLWELL, G.P., HAMDAN, M.A.M.S., ROBBINS, M.P. 1987. Molecular biology of induced resistance. In: Genetics and Plant Pathogenesis. (P.R. Day and G.J. Jellis, eds.) Blackwell Scientific, Oxford, pp. 245-259.

2.   DYER, W.E., HENSTRAND, J.M., HANDA, A.K., HERRMANN, K.M. 1989. Wounding induces the first enzyme of the shikimate pathway in *Solanaceae*. Proc. Natl. Acad. Sci. USA 86:7370-7373.

3.   HOCH, H.C., PRATT, C., MARX, G.A. 1980. Subepidermal air spaces: basis for the phenotypic expression of the *Argenteum* mutant of *Pisum*. Amer. J. Bot. 67:905-911.

4.   MARX, G.A. 1978. *Argenteum*: a mutant under nuclear and extra nuclear control. Pisum Newsl. 10:34-37.

5.   HRAZDINA, G., MARX, G.A., HOCH, H.C. 1982. Distribution of secondary plant metabolites and their biosynthetic enzymes in pea (*Pisum sativum* L.) leaves. Plant Physiol. 70:745-748.

6.   JAHNEN, W., HAHLBROCK, K. 1988. Cellular localization of nonhost resistance reactions of parsley (*Petroselinum crispum*). Planta 173:197-204.

7. SCHMELZER, E., JAHNEN, W., HAHLBROCK, K. 1988. *In situ* localization of light-induced chalcone synthase MRNA, chalcone synthase and flavonoid end products in epidermal cells of parsley leaves. Proc. Natl. Acad. Sci. USA 85:2989-2993.

8. KNOGGE, W., WEISSENBÖCK, G. 1986. Tissue-distribution of secondary phenolic biosynthesis in developing primary leaves of *Avena sativa* L. Planta 167:196-205.

9. DENNIS, D.T., MIERNYK, J.A. 1982. Compartmentation of nonphotosynthetic carbohydrate metabolism. Annu. Rev. Plant Physiol. 33:27-50.

10. JENSEN, R.A. 1986. Tyrosine and phenylalanine biosynthesis: relationship between alternative pathways, regulation and subcellular location. Rec. Adv. Phytochem. 20:57-82.

11. CARLSON, T.E., WIDHOLM, J.M. 1978. Separation of two forms of anthranilate synthase from 5-methyltryptophan susceptible and resistant cultured *Solanum tuberosum* cells. Physiol. Plant. 44:251-255.

12. MOUSDALE, D.M., CAMPBELL, M.S., COGGINS, J.R. 1987. Purification and characterization of bifunctional dehydroquinase-shikimate:NADP oxidoreductase from pea seedlings. Phytochemistry 26:2665-2670.

13. MOUSDALE, D.M., COGGINS, J.R. 1985. Subcellular localization of the common shikimate-pathway enzymes in *Pisum sativum* L. Planta 163:241-249.

14. D'AMATO, T.A., GANSON, R.J., GAINES, R.A., JENSEN, R.A. 1984. Subcellular localization of chorismate mutase isoenzymes in protoplasts from mesophyll and suspension cultured cells of *Nicotiana silvestris*. Planta 162:104-108.

15. SMART, C.C, AMRHEIN, N. 1987. Ultrastructural localization by protein A-gold immunocytochemistry of 5-enolpyruvylshikimic acid 3-phosphate synthase in a plant cell culture which overproduces the enzyme. Planta 170:1-6.

16. SAUNDERS, J.A., MCCLURE, J.W. 1976. The distribution of flavonoids in chloroplasts of twenty five species of vascular plants. Phytochemistry 15:809-816.

17. WEISSENBÖECK, G., FLEING, I., RUPPEL, H.G. 1972. Untersuchungen zur Lokalisation von Flavonoiden in Plastiden. I. Flavonoide in Etioplasten von *Avena sativa* L. Z. Naturforschg. 27:1216-1224.

18. HRAZDINA, G., ALSCHER-HERMAN, R., KISH, V.M. 1980. Subcellular localization of flavonoid synthesizing enzymes in *Pisum*, *Phaseolus*, *Brassica* and *Spinacia* cultivars. Phytochemistry 19:1355.

19. SAUNDERS, J.A., CONN, E.E., LIN, C.H., SHIMADA, M. 1977. Localization of cinnamic acid 4-monooxygenase and the membrane-bound enzyme system for dhurrin biosynthesis in *Sorghum* seedlings. Plant Physiol. 60:629-634.

20. HAGMANN, M-L., HELLER, W., GRISEBACH, H. 1983. Induction and characterization of a microsomal flavonoid 3'-hydroxylase from parsley cell cultures. Eur. J. Biochem. 134:547-554.

21. URETA, T. 1978. The Role of Isozymes in Metabolism: A Model of Metabolic Pathways as the Basis for The Biological Role of Isozymes. Current Topics in Cellular Regulation 13:233-258.

22. GORRINGE, D.M., MOSES, V. 1978. A multienzyme aggregate with glycolytic activity from *Escherichia coli*. Biochem. Soc. Transactions 6:167-169.

23. AHMED, S.I., GILES, N.H. 1969. Organization of enzymes in the common aromatic synthetic pathway: Evidence for aggregation in fungi. J. Bact. 99:231-237.

24. BERLYN, M.B., GILES, N.H. 1969. Organization of enzymes in the polyaromatic synthetic pathway: Separability in bacteria. J. Bact. 99:222-230.

25. DUNCAN, K., EDWARDS, R.M., COGGINS, J.R. 1987. The pentafunctional *arom* enzyme of *Saccharomyces cerevisisiae* is a. mosaic of monofunctional domains. Biochem. J. 246:375-386.

26. KOSHIBA, T. 1978. Purification of two forms of the associated 3-dehydroquinate hydro-lyase and shikimate:NADP$^+$ oxidoreductase in *Phaseolus mungo* seedlings. Biochem. Biophys. Acta 522:10-18.

27. POLLEY, L.D. 1978. Purification and characterization of 3-dehydroquinate hydrolase and shikimate oxidoreductase. Evidence for a bifunctional enzyme. Biochim. Biophys. Acta 526:259-266.

28. GAERTNER, F.H., DEMOSS, J.A. 1969. Purification and characterization of a multienzyme complex in the tryptophan pathway of *Neurospora crassa*. J. Biol. Chem. 244:2716-2725.

29. HYDE, C.C., AHMED, S.A., PADLAN, E.A., MILES, E.W., DAVIES, D.R. 1988. Three-dimensional structure of the tryptophan synthase $\alpha^2\beta^2$ multienzyme complex from *Salmonella typhimurium*. J. Biol. Chem. 263:17857-17871.

30. DUNN, M.F., AGUILAR, V., BRZOVIC, P., DREWE, JR., W.F., HOUBEN, K.F., LEJA, C.A., ROY, M. 1990. The tryptophan synthase bienzyme complex transfers indole between the α- and β-sites via a 25-30 Å long tunnel. Biochemistry 29:8598-8607.

31. AMRHEIN, N., ZENK, M. H. 1971. Untersuchungen zur Rolle der Phenylalanine Ammonia Lyase (PAL) bei der Regulation der Flavonoidsynthese in Buchweizen (*Fagopyrum Esculentum* Moench). Z. Pflanzenphysiol. 64:145-168.

32. CZICHI, U., KINDL, H. 1975. Formation of *p*-coumaric acid and *o*-coumaric acid from L-phenylalanine by microsomal membrane fractions from potato: evidence of membrane-bound enzyme complexes. Planta 125:115-125.

33. CZICHI, U., KINDL, H. 1977. Phenylalanine ammonia lyase and cinnamic acid hydroxylase as assembled consecutive enzymes on microsomal membranes of cucumber cotyledons: cooperation and subcellular distribution. Planta 134:133-143.

34. HRAZDINA, G., WAGNER, G.J. 1985. Metabolic pathways as enzyme complexes: evidence for the synthesis of phenylpropanoids and flavonoids on membrane-associated enzyme complexes. Arch. Biochem. Biophys. 237:88-100.

35. CZICHI, U., KINDL, H. 1975. A model of closely assembled consecutive enzymes on membranes: Formation of hydroxycinnamic acids from *L*-phenylalanine of thylakoids of *Dunaliella marina*. Hoppe-Seyler's Z. Physiol. Chem. 475-485.

36. STAFFORD, H.A., 1981. Compartmentation in natural product biosynthesis by multienzyme complexes. In: The Biochemistry of Plants, Vol. 7, (E.E.Conn, ed), Academic Press, New York, pp.117-137.

37. HRAZDINA, G., WAGNER, G.J., SIEGELMAN, H.W. 1978. Subcellular localization of enzymes of anthocyanin biosynthesis in protoplasts. Phytochemistry 17:53-56.

38. WAGNER, G.J. HRAZDINA, G. 1984. Endoplasmic reticulum as a site of phenylpropanoid and flavonoid metabolism in *Hippeastrum*. Plant. Physiol. 74:901-906.

39. LORD, J.M., KAGAWA, T., MOORE, T.S., BEEVERS, H. 1973. Endoplasmic reticulum as the site of lecithin formation in castor bean endosperm. J. Cell Biol. 57:659-667.

40. HRAZDINA, G., WAGNER, G.J. 1985. Compartmentation of plant phenolic compounds; sites of synthesis and accumulation. In: Annual Proceedings of the Phytochemical Society of Europe. Vol. 25. (C.F.

Van Sumere, P.J. Lea, eds.) Clarendon Press, Oxford, pp. 120-133.

41. PARHAM, R.A., KAUSTINEN, H.M. 1977. On the site of tannin synthesis in plant cells. Bot. Gaz. 138:465-467.

42. HRAZDINA, G., Zobel, A.M., HOCH, H.C. 1987. Biochemical, immunological and immunocytochemical evidence for the association of chalcone synthase with endoplasmic reticulum membranes. Proc. Natl. Acad. Sci. USA 84:8966-8970.

43. BEERHUES, L., WIERMANN, R. 1988. Chalcone synthases from spinach (*Spinacia oleracea* L.). Planta 173:532-543.

44. SUN, Y., WU, Q., VAN ETTEN, H.D., HRAZDINA, G. 1991. Stereo-isomerism in plant disease resistance: Induction and isolation of the 7,2'-dihydroxy-4',5'-methylenedioxyisoflavone oxidoreductase, an enzyme introducing chirality during synthesis of isoflavonoid phytoalexins in pea (*Pisum sativum* L). Arch. Biochem. Biophys. 284:167-173.

45. LIPMAN, T. 1926. Die Anthocyanophore der Erythrea-Arten. Beitr. Bot. Zentralbl. Abt. I 43:127-132.

46. MOLLISCH, H. 1923. Mikrochemie der Pflanzen. Fischer Verlag, Jena.

47. MATILE, P. 1978. Biochemistry and function of vacuoles. Annu. Rev. Plant Physiol. 29:193-213.

48. WAGNER, G.J. SIEGELMAN, H.W. 1975. Large scale isolation of intact vacuoles and isolation of chloroplasts of mature plant tissues. Science 190:1298-1304.

49. SASSE, F., BECKS-HUSEMANN, D., BARZ, W. 1979. Isolation and characterization of vacuoles from cell suspension cultures of *Daucus carota*. Z. Naturforsch. 34c:848-853.

50. MOSKOWITZ, A.H., HRAZDINA, G. 1981. Contents of vacuoles isolated from grape berry subepidermal tissues. Plant Phys.. 65:97-105.

51. MATERN, U., HELLER, W., HIMMELSPACH, K. 1983. Conformational changes of apigenin 7-O-(6-O-malonylglucoside), a vacuolar pigment from parsley, with solvent composition and proton concentration. Eur. J. Biochem. 133:439-448.

52. SAUNDERS, J.A., CONN, E.E. 1978. Presence of the cyanogenic glucoside dhurrin in isolated vacuoles from sorghum. Plant Physiol. 61:154-157

53. OBA, K., CONN, E.E., CANUT, H., BOUDET, A.M. 1981. Subcellular localization of 2-(β-D-glucosyl-oxy)cinnamic acids and the related β-glucosidase in leaves of *Melilotus alba* Dear. Plant Physiol. 68:1359-1363.

54. GROB, K., MATILE, P. 1979. Vacuolar location of glucosinolates in horseradish root cells. Plant Sci. Lett. 14:327-335.

55. LEIGH, R.A., BRANTON, D. 1976. Isolation of vacuoles from root storage tissue of *Beta vulgaris*. Plant. Physiol. 58:656-662.

56. DAVIS, R.H. 1967. Channeling in *Neurospora* metabolism. In: Organizational Biosynthesis. (H.J. Vogel, J.O. Lampen, and V. Bryson, eds.) Academic Press, New York, pp. 303-333.

Chapter Two

IMMUNOLOCALIZATION OF FLAVONOID CONJUGATES AND THEIR ENZYMES

Ragai K. Ibrahim

Plant Biochemistry Laboratory
Department of Biology, Concordia University
Montreal, Canada H3G 1M8

Phenolic Metabolism in Plants, Edited by H.A. Stafford
and R.K. Ibrahim, Plenum Press, New York, 1992

INTRODUCTION

Flavonoid compounds are among the most widely distributed natural plant products. They occur in several conjugated forms, most commonly as glycosylated and methylated derivatives, although other flavonoids may be acylated, prenylated or sulfated.[1,2] A variety of flavonoids have been assigned different roles as antimicrobial compounds, stress metabolites or signalling molecules.[3-7]

The spatial organization of secondary metabolism is the result of the separation of precursors, metabolic intermediates and products from the enzymes involved in their biosynthesis. Such compartmentation is achieved at the cellular level by the membranes which surround the organelles, and those which separate the cytoplasmic regions into different microcompartments.[8,9] In addition, the association of biosynthetic enzymes as loose aggregates[10,11] or clusters,[12] as well as the interaction between constituent enzymes, permits direct transfer of metabolites from one enzyme to another, and channeling of intermediates and final products to the different sites of accumulation.[10-15]

Flavonoid conjugates that are hydrophilic in nature usually accumulate in the central vacuoles of epidermal and subepidermal cells,[9,16,17] whereas lipophilic compounds are found in epidermal glandular cells,[18,19] on plant surfaces[20,21] or are secreted from roots as chemical signals.[7,22] These sites of accumulation have been considered adaptations to various ecological conditions, or in reponse to symbiotic association with plant roots.

TISSUE/CELLULAR DISTRIBUTION

Flavonoid conjugates

Several studies have indicated a high degree of compartmentation of flavonoid compounds and the enzymes involved in their biosynthesis (see Chapter 1 by Hrazdina). Many secondary metabolites, especially flavonoid compounds, accumulate in the central vacuoles of epidermal and subepidermal cells of leaves,[9,14,16,23,24] and shoots,[35] or on the external surfaces of plant organs.[18,21] The methods used for localization studies include light, fluorescence, or electron microscopy, physical separation of tissues for chromatographic analysis, isolation of organelles for labeling or enzymic studies and the preparation of protoplasts for the isolation of vacuoles. Table 1 shows some examples of the sites of flavonoid localization that exhibit tissue-specific distribution. Two examples are presented here to demonstrate some of the problems in this phenomenon.

In oat primary leaves, isovitexin glycosides were predominantly present in the lower epidermis, with 30% of the total found in the mesophyll tissue.[31] However, 80-90% of activity of the enzymes involved in their biosynthesis—chalcone synthase (CHS), chalcone isomerase (CHI), isovitexin arabinosyltransferase (GT) and vitexin rhamnoside 7-O-methyltransferase (OMT)—was located in the mesophyll.[31] These results implied that the entire pathway for flavonoid biosynthesis is located in the leaf mesophyll, and that C-glycoflavones are transported to the epidermis either apoplastically (across the cell walls) or symplastically (across the plasmodesmata).[31] Re-examination of this and three other monocotyledonous species, using monoclonal anti-CHS antibody, indicates that the amount of immunodetectable CHS is equally distributed in both epidermal layers and the mesophyll tissue, and coincides with flavonoid accumulation in the primary leaves of barley, oats and rye. In maize, on the other hand, these amounts are 3- to 4-fold higher in the upper epidermis than in the lower epidermis or mesophyll (G. Weissenböck, personal communication). These results seem to indicate that each tissue type is autonomous in flavonoid biosynthesis and accumulation. In contrast, soybean epidermis contains appreciable levels of flavonol glucosides, whereas mesophyll cells contain neither flavonoids nor exhibit any activity for PAL, CHS or isoflavone GT.[32] However, following treatment with acifluorfen (a herbicide) mesophyll cells were found to contain 12-20% of the total isoflavone glucosides, whereas the epidermis accumulated isoflavone aglycones and pterocarpans.[32] Although the foregoing represents an example of elicitor-induced metabolites, there appears to

Table 1. Sites of localization of some flavonoid compounds.

Flavonoid	Plant (tissue)	Site of localization[a]	Reference
Flavonol & anthocyanin glycosides	*Vitis sp.* (berry)	Epid. & subepidermal vacuoles (C)	23
	Pisum sativum (leaf)	Epid. vacuoles (C)	24
Flavonols & anthocyanins	*Sinapis alba* (colytedon)	Upper epid (flavonols) lower epid. (anthocy.)(C)	25
Kaempferol 3-glc-7-rhamn.	*Vicia faba* (leaf)	Guard cell protoplast & epid. vacuoles (Ms)	26
Kaempferol & quercetin 3-glycosides	*Picea abies* (needle)	Cell-wall bound (C)	27
Methylated flavonols	*Aesculus hippo- castaneum* (bud)	Surface exudate through ER (EM)	18
Polymethylated flavonol gluc.	*Chrysosplenium americanum* (leaf)	Cell wall, vesicles (C, EM)	28
Flavonoid aglycones	Several spp. (stems, leaves)	Surface exudate (C)	21
Apigenin (6"-malonyl) 7-glucoside	*Petroselinum sp.* (cell culture)	Protoplast vacuole (M, Ms)	29
Isovitexin glycosides	*Secale cereale* (1[ary] leaf)	Epidermal vacuole (C)	30
C-Glycoflavones	*Avena sativa* (1[ary] leaf)	70% epid. protoplast, 30% mesophyll (C)	31
Isoflavone gluc., pterocarpans	*Glycine max* (leaf)[b]	Mesophyll protoplast (C)	32
Proanthocyanidins (cond. tannins)	Several gymnosp. (cell cultures)	Vesicles, vacuoles (EM, H)	33,34
	Sambucus racemosa (promeristem)	Tannin cells (EM)	35

[a] Methods used:C, chromatography; EM, electron microscopy; FM, fluorescence microsocpy; Ms, microspectrophotometry; H, histochemistry
[b] After treatment with acifluorfen.

be basic differences in tissue distribution of flavonoid compounds and their enzymes in a cereal[31] and a legume.[32]

On the other hand, lipophilic metabolites (polymethylated flavonoids)[18,28] and highly toxic compounds such as proanthocyanidins,[33,34] have been shown to be sequestered in membrane vesicles that are possibly derived from the endoplasmic reticulum membranes (but not the Golgi). These vesicles either secrete their metabolites to the external surface,[18,28] fuse with the large central vacuole[33,34] or coalesce to form specialized cells.[35]

Flavonoid enzymes

The enzymes involved in flavonoid biosynthesis have previously been reported to be mostly cytosolic[31,32,36-38] (Table 2), although a small portion of CHS, CHI and GT has been reported to be membrane-associated.[14] Recently, however, experiments have indicated that a significant part of the above enzyme activities was associated with the 'particulate cytoplasm' of *Hippeastrum* protoplasts.[39] The latter fraction was characterized as membranes of the endoplasmic reticulum (ER) by the EDTA shift method, marker enzymes and structural latency experiments.[39] Similar results have been reported with buckwheat (*Fagopyrum esculentum*) hypocotyls, where the 100,000xg ER pellet was recovered in the exclusion volume after gel filtration, and shown to contain significant amounts of the above enzymes after trypsin digestion.[40] In contrast, oxidative enzymes such as polyphenol oxidase[41] and peroxidase,[42] have been reported to be associated with walls and vacuolar compartments, and a dihydroflavonol glucoside β-glucosidase to be bound to the cell wall.[43]

Limitations of Localization Methods

Whereas the methods used for localization studies have yielded valuable information, some of the reported results should be interpreted with caution dueto inherent problems with some techniques that may have resulted in artifacts. The limitations of, and artifacts which may arise from, the use of such methods have recently been summarized.[9,14] Furthermore, during tissue or cell disruption, membranes of chloroplasts or of other organelles may bind vacuolar flavonoids[44] as well as the enzymes involved in their biosynthesis.[45] The latter problem has resulted in the artifactual reporting of the association of flavonoid OMTs with spinach chloroplast membranes.[46] In addition, membrane-associated enzymes may become displaced during tissue processing and consequently recovered in the cytosolic fraction.[11,14] Such artifacts suggest that the a re-evaluation of previous reports of the localization of plant secondary metabolites and their biosynthetic enzymes is required.[14,47]

IMMUNOLOGICAL TECHNIQUES

Introduction

The main thrust of this section is to foster the application of immunological methods to localization studies. The binding reaction between an antibody and its corresponding antigen, whether a macromolecule or hapten, is sufficiently specific to allow the use of antibodies as molecular probes in the subcellular localization of secondary metabolites and their enzymes.[48]

The antigenic determinant of a protein (epitope) is the particular site on the surface of the antigen responsible for binding the antibody. Experiments with synthetic peptides, as well as with protein antigens, have revealed that the antigenic site usually consists of only 3-8 amino acid residues. The latter may form part of a continuous stretch of a peptide sequence (continuous antigenic site). Alternatively, the antigenic sites may be separated from each other in the primary protein sequences, although they are brought together in the native folded form by protein conformation (discontinuous topographic antigenic site). Most recent studies suggest that most, if not all, determinants are discontinuous to some extent.[49]

A conventional antiserum (polyclonal) not only has antibodies to several determinants, but also a family of antibodies of different structure and avidity which compete for each individual determinant of the antigen molecule. On the other hand, monoclonal antibodies produced by hybridoma technology are homogeneous because they are selected for their ability to bind a single, unique determinant.

The ease of production of antibodies to small molecules (<1000 Da), especially natural products,[50] makes it possible to apply these antibodies to study the localization of their antigens. The successful use of antibodies in immunological studies depends on the specificity of the binding reaction which, in itself, is determined by the strategy used to isolate and purify the antigen in question. Recent technical innovation in fluorescence and electron microscopy, as well as the embedding materials and cryomicrotomy, have brought the techniques of immunocytochemistry to the point of routine laboratory application.

Table 2. Tissue/cell localization of flavonoid enzymes.[a]

Enzyme	Plant (tissue)	Site of localization	Reference
CHI, GT	*Hippeastrum* & *Tulipa* (protopl.)	Cytosol	36
CHS, OMT, GT	Several spp. (leaves)	Cytosol, homogenate	37
CHS, GT, OMT	*Pisum sativum* (leaves)	CHS, flv.3GT (epidermis); flv. 7GT, OMT (mesophyll)	24
PAL, CHS, GT	*Hippeastrum sp:* (petal protoplasts)	ER membranes[b]	39,40
Anthocy. OMT	*Petunia hybrida* (flower)	Cytosol	38
CHS, CHI, *C*-glyco-flavone, GT & OMT	*Avena sativa* (1[ary] leaf)	Mesophyll protoplasts	31
CHS, isoflv. GT	*Glycine max* (leaves)[c]	Mesophyll protoplasts (25% of total)	32
Polyphenol oxidase	Several spp.	Cytosol, chloroplast, vacuole	41
DHF-glucoside β-glucosidase	*Petunia hybrida* (flowers)	Cell wall	42

[a]CHI, chalcone isomerase; CHS, chalcone synthase; DHF, dihydroflavonol;GT, gluco/glycosyltransferase; OMT, *O*-methyltransferase; PAL, phenylalanine ammonia-lyase.

[b] Gradients lacking Mg^{2+} and containing 10 mM EDTA.

[c] After treatment with acifluorfen.

Preparation of Antigen

Despite the fact that enzymes of seconday metabolism are usually present in extremely low abundance, the recent introduction of fast protein liquid chromatography (FPLC) has revolutionized protein separation technology of these enzymes.[51,52] Since the immunization of animals requires only a few milligrams of pure antigen, scaled-up conventional column chromatographic techniques are commonly used in the preliminary steps of protein purification.

These include a combination of gel permeation, ion exchange, hydroxyapatite and hydrophobic interaction columns. Affinity and dye-ligand chromatography offer tremendous possibilities in separations which are difficult to achieve with less specific procedures, especially when adapted to FPLC to take advantage of the more exacting, automated gradient elution and high resolution in a minimum of time.[53] Immunoaffinity purification represents another form of affinity chromatography which can be achieved by linking the purified antibodies to an insoluble matrix such as sepharose or polyacrylamide.[54] The use of chromato-focusing offers a powerful tool in protein purification because it achieves the separation of isoenzyme forms.[55] High performance capillary electrophoresis[56] may be used as the final step in protein purification, especially when small fractions need to be collected for analysis.

The importance of establishing a high degree of purity of the isolated protein in order to insure purity of the resulting polyclonal antibody cannot be overemphasized. This is usually carried out by electrophoresis of the native or the SDS-denatured protein on polyacrylamide gels (PAGE), with high-resolution isoelectric focusing or better still by two-dimensional PAGE in conjunction with specific stains. The recent introduction of compact, automated electro-phoresis systems, such as Pharmacia's PHAST system, allows for the routine analysis of minute samples with high resolution, in a minimum of time.

One of the major limitations in the production of highly specific antibodies is the amount of available protein present in a highly purified state. This, however, can often be achieved by one- or two-dimensional gel electropho-resis, followed by electro-blotting the protein to nitrocellulose membranes.[57] The blotted protein is then dissolved in DMSO, mixed with adjuvant and injected into a suitable animal.[58] Alternatively, it is possible to elute proteins which have been electroblotted to immobilon membranes,[59] into a detergent solution and thus avoid injecting the animal with polyacrylamide gel debris. Recent advances in protein chemistry make it possible to obtain NH_2-terminal sequence information from as little as 10 picomole samples, sufficient to allow the preparation of a synthetic oligopeptide. The latter may be coupled to a carrier protein for the production of monospecific antisera.[60]

Preparation of Hapten-Protein Conjugates

Small molecules are rendered antigenic by covalent linking to a larger molecule, usually a protein. Conjugation of haptens to proteins is usually achieved by linkages through carboxylic, alcoholic or phenolic hydroxyl groups. These and other group-directed reagents have recently been reviewed[61] and details of the methodology and applications have been described.[50] The conjugate is used to immunize animals, and the antisera thus generated contain a mixture of antibodies, some of which will recognize the hapten even when the latter is free in solution. Antibodies to hapten-protein conjugates are useful probes to study metabolite distribution and quantification using different types of immuno-assays.[50] However, they have seldom been applied to intracellular localization of their antigens.

Production of Antisera

The process of eliciting monoclonal and polyclonal antibodies has been extensively reviewed.[62-65] The choice between monoclonal and polyclonal antibodies depends on the desired application, and requires an understanding of the differences between conventional (polyclonal) and monoclonal serology. Despite their extremely high specificity and high affinity, the production of monoclonals is very costly and time-consuming. In addition, their strict site-specificity may be considered a drawback in immunolocalization studies, where the particular antigenic sites of a protein may be denatured during tissue processing or rendered unavailable due to protein-protein interactions within the cell.[66] On the other hand, the specificity of polyclonal antibodies is the consensus of hundreds of different epitopes which bind to antigenic determinants covering most of the accessible surface of the antigen. Small changes in structure of the antigen have little or no effect on polyclonal antibody binding ability to the antigen. Furthermore, since tissue processing tends to cause some denaturation changes in antigenic structure, the use of polyclonal antibodies is preferred in localization studies.

Most plant proteins, especially enzymes of secondary metabolism, are absent in animals and therefore, exhibit strong antigenicity when injected in rabbits, goats, sheep, rats or mice. Immunization of rabbits is usually carried out with 100 to 200 μg aliquots of protein. The first injection is usually emulsified with Freund's complete adjuvant, and booster injections, mixed with Freund's incomplete adjuvant, are commonly applied until an adequate antibody titer is reached. The antiserum is purified by precipitation with solid ammonium sulfate (40-50% saturation), desalted by gel filtration, then chromatographed on DEAE-cellulose (Mono Q) or protein A sepharose columns. The specificity of the eluted antibodies is determined by enzyme-linked immunosorbent assay (ELISA)[67] or radioimmunoassay (RIA).[50] Contaminating antibodies in polyclonal antisera may be absorbed with either the protein impurities obtained after acrylamide gel separation, with the carrier protein of the conjugate, or by using immunoaffinity chromatography. Specificity of of purified IgGs should be assessed by immunoblotting (Western blots) following native or SDS-PAGE.[68] It should be noted that monospecificity of antibodies is an absolute requirement for reliable localization studies, although it is often difficult to prove that an antibody does not cross-react with other proteins.[69]

IMMUNOCYTOCHEMISTRY

Immunocytochemistry involves the detection of antigens in cells and tissues through binding reactions using antibodies labeled with either fluorescent indicator (immunofluorescence), colloidal gold particles (immunogold labeling), or with enzyme markers such as alkaline phosphatase (immunohistochemistry). These techniques involve common preparative steps, although they differ with respect to the type of immunodetection systems. The latter have been discussed in several reviews and monographs.[69-73]

Tissue Fixation and Embedding

Tissue fixation is absolutely required in order to avoid extraction or displacement of the antigen. The cross-linking reagents, formaldehyde (FA) and glutaraldehyde (GA) are commonly used fixatives. While FA preserves most antigenic sites, its reversibility may adversely affect ultrastructural preservation. On the other hand, while GA offers better structural preservation, it only maintains half as many antigenic sites when compared with FA fixation. Since GA permeates more slowly than FA, both fixatives are usually combined in immunocytochemistry. Although osmium post-fixation is essential for membrane preservation, it results in an irreversible masking of epitopes, and its removal by periodic acid (or sodium metaperiodate) destroys some antigenic sites. It may appear, therefore, that some compromises have to be made in order to achieve successful results.[69,73]

Immunofluorescence

The indirect immunofluorescence technique is commonly used after the tissue is cryoprotected, embedded in glycol methacrylate (-30°C) and sectioned. The tissue is first incubated with the primary antibody, then washed to remove unbound antibodies. Specifically bound antibodies are visualized by treatment with a fluorochrome (usually fluorescein isothiocyanate, FITC) conjugated to a second antibody.[74,75]

Immunogold Labeling

Post-embedding labeling is usually preferred to the pre-embedding labeling technique, since the former employs the classical procedures of electron microscopy.[76] In this technique, fixed tissues are embedded in plastic resin and ultrathin sections are first incubated with the unlabeled primary antibody, followed by incubation with either a second antibody or protein A, both of which are conjugated with colloidal gold particles. The choice of labeling technique and embedding resins determines the processing protocols and the consequent compromises between ultrastructural preservation and retention of antigenicity.[69,73] The advantages and drawbacks of both techniques, as well as the characteristics of the different embedding resins have previously been summarized.[76]

THE MULTIENZYME SEQUENCE INVOLVED IN THE BIOSYNTHESIS OF *CHRYSOSPLENIUM* FLAVONOIDS

The *Chrysosplenium* System

Chrysosplenium americanum (Saxifragaceae) is a tiny semiaquatic weed which accumulates six, tri- to pentamethylated flavonol glucosides (Figs. 1-3). Two of these glycosides (A & B) are derived from 2'-hydroxyquercetin (Fig. 2), two others (C & D) are 6-hydroxyquercetin (quercetagetin) derivatives, and the remaining two (E & F) are 6,2'-substituted derivatives (Fig. 3).[77] Whereas the first pair of metabolites is glucosylated at the less common 2'-posi-tion, all the other compounds are 5'-glucosides. However, of the six metabolites which accumulate in *Chrysosplenium*, compounds A, E and F occur in higher amounts than C, D or E. In addition, none of the partially methylated aglycones accumulate in this tissue, although they can be synthesized enzymatically *in vitro*.[78]

Tracer experiments using [2-^{14}C] cinnamate administered to young *Chrysosplenium* shoots resulted in labeling of the six flavonoid metabolites within a 5-10 min pulse (unpublished data). None of the early methylated intermediates accumulated any label, thus suggesting efficient channeling of substrates and intermediates to final products. Furthermore, partially purified enzyme preparations catalyzed the methylation of quercetin to its 3-mono-, 3,7-di- and 3,7,4'-trimethyl derivatives (Fig. 1),[78] as well as the glucosylation of partially methylated intermediates to their corresponding glucosides.[79] These results indicated the existence of the enzyme complement involved in flavonoid biosynthesis in these shoots.

O-Methyltransferases

Six enzymes involved in the methylation sequence have been partially purified (400 to 650-fold) from *C. americanum* shoots by a combination of conventional and high performance columns.[80,81] The highly purified O-methyltransferases (OMTs) exhibited strict substrate- and position specificities for positions 3 of the parent aglycone, quercetin (3-OMT), 7 of 3-methyl-quercetin (7-OMT), 4' of 3,7-dimethylquercetin (4'-OMT) and 6 of 3,7,4'-trimethylquercetin (6-OMT).[80] In contrast with the four latter enzymes which accept flavonol aglycones as substrates, further methylation at the 2'- and 5'-

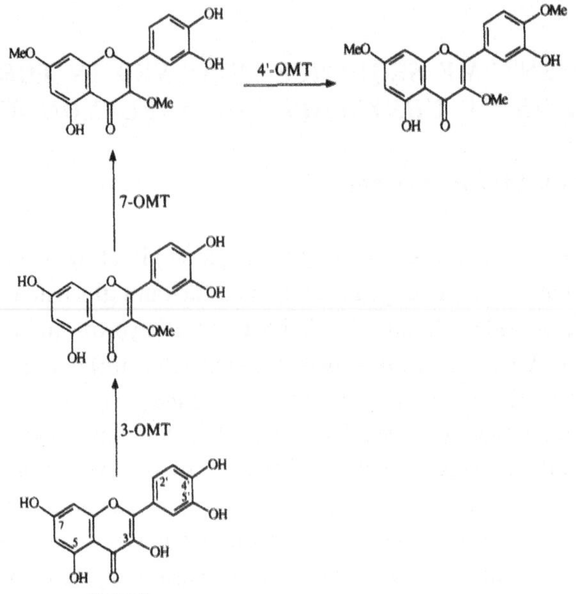

Fig. 1. Initial steps in the pathway of polymethylated flavonol glucoside biosynthesis showing the stepwise O-methylation of quercetin to 3-methyl-, 3,7-dimethyl- and 3,7,4'-trimethyl derivatives by the 3-, 7-, and 4'-O-methyltransferases (OMT), respectively. Due to page size, Figs. 1-3 are drawn in tandem to complete the whole pathway. Note that the *o*-dihydroxy β-ring for quercetin, usually numbered 3,4'dihydroxy, is shown as 4',5' in order to indicate the position of the new 2'-OH group in a *para* position to that of the 5'-OH group; the 3'/5' positions can be considered equivalent.

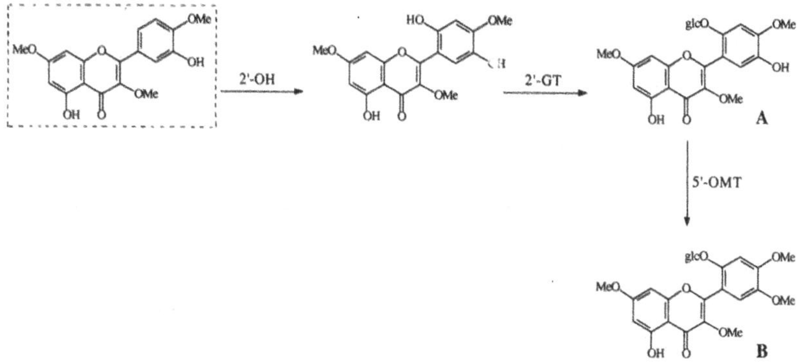

Fig. 2. The 2'-branch pathway for the biosynthesis of flavonoids A and B. The branch-point intermediate, 3,7,4'-trimethylquercetin (dotted box) is first hydroxylated at the 2'-position by a 2'-hydroxy- lase (2'-OH), glucosylated at that position by a 2'-glucosyltrans- ferase (2'-GT, compound A), and finally methylated by a 5'-O- methyltransferase (5'-OMT, compound B).

positions takes place at the glucoside level by a 2'-/5'-OMT. The latter enzyme utilizes 5,2',5'-trihydroxy-3,7,4'-tetramethoxyflavone 2'-glucoside (Fig. 2) and 5,2',5'-trihydroxy-3,6,7,4'-trimethoxyflavone 5'-glucoside (Fig. 3) as substrates, equally well, for methylation at the 5'- and 2'-positions, respectively.[81] In view of the strict position specificity of earlier enzymes in the pathway, it can be assumed that the two later methylation steps are mediated by two distinct OMTs which remain to be resolved. That the 2'-/5'-OMT is a bifunctional enzyme of the methylation sequence can not be excluded, since both methylations constitute the final steps in flavonoid biosynthesis in this tissue (Figs. 2,3). These enzymatic steps indicate a coordinated sequence of methyl transfers from S-adenosyl-L-methionine (SAM) to quercetin;→ 3-methylquercetin→ 3,7-dimethylquercetin→ 3,7,4'-trimethylquercetin. After hydroxylation of the latter at the 2'-position and its subsequent glucosylation (compound A), it is further methylated to its 5'-methyl derivative to give compound B (Fig. 2). 3,7,4'-trimethyl-quercetin is also hydroxylated at position 6, and glucosylated at 5' (compound C); further methylated at the 6-position, then glucosylated at position 5' (compound D). The latter compound, after hydroxylation of the 2'-OH, is glucosylated at the same position (compound E), then finally methylated at position 5' (Fig. 3) to give rise to the pentamethyl derivative (compound F).

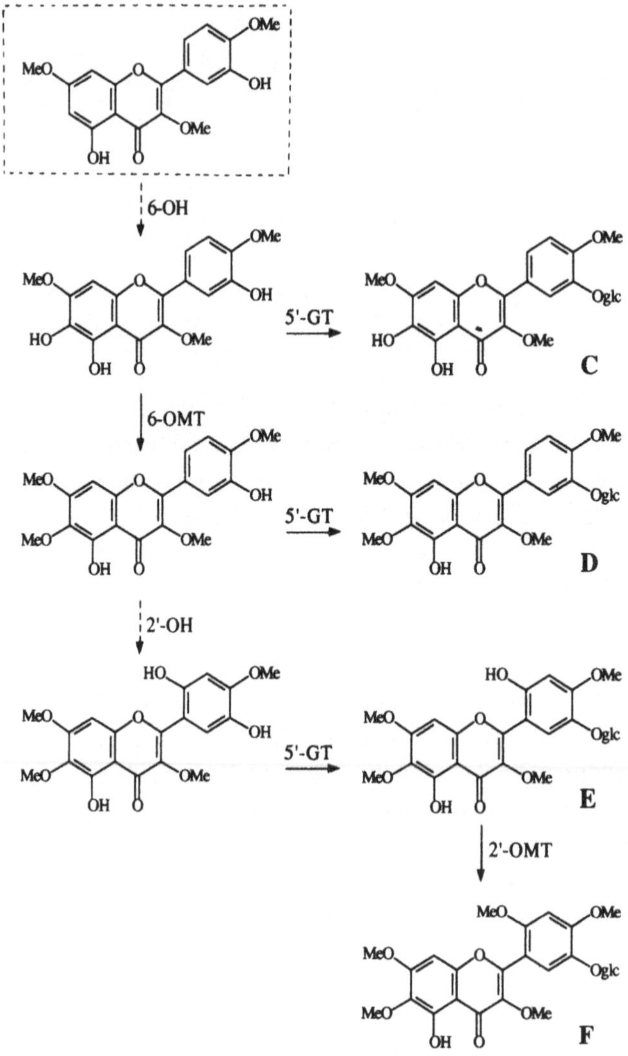

Fig. 3. The 6- and 6,2'-branch pathways for the biosynthesis of metabolites C-F. 3,7,4'-Trimethylquercetin is first hydroxylated at position 6, methylated at the same position, then hydroxylated at the 2'-position. Each of the latter intermediates is glucosylated at position 5' and accumulate as compounds C-E, respectively. Compound E is finally methylated at position 2' to give rise to the pentamethyl derivative, compound F. OH, hydroxylase; GT, *O*-glucosyltransferase; OMT, *O*-methyltransferase (see Fig. 1 for numbering system).

Kinetic analysis of *Chrysosplenium* OMTs[82,83] showed K_m values of 2-18 µM for the flavonoid substrates and 50-130 µM for the methyl donor, SAM. The K_i values for the methylated products and S-adenosyl-L-homocysteine (SAH) were 10-167 µM and 4.4-16 µM, respectively. Of all the enzymes of the methylation sequence, only the 6-OMT exhibited an absolute requirement for Mg^{2+} and was severely inhibited by its respective flavonoid substrate at concentrations close to K_m. The tendency towards decreasing Km values for the flavonoid substrates and K_i values for the methylated products along the pathway, in addition to the low K_i values for SAH, suggest that earlier enzymes in the pathway may regulate the rate of synthesis of the final products.[82,83]

O-Glucosyltransferases

Previous studies have demonstrated the existence, in *C. americanum*, of a novel ring B-specific glucosyltransferase (GT) which required two para-oriented substituents for optimal activity.[79] It catalyzed the glucosylation of 5,2',5'-trihydroxy-3,7,4'-trimethoxy-flavone (Fig. 2) and its 6-methoxy analogs (Fig. 3) to form their 2'- and 5'-glucosyl derivatives, respectively. This enzyme was later resolved into two, distinct 2'- and 5'-GT activities by affinity chromatography on UDP glucuronic acid-agarose followed by Brown 10 dye ligand columns.[84] The enzyme protein was eluted from the latter column using a linear pH-salt gradient and resulted in the separation of the 2'- and 5'-glucosylating activities at pH values of 7.8 and 7.3, respectively. The fact that each of the purified enzymes gave a single product when assayed against the respective aglycone indicates that glucosylation of the 2'- and 5'-positions of *Chrysosplenium* flavonoids is catalyzed by two distinct GTs.[84]

Further evidence for the existence of two position-specific GTs was obtained from immunological studies.[66] A murine monoclonal antibody to the partially purified flavonoid GT[79] was produced by an *in vitro* immunization technique[85] of Balb/c mice spleen cells, followed by fusion with mouse myeloma cells. A highly immunoreactive IgM-secreting clone displayed >50% inhibition of the 2'- but not the 5'-GT activity. Moreover, the native form of this enzyme was essential for antibody recognition on Western blot analysis.[66]

Except for the strict substrate and position specificity, both GTs exhibited similar properties. Their K_m values for both the flavonoid substrates and the glucosyl donor, UDPG, were 5-10 µM and 250 µM, respectively. The high K_i values for the glucosylated products (1 mM), as compared with those for the substrates and cosubstrate (20-25 µM) indicate that glucosylation of *Chrysosplenium* flavonoids is not inhibited by the products formed and is consistent with their accumulation in this tissue.[86]

The above account indicates that the enzymatic synthesis of poly-methylated flavonol glucosides in *Chrysosplenium* is regulated by the strict substrate and position specificities of the enzymes involved and their ordained sequence in the pathway. This suggests that the multienzyme system for flavonoid biosynthesis in this tissue is highly organized as a linear sequence of enzymes, similar to those described by Stafford[15] for other flavonoids. Such a multienzyme sequence permits effective channeling of substrates and products, prevents accumulation of the early methylated intermediates, and facilitates packaging and transport of the final metabolites.

IMMUNOLOCALIZATION OF POLYMETHYLATED FLAVONOL GLUCOSIDES

In contrast with the voluminous literature on immunolocalization of the enzymes of primary metabolism, storage proteins and lectins[73-75], only a few studies have dealt with plant secondary metabolites and their enzymes. These studies have been recently summarized (Table 3).[76] This section describes the immunolocalization of polymethylated flavonol glucosides in *Chryso-splenium* leaves, as well as the first and later enzymes involved in their biosynthesis (CHS and GT) in order to formulate a compartmentation model for this biosynthetic pathway.

Chrysosplenium flavonoids are mostly lipophilic compounds due to the presence of 3 to 5 methoxyl groups, although the presence of a glucosyl residue may render them partly hydrophilic. It is not unexpected, therefore, that these compounds may be excreted on the external surfaces of leaves, or may accumulate in the cellular vacuole. Three different approaches have been used to study their localization *in situ*.

Electron Microscopic Observations

Histochemical studies, using caffeine as a prefixative and staining reagent, have indicated that the walls of epidermal and mesophyll cells were impregnated with electron-dense deposits indicative of the site of flavonoid accu-mulation. In addition, various membrane profiles and associated vesicles in the periplasmic area were electron-dense and appeared fused with the plasma-lemma (Fig. 4a), suggesting the secretory nature of these cells.[28] The Golgi bodies did not acquire the caffeine stain (Fig. 4b), suggesting that they are not involved

Table 3. Immunolocalization of phenolic metabolites and their enzymes.

Metabolite/enzyme[a]	Plant tissue studied	Site of localization[b]	Reference
Polymethylated flavonol glucosides	*Chrysosplenium americanum* (leaf)	Cell wall (IF, IG)	87,88
CHS	*Fagopyrum esculentum* (hypocotyl)	ER membranes (IG)	89
	Spinacea oleracea (leaf)	Cytosol (IF, IG)	90
	Larix sp. (needle)	Epidermis, mesophyll, vascular cells (IF)	91
CHS, CHS-mRNA	*Petroselinum sp.* (leaf)	Epidermis (IH, RH)	92
PAL, CHS	*Tulipa sp.* (anther)	Tapetum layer (IH)	93
Isoflavone glucoside β-glucosidase	*Cicer arietinum* (seedling)	Parenchyma cytosol (IF)	94
Coniferyl alcohol GT	*Picea abies* (hypocotyl)	Cytosol of epid., subepid. & vascular cells (IF)	95
Coniferin β-glucosidase	*Picea abies* (hypocotyl)	Cell walls of epid.& vasc. tissue (IF)	96

[a]CHS, chalcone synthase; GT, glucosyltransferase; PAL, phenylalanine ammonia-lyase. [b]IF, immunofluorescence; IG, immunogold labeling; IH, immunohisto-chemistry; RH, RNA hybridization.

in the secretion of flavonoids. In addition, the cell wall deposits could be leached out after dipping the leaves in organic solvents for 1-2 sec intervals. HPLC analysis of these effusates indicated the recovery of the major flavonoid constituents of this tissue. These observations are consistent with the lipophilic nature of highly methylated flavonol glucosides and their association with the cell wall material, although the nature of this association remains to be determined.[28]

Anti-trimethylflavonol Glucoside Antibodies

5,2',5'-Trihydroxy-3,7,4'-trimethoxyflavone-2'-glucoside (Fig. 2), one of the major end metabolites in this tissue, was conjugated to bovine serum albumin (BSA) by the diazo reaction, and the conjugate was used to immunize rabbits.[97] Polyclonal antibodies raised against the latter conjugate were first preabsorbed with the BSA-p-aminohippuric acid complex, then characterized by counterelectrophoresis and ELISA. These antibodies were found to be specific for the 2'-glucosides of tri- (compound A) and tetra- (compound B) methoxy-flavones, with some cross reactivity (55%) observed against the pentamethoxy-flavone 5'-glucoside (compound F). However, the purified antibodies did not recognize the parent aglycone, quercetin, nor any of its partially methylated

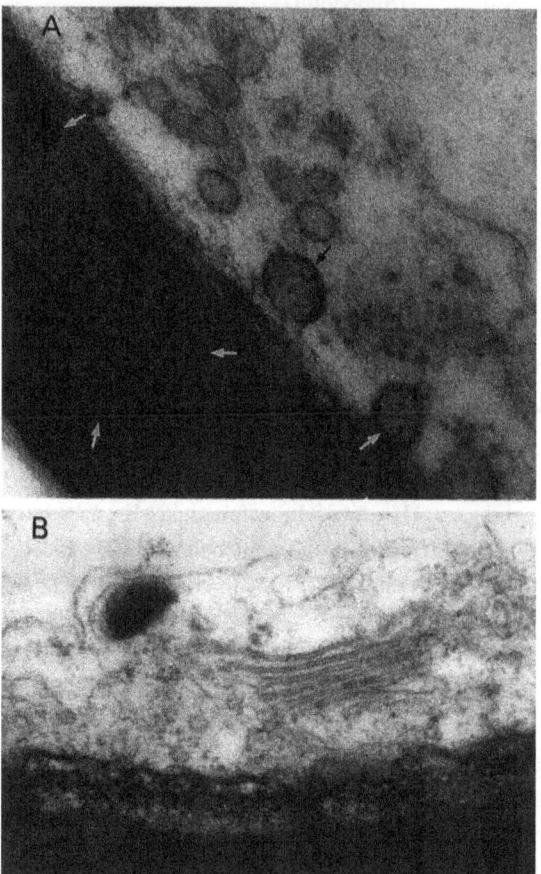

Fig. 4. Electron micrographs of caffeine-treated leaf sections showing (A) electron-dense deposits on the cell wall, and in vesicles fused with the plasmalemma, 50,000X and (B) the Golgi body did not acquire the caffeine stain, 45,000X.

(3,7,4'-tri- or 3,7,3',4'-tetramethyl) derivatives.[97] These antibodies were used for immunocytochemical localization of flavonoids in *Chrysosplenium* leaves.

Immunofluorescence Localization

An indirect immunofluorescence technique utilizing FITC-labeled goat anti-rabbit antibody was used with leaf epidermis, frozen leaf sections and mesophyll protoplasts.[87] Specific immunofluorescence labeling of flavonoids was observed mainly in the walls of epidermal and mesophyll cells, but not in the cellular vacuoles. The uneven distribution of fluorescence in cell walls suggests an irregular deposition of flavonoid glucosides, and is comparable to their clustered appearance in EM observations (Fig. 4a).[28] A small proportion of protoplasts (about 10%) exhibited weak immuno-fluorescence in the vacuoles, suggesting a minor role of this compartment in the accumulation of the trimethylated product.[87]

Immunogold Labeling

Further evidence for the site of flavonoid accumulation was obtained using the protein A-gold, postembedding technique on thin sections with transmission electron microscopy.[88] Antibody-specific labeling was observed mainly in epidermal cell walls. There was a significant amount of labeling associated with the plasmalemma, but none with other organelles such as the ER membranes, Golgi bodies or chloroplasts.[88] The corroboration of the results obtained from both immunocytochemical studies, as well as the histochemical observations, provide strong evidence for the localization of flavonoids within the cell walls of *Chrysosplenium* tissues.

IMMUNOLOCALIZATION OF CHALCONE SYNTHASE AND FLAVONOL GLUCOSYLTRANSFERASE

In spite of the fact that most of the enzymes of flavonoid biosynthesis are usually recovered in the supernatant fraction, this does not mean that they are actually 'cytosolic' since their solubilization may be due to artifacts of the isolation procedures as discussed above.[15,39,40] The enzymes involved in the multiple methylations and glucosylations of *Chrysosplenium* flavonoids have been shown to operate in a sequentially coordinated, tightly regulated manner. The components of such a multienzyme sequence are expected to be organized in a linear pathway as a membrane-associated aggregate or complex of the type proposed by Stafford.[15] In order to localize such a pathway, we have used antibodies raised against CHS and flavonol GT, the first and terminal enzymes involved in the biosynthesis of most flavonoid compounds. In *Chrysosplenium*, however, the latter enzyme is actually the second to last step in the pathway of polymethylated flavonoid glucosides.

Localization of Chalcone Synthase

Anti-CHS polyclonal antibodies were raised against the SDS-PAGE-purified CHS from primary leaves of rye.[98] These antibodies were purified by chromatography on a Mono Q column, and characterized by ELISA and immunotitration, and Western blotting. The fractions containing anti-CHS-specific IgGs were pooled and used for enzyme localization in *Chrysosplenium* leaves.

In order to preserve the structural integrity and protein antigenicity, a mild fixation technique was employed using 2% paraformaldehyde and 0.1% glutaraldehyde in 0.1M cacodylate buffer, pH 7.4 and microwave irradiation for 15 seconds.[99] Localization of CHS was carried out on Lowicryl K4M- and LR White-embedded tissues.

In Lowicryl, CHS was found associated with the cytoplasmic membranes of young epidermal cells prior to vacuole formation (Fig. 5a), and in membrane vesicles of vacuolated cells (Fig. 5b). The absence of gold label in LR White-embedded tissue (not shown) indicates that CHS is weakly associated with membranes, so that its epitopes may have been destroyed during embedding in this resin. These results in Lowicryl are similar to those obtained with buckwheat CHS,[89] in contrast to the cytoplasmic localization of CHS in spinach.[90]

Localization of Flavonol Glucosyltransferase

Since glucosylation of polymethylated flavonols is catalyzed by two GTs with similar molecular weight, a protocol was designed to copurify both enzyme activities to apparent homogeneity. This was achieved by successive chromatography of protein extracts on Sephacryl S-200, UDP-glucuronic acid-agarose, Mono P, Superose 12 and Mono Q columns.[100] This protocol resulted in a 3500-fold increase in specific activity of the GTs, and the apparently homogeneous protein was then used to raise antibodies in rabbits.

The immune serum exhibited positive reaction when tested by ELISA and was further purified by chromatography on a Mono Q column. Anti-GT-specific antibodies were characterized by inhibition studies and Western blot analysis before being used for localization studies.[100]

In situ localization of the flavonol GT was performed by applying a postembedding, immunogold labeling technique on ultrathin sections of Lowicryl K4M- and LR White-embedded leaf tissues. Specific gold labeling was observed associated with cytoplasmic membranes (Fig. 6a) and vesicle-like

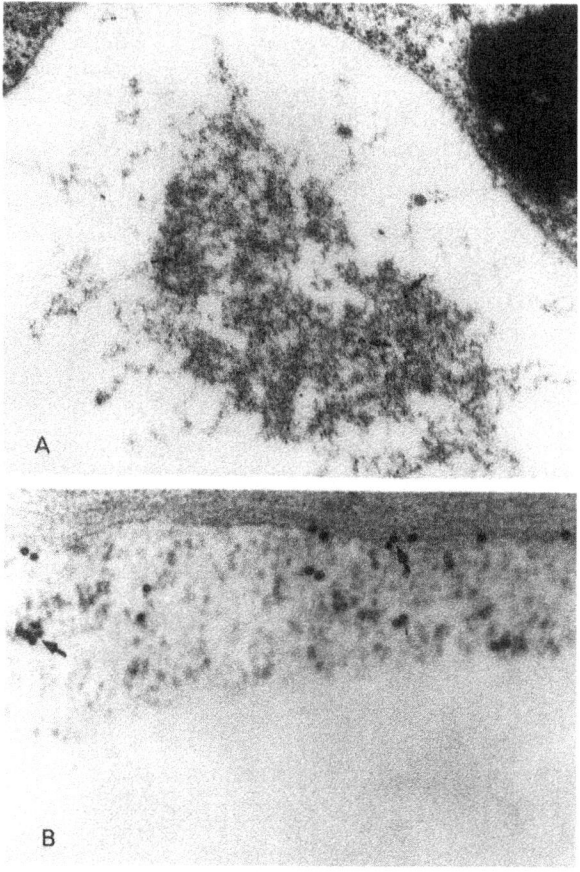

Fig. 5. Immunogold localization of chalcone synthase (CHS) in Lowicryl-embedded leaf sections using anti-CHS-specific IgGs and protein A-gold. Note that the gold labeling is associated with the cytoplasmic membranes (arrows) of vacuolating epidermal cells (A), 25,000X, and in membrane vesicles (arrows) of mature cells (B), 75,000X.

structures (Fig. 6b) in the periplasmic area.[100] These membranes and vesicles most probably arise from the ER, since there was no labeling associated with the Golgi body. It should be pointed out that embedding leaf samples in Lowicryl K4M (at -20°) results in poorly preserved membranes, although they retain a high degree of antigenicity and exhibit low background labeling. LR White-embedded tissues (at 60°) on the other hand, exhibit better membrane preservation, but protein antigenicity is greatly reduced. Since the main objective of this

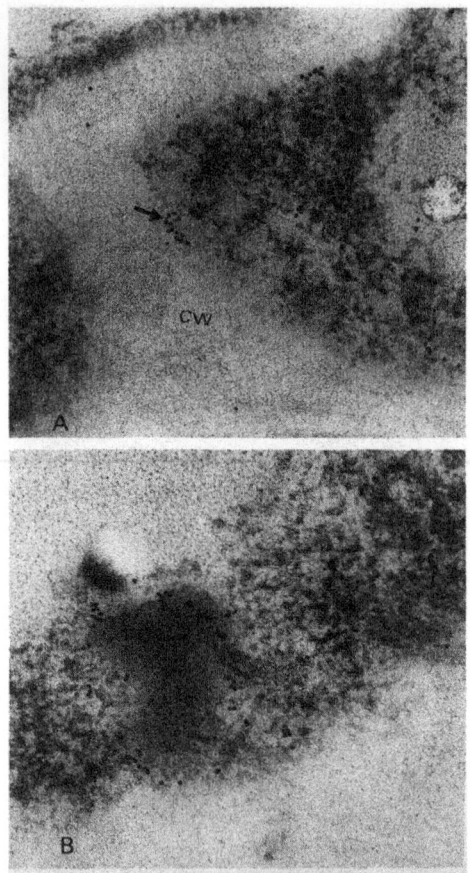

Fig. 6. Immunogold localization of flavonol glucosyltransferase (GT) in Lowicryl-embedded leaf sections using anti-GT-specific IgGs and goat anti-rabbit IgG-gold. Note gold labeling association with the cytoplasmic membranes (arrow) of the periplasmic area (A), 45,000X and in membrane vesicles (B), 50,000X. cw, cell wall; ve, vesicle.

work was to localize an enzyme of very low abundance, a compromise was made to retain antigenicity at the expense of contrasting membrane ultrastructure.[100] The density of immunogold labeling associated with different cellular compartments of Lowicryl-embedded tissues was determined by morphometric analysis (gold particles/μm^2) and amounted to 140 in periplasmic membranes, 750 in membrane vesicles, 2.5 on the cell wall, and 3.8 in the vacuole, as compared with an average of 0.7-0.9 in control sections treated with nonimmune serum. There was no labeling associated with other organelles such as mitochondria, chloroplasts or Golgi bodies. Furthermore, elimination of anti-GT IgGs by pre-absorption with a purified enzyme preparation resulted in a negligible background labeling similar to that observed with the nonimmune IgG control.[100]

These results strongly suggest that both CHS and the flavonol GT are membrane associated enzymes. The nature of this association is so loose that it results in their displacement from their *in situ* compartment and their recovery in the cytosolic fraction.

MODEL FOR THE COMPARTMENTATION OF FLAVONOID ENZYMES AND THEIR METABOLITES

Compartmentation of Flavonoid Enzymes

The concept that flavonoid enzymes are freely soluble in the cytoplasm is no longer valid, as is also the case with the easily solubilized glycolytic [101] and Calvin cycle[102] enzymes. Instead, it has been postulated that biosynthetic pathways are organized as multienzyme aggregates or complexes that permit enzyme-enzyme interactions, prevent diffusion of intermediates, and allow their efficient channeling. Examples of channeling in multienzyme complexes have been described in earlier[10,11] and recent[14] literature. More recently, Stafford[15] has presented an elegant discussion of the organization of flavonoid enzymes as membrane-associated linear sequences (see also Hrazdina, this volume). The latter concept would allow endogenous regulators to control the entire biosynthetic pathway at the level of the multienzyme complex rather than the initial enzyme, chalcone synthase.[15]

The elaborate network of endoplasmic reticulum is known to play an important role in the synthesis of proteins on associated ribosomes, and in the packaging of metabolic products.[103] This organelle would be an appropriate compartment for a multienzyme aggregate involved in flavonoid synthesis, and for the transport of the final products towards the different sites of accumulation.

Most of the models depicting the organization of the enzymes involved in flavonoid biosynthesis have dealt with simple metabolites which may require one or two enzymatic steps after aglycone formation.[14,15] The enzymatic synthesis of polymethylated flavonol glucosides in *Chrysosplenium*, on the other hand, is catalyzed by ten substrate-specific and position-oriented enzymes, starting with the aglycone quercetin. These consist of six OMTs, two hydroxylases and two GTs (Figs. 1-3). It is reasonable to assume, therefore, that the ER membrane would be a suitable site for such an elaborate multienzyme system, if it is to operate efficiently as a unidirectional linear sequence. The fact that neither of the early methylated intermediates nor any of the highly methylated aglycones accumulate in this tissue indicates efficient channeling and strongly supports the above concept.

In 1987, we proposed a model for the compartmentation of the later enzymes involved in the biosynthesis of polymethylated flavonol glucosides in *Chrysosplenium*.[104] This model which is based on isotopic, enzymic and kinetic evidence, and further corroborated with immunocytochemical observations, assumes a linear sequence of the early enzymes involved in flavonol methylation (3-, 7-, and 4'-OMTs; Fig. 1) along the cytoplasmic side of the endoplasmic reticulum (Fig. 7). The product of these methylation steps, 3,7,4'-trimethylquercetin, is a branch-point intermediate for two alternative pathways. One of these (Fig. 2) leads to the formation of tri- and tetramethoxyflavone 2'-glucosides (compounds A & B) and the other (Fig. 3), gives rise to tetra- and pentamethoxyflavone 5'-glucosides (compounds C-F). Since the 2'- and 6-hydroxylases which act on the branch-point intermediate are expected to be firmly bound to the membrane, it is reasonable to assume that the 6-OMT would be located in the vicinity of the 6-hydroxylase (Fig. 7). This would permit direct transfer of the hydroxylated intermediate for subsequent methylation at the 6-position. It should also be noted that, of all the OMTs involved in the methylation sequence, the 6-OMT is tightly regulated by Mg^{2+} and its substrate concentration. Since the end metabolites in this pathway accumulate as flavonoid glucosides, it is expected that the 2'- and 5'-GTs would be accessible to their corresponding substrates. However, the fact that the remaining (2' & 5') methylations in the two-branch pathway take place at the level of glucosylated substrates, implies their requirement for a different microenvironment. It seems logical, therefore, that glucosylation of intermediates at the 5'- and 2'-positions and their subsequent methylation at the respective 2'- and 5'-positions take place on the other (lumen) side of the membrane (Fig. 7). Furthermore, the fact that the two later metabolites (E and F) are 6,2'-substituted derivatives, suggests the association of another 2'-hydroxylase with this branch of the pathway, since

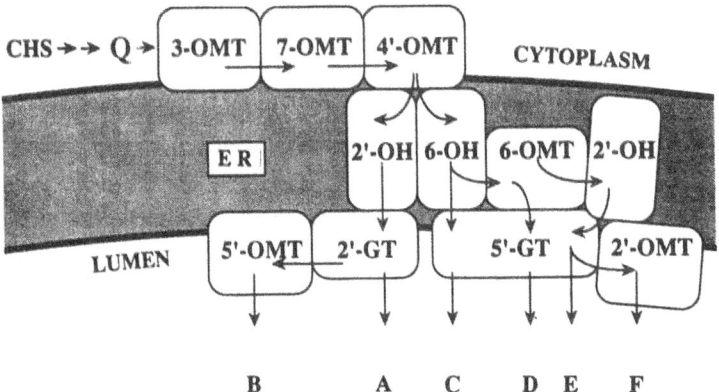

Fig. 7. Compartmentation model for the enzymes involved in the later steps of polymethylated flavonol glucoside biosynthesis (see text for details). A-F, final metabolites; CHS, chalcone synthase; ER, endoplasmic reticulum; OH, hydroxylase; GT, O-glucosyltransferase; OMT, O-methyltransferase; Q, quercetin.

6-substituted intermediates would not have access to the analogous enzyme in the former branch (Fig. 7). This model would permit effective discrimination among the 2'-, 6-, and 6,2'-substituted intermediates and direct access to their corresponding enzymes. It also allows the regulation of synthesis of each pair of final metabolites and their eventual sequestering in vesicles derived from the membrane. In addition, such a model postulates that the five enzymes involved in the formation of quercetin, starting with chalcone synthase and including flavonol synthase, form the proximal segment of this linear pathway (Fig. 7). It can also be envisaged that the latter sequence may be compartmented in a separate aggregate on an adjacent ER membrane.

It is evident, therefore, that regulation of the biosynthesis of polymethylated flavonol glucosides in *Chrysosplenium* may be achieved by the spatial organization of the component enzymes in this aggregate. The association of CHS with the ER membranes (Fig. 5) and of flavonol GT with these membranes and associated vesicles (Fig. 6), as demonstrated by immuno-gold labeling, strongly supports the above model which resembles those recently described.[14,15] The association of both of these enzymes with either the membranes or vesicles is considered to be sufficiently weak, thus resulting in their solubilization during isolation procedures. However, immunolocalization of the first and last enzymes in the methylation sequence should provide further evidence for the compartmentation of this elaborate pathway.

Compartmentation of the Flavonoid Metabolites

The proposed working model for enzyme compartmentation described above allows for sequestering and accumulation of the highly lipophilic metabolites in the lumen of the ER membranes (Fig. 7). Metabolite accumulation would result in the swelling of certain regions of these membranes and the formation of membrane vesicles. These vesicles would eventually move towards and fuse with the membrane at the cell surface (plasmalemma) releasing their flavonoid content within (or outside) the cell wall by exocytosis. In fact, electron microscopic studies of flavonoid accumulation in *Chrysosplenium* leaf tissue (Fig. 4a) revealed the presence of electron-dense deposits in various membrane profiles and associated vesicles that were fused with the plasmalemma.[28] Electron-dense deposits were also associated with the cell wall. These could be readily leached with organic solvents and identified as the major flavonoid metabolites of this tissue. In addition, by using an antibody against one of the major partially methylated flavonoid glucosides, it was possible to localize the final metabolites within the cell walls by immunofluorescence[87] and immunogold labeling.[88] It is conceivable that similar events take place in other systems where metabolites are secreted near or on the outer surface of cell walls.[18-21] However, none of these localization studies gave any indication of the participation of the Golgi bodies in the synthesis or transport of flavonoids.[18,28,88]

Except for the microscopic observations on the vacuolar localization of proanthocyanidins,[33-35] and the presumed packaging of anthocyanins in cyanoplasts,[105] there is a lack of documentation of the events leading to metabolite accumulation in the cellular vacuole.

For example, the monoterpenoid indole alkaloid, vindoline, in *Catharanthus roseus* is believed to accumulate in cellular vacuoles. Using a specific antivindoline antibody, indirect immunofluorecence and immunogold labeling techniques, we have recently demonstrated its localization in the vacuoles of mesophyll protoplasts.[106] In fact, immunogold localization of vindoline could only be achieved using cryotechniques.[107,108] Immunogold labeling was observed associated with numerous membrane vesicles which seemed to coalesce before fusion with the central vacuole (Fig. 8). These vesicles presumably represent the vindoline pool en route to the site of accumulation. However, not

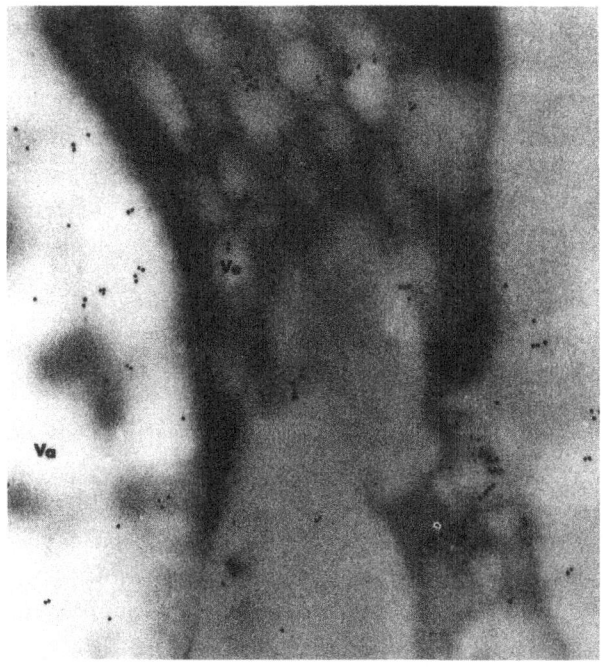

Fig. 8. Immunogold labeling of vindoline in the central vacuole (va) and membrane vesicles (ve, arrow heads) of cryofixed, cryosubstituted mesophyll protoplasts of *Catharanthus roseus*, 45,000X.

all the vesicles may necessarily be implicated in this process, since such an endomembrane system is known to be involved in the secretion of other metabolites or of enzymes associated with detoxification or degradative processes.[109] Although the localization of the later enzymes in vindoline biosynthesis remain to be determined, the observations on metabolite accumulation (Fig. 8) provide the first evidence for the involvement of membrane vesicles in the transport of secondary metabolites towards the central vacuole.[106]

CONCLUSION

The variety of substitution patterns of *Chrysosplenium* flavonoids and the organization of the enzymes involved in their biosynthesis, as well as the secretory nature of this tissue, provide a model system and, indeed, a 'treasure chest' for further biochemical and localization studies. Immunolocalization of the enzymes involved in the early and later stages of flavonoid biosynthesis indicates their association with cytoplasmic membranes and membrane vesicles, and suggests that the multienzyme system in this tissue operates as a linear uni-directional sequence. Immunolocalization of the final metabolites corroborates electron microscopic observations of their accumulation in membrane vesicles prior to secretion within the cell wall. The complementarity and sensitivity of these techniques can be successfully applied to the localization of other secondary metabolites and their biosynthetic enzymes.

Immunocytochemical techniques provide a powerful tool for the precise and reproducible localization of most hapten and protein antigens. In spite of the paucity of localization studies involving plant secondary metabolism, it is hoped that the next few years will witness a proliferation of these studies. The growing interest in modification of secondary metabolism by genetic manipula-tion requires a better understanding of the compartmentation of both biosynthetic enzymes and their catalytic products, as well as the verification of proposed pathways where different metabolic steps take place in different subcellular compartments.

Some inconsistencies may arise as to the precise site of localization of a given protein antigen in different species. Such problems can be alleviated by encouraging the exchange of antibodies raised in different laboratories and their use with as many tissue systems as possible. It is also possible that compart-mentation patterns of enzymes or metabolites may be different, not only in different species, but also with age, type of organ and tissue distribution within the plant.

The importance of comparing various fixation methods and embedding resins of different polarities cannot be overemphasized. This approach often permits better interpretation of the results, since different resins affect cellular ultrastructure and protein antigenicity to various degrees. Finally, it is hoped that further advances in immunocytochemical technology will bring about improved fixation methods and better embedding resins with the dual properties of good ultrastructural preservation and retention of antigenicity.

ACKNOWLEDGEMENTS

I wish to thank my past and present collaborators for their valuable contributions to this work; Catherine Avezard and Jacynthe Seguin for their excellent technical assistance; Professor G. Weissenböck for the anti-chalcone synthase antibody; Dr. J. Grandmaison for enlightening discussions and Dr. E. Bleichert for careful reading of the manuscript. The work cited from the author's laboratory was supported by grants from the Natural Sciences and Engineering Research Council of Canada and the Department of Higher Education, Government of Québec.

REFERENCES

1. HARBORNE, J.B., ed. 1988. The Flavonoids: Advances in Research Since 1980. Chapman & Hall, London.

2. BARRON, D., VARIN, L., IBRAHIM, R.K., HARBORNE, J.B., WILLIAMS, A.C. 1988. Sulphated flavonoids—an update. Phytochemistry 27:2375-2395.

3. HARBORNE, J.B. 1988. Phenolics in the environment—an overview of recent progress. Bull. Liaison Groupe Polyphenols 3-11.

4. HARBORNE, J.B. 1988. Introduction to Ecological Biochemistry, Third edition. Academic Press, London.

5. DIXON, R.A. 1986. The phytoalexin response: elicitation, signalling and the control of host gene expression. Biol. Rev. 61:239-291.

6. SMITH, D.A., BANKS, S.W. 1986. Biosynthesis, elicitation and biological activity of isoflavonoid phytoalexins. Phytochemistry 25:979-995.

7. LONG, S.R. 1989. Rhizobium-legume nodulation: life together in the under-ground. Cell 56:203-214.

8. SERERE, P.A., MOSBACH, K. 1974. Metabolic compartmentation: symbiotic, organellar, multienzymic and microenvironmental. Annu. Rev. Microbiol. 28:61-83.

9. LUCKNER, D., DIETTRICH, B., LERBS, W. 1980. Cellular compartmentation and channeling of secondary metabolism in micro-organisms and higher plants. In: Progress in Phytochemistry, Vol. 6. (R. Reinhold, J.B. Harborne, T. Swain, eds.) Pergamon Press, Oxford, pp. 103-142.

10. STAFFORD, H.A. 1974. The metabolism of aromatic compounds. Annu. Rev. Plant Physiol. 25:459-486.

11. STAFFORD, H.A. 1981. Compartmentation in natural product biosynthesis by multienzyme complexes. In: The Biochemistry of Plants, Vol. 7 (E.E. Conn, ed.) Academic Press, New York, pp. 117-137.

12. WOMBACHER, H. 1983. Molecular compartmentation by enzyme cluster formation. Mol. Cell. Biochem. 56:155-164.

13. ROOS, W., LUCKNER, M. 1986. The spatial organization of secondary metabolism in microbial and plant cells. In: Cell Metabolism—Growth and Environment, Vol. 1. (T. Subramanian, ed.) CRC Press, Boca Raton, FL, pp. 46-73.

14. HRAZDINA, G., WAGNER, G.J. 1985. Compartmentation of plant phenolic compounds: sites of synthesis and accumulation. In: Annu. Proc. Phytochemical Society of Europe, Vol. 25 (C.F. Van Sumere, P.J. Lea, eds.). Clarendon Press, Oxford, pp. 119-129.

15. STAFFORD, H.A. 1990. Flavonoid Metabolism. CRC Press, Boca Raton, FL.

16. WAGNER, G.J. 1982. Compartmentation in plant cells: the role of the vacuole. Rec. Adv. Phytochem. 16:1-46.

17. MATILE, Ph. 1984. Das toxische Kompartiment der Pflanzenzelle. Naturwissen-schaften 71:18-24.

18. CHARRIÉRE-LADREIX, Y. 1975. La sécrétion lipophile des bourgeons d'*Aesculus hippocastaneum*: modifications structurales des trichomes au cours du processus glandulaire. J. Microsc. 24:75-90.

19. CHARRIÉRE-LADREIX, Y. 1979. Intracellular distribution of flavonoids in glandular cells. In: Regulation in Secondary Product and Plant Hormone Metabolism. (M. Luckner, K. Schreiber eds.) FEBS Symp. Vol. 55, Pergamon Press, Oxford, pp. 101-109.

20. WOLLENWEBER, E. 1984. The systematic implication of flavonoids secreted by plants. In: Biology and Chemistry of Plant Trichomes. (E. Rodriguez, P. Healy, I. Metha, eds.) Plenum Press, New York, pp. 53-69.

21. WOLLENWEBER, E. 1989. Exkret-Flavonoide bei Bittenpflanzen und Farnen. Naturwissenschaften 76:458-463.

22. ROLFE, B.G., GRESSHOFF, P.M. 1988. Genetic analysis of legume nodule initiation. Annu. Rev. Plant Physiol. 39:297-319.

23. MOSKOWITZ, A.H., HRAZDINA, G. 1981. Vacuolar contents of fruit subepidermal cells from *Vitis sp.* Plant Physiol. 68:686-692.

24. HRAZDINA, G., MARX, G.A., HOCH, H.C. 1982. Distribution of secondary plant metabolites and their biosynthetic enzymes in pea leaves. Plant Physiol. 70:745-748.

25. BEGGS, C.J., KUHN, K., BÖCKER, R., WELLMANN, E. 1987. Phytochrome-induced flavonoid biosynthesis in mustard *Sinapis alba* cotyledons: enzymic control and differential regulation of anthocyanin

and quercetin formation. Planta 172:121-126.

26. SCHNABL, H., WEISSENBÖCK, G., SCHARF, H. 1988. *In vitro* microspectro-photometric characterization of flavonol glycosides in *Vicia faba* guard and epidermal cells. J. Exp. Bot. 37:61-72.

27. STRACK, D., HEILEMANN, J., KLINKOTT, J.S. 1988. Cell wall-bound phenolics from spruce needles. Z. Naturforsch. 43c:37-41.

28. CHAREST, P.M., BRISSON, L., IBRAHIM, R.K. 1986. Ultrastructural features of flavonoid accumulation in leaf cells of *Chrysosplenium americanum*. Protoplasma 134:95-101.

29. MATERN, U., HELLER, W., HIMMELSPACH, K. 1983. Configurational changes of apigenin 7-O-(6"-O-malonyl) glucoside, a vacuolar pigment from parsley, with solvent composition and proton concentration. Eur. J. Biochem. 133:439-448.

30. STRACK, D., MEURER, B., WEISSENBÖCK, G. 1982. Tissue-specific kinetics of flavonoid accumulation in primary leaves of rye. Z. Pflanzenphysiol. 108:131-141.

31. KNOGGE, W., WEISSENBÖCK, G. 1986. Tissue distribution of secondary phenolic biosynthesis in developing primary leaves of *Avena sativa*. Planta 167:196-205.

32. COSIO, E.G., WEISSENBÖCK, G., McCLURE, J.W. 1985. Acifluorfen-induced isoflavonoids and enzymes of their biosynthesis in mature soybean leaves. Plant Physiol. 78:14-19.

33. PARHAM, R.A., KAUSTINEN, H.M. 1977. On the site of tannin synthesis in plant cells. Bot. Gaz. 138:465-467.

34. STAFFORD, H.A., LESTER, H.H., WEIDER, R. 1987. Histochemical assay of proanthocyanidin heterogeneity in cell cultures. Plant Sci. 52:99-104.

35. ZOBEL, A.M. 1986. Localization of phenolic compounds in tannin-secreting cells from *Sambucus racemosa* shoots. Ann. Bot. 57:801-810.

36. HRAZDINA, G., WAGNER, G.J., SIEGELMAN, H.W. 1978. Subcellular localization of enzymes of anthocyanin synthesis in protoplasts. Phytochemistry 17:53-56.

37. HRAZDINA, G., ALSCHER-HERMAN, R., KISH, V.M. 1980. Subcellular localization of flavonoid synthesizing enzymes in *Pisum*, *Phaseolus*, *Brassica* and *Spinacea* cultivars. Phytochemistry 19:1355-1359.

38. JONSSON, L.M.V., DONKER-KOOPMAN, W.E., UITSLAGER, P., SCHRAM, A.W. 1983. Subcellular localization of anthocyanin methyltransferases in flowers of *Petunia hybrida*. Plant Physiol.

72:187-290.

39.   WAGNER, G.J., HRAZDINA, G. 1984. Endoplasmic reticulum as a site
       of phenylpropanoid and flavonoid metabolism in *Hippeastrum*. Plant
       Physiol. 74:901-906.

40.   HRAZDINA, G., WAGNER, G.J. 1985. Metabolic pathways as enzyme
       complexes: evidence for the synthesis of phenylpropanoids and
       flavonoids on membrane-associated enzyme complexes. Arch.
       Biochem. Biophys. 237:88-100.

41.   MAYER, A.M. 1987. Polyphenol oxidases in plants—recent progress.
       Phytochemistry 26:11-20.

42.   SCHLOSS, P., WALTER, C., MÄDER, M. 1987. Basic peroxidases in
       isolated vacuoles of *Nicotiana tabacum*. Planta 170:225-229.

43.   SCHRAM, A.W., AL, E.J.M., DOUMA, N., JONSSON, L.M.V., DE
       VLAMING, P., KOOI, A., BENNINK, G.J.H. 1982. Cell wall
       localization of dihydroflavonol glucoside β-glucosidase in flowers of
       Petunia hybrida. Planta 155:162-165.

44.   CHARRIÉRE-LADREIX, Y., TISSUT, M. 1981. Foliar flavonoid distri-
       bution during *Spinacea* chloroplast isolation. Planta 151:309-313.

45.   THRESH, K., IBRAHIM, R.K. 1985. Are spinach chloroplasts involved
       in flavonoid *O*-methylation? Z. Naturforsch. 40c:331-335.

46.   CHARRIÉRE-LADREIX, Y., DOUCE, R., JOYARD, J. 1981. Charac-
       terization of *O*-methyltransferase activities associated with spinach
       chloroplast fractions. FEBS Lett. 133:55-58.

47.   IBRAHIM, R.K., KHOURI, H.E., BRISSON, L., LATCHINIAN, L.,
       BARRON, D., VARIN, L. 1986. Glycosylation of phenolic
       compounds. Bull. Liaison Groupe Polyphénols 13:3-14.

48.   KNOX, R.B. 1982. Immunology and the study of plants. In: Antibody as
       a Tool. (J.J. Marchalonis, G.W. Warr, eds.) John Wiley & Sons,
       New York, pp. 293-346.

49.   BARLOW, D.J., EDWARDS, M.S., THORNTON, J.M. 1986.
       Continuous and discontinuous protein antigenic determinants. Nature
       322:747-748.

50.   ROBINS, R.J. 1986. The measurement of low-molecular-weight
       nonimmuno-genic compounds by immunoassay. In: Immunology in
       Plant Sciences. (H.F. Linskins, J.F. Jackson, eds.) Springer-Verlag,
       Berlin, pp. 86-141.

51.   JANSON, J.C., RYDÉN, L. 1989. Protein Purification: Principles, High
       Resolution Methods and Applications. VCH Publishers, New York.

52. DEUTSCHER, M.P., ed. 1990. Guide to Protein Purification. Methods Enzymol. 182:1-818.

53. LATCHINIAN, L., KHOURI, H.E., IBRAHIM, R.K. 1987. Fast protein affinity chromatography of two flavonol O-glucosyl-transferases J. Chromatogr. 388:235-242.

54. FORNSTEDT, N. 1984. Affinity chromatographic studies of antigen—antibody dissociation. FEBS Lett. 177:195-199.

55. KHOURI, H.E., IBRAHIM, R.K. 1987. Resolution of five position-specific flavonoid O-methyltransferases by fast protein liquid chromatofocusing. J. Chromatogr. 407:291-297.

56. KRAGER, B.L. 1989. High performance capillary electrophoresis. Nature 339:641-642.

57. CHILES, T.C., O'BRIEN, T.W., KILBERG, M.S. 1987. Production of monospecific antibodies to a low-abundance hepatic membrane protein using nitrocellulose immobilized protein as antigen. Anal. Biochem. 163:136-142.

58. KUNSDEN, K.A. 1985. Proteins transferred to nitrocellulose for use as immunogens. Anal. Biochem. 147:285-288.

59. SZEWCZYK, B., SUMMERS, D.F. 1988. Preparative elution of proteins blotted to immobilon membranes. Anal. Biochem. 168:48-53.

60. AEBERSOLD, S.H., LEAVITT, J., SAAVEDRA, R.A., HOOD, L.E. 1987. Amino acid sequence analysis of proteins separated by one-dimensional and two-dimensional gel electrophoresis after *in situ* protease digestion on nitrocellulose. Proc. Natl. Acad. Sci. USA 84:6970-6974.

61. ERLANGER, B.F. 1980. The preparation of antigen hapten-carrier conjugates: a survey. Methods Enzymol. 70:85-103.

62. CAMPBELL, A.M. 1984. Monoclonal antibody technology. In: Laboratory Techniques in Biochemistry and Molecular Biology. (R.H. Burdon, P.H. van Knippenberg, eds.) Vol. 13. Elsevier, Amsterdam, pp. 1-259.

63. GALFRÉ, G., BUTCHER, G.W. 1986. Making antibodies. In: Immunology in Plant Science. (T.L. Wang, ed.) Cambridge University Press, Cambridge, pp. 1-25.

64. GODING, J.W. 1986. Monoclonal Antibodies: Principles and Practice. Academic Press, New York.

65. BREWIN, N.J., DAVIES, D.D., ROBINS, R.J. 1987. Immuno-cytochemistry for enzymology. In: The Biochemistry of Plants. (D.D. Davies, ed.) Vol. 13. Academic Press, New York, pp. 1-31.

66.   LATCHINIAN, L., IBRAHIM, R.K. 1989. Characterization of a monoclonal antibody specific to a flavonol 2'-O-glucosyltransferase. Biochem. Cell Biol. 67:210-213.

67.   ENGVALL, E. 1980. Enzyme immunoassay—ELISA and EMIT. Methods Enzymol. 70:419-438.

68.   TOWBIN, H., STAEHELIN, T., GORDON, J. 1979. Electrophoretic transfer of proteins from polyacrylamide gels to nitrocellulose sheets: procedure and some applications. Proc. Natl. Acad. Sci. USA 76:4350-4354.

69.   LARSSON, L.I. 1988. Immunocytochemistry: Theory and Practice. CRC Press, Boca Raton, FL.

70.   MARCHALONIS, J.J., WARR, G.W. 1982. Antibody as a Tool. John Wiley & Sons, New York.

71.   LINSKINS, H.F., JACKSON, J.F. 1986. Immunology in Plant Sciences. Springer-Verlag, Berlin.

72.   WANG, T.L. 1986. Immunology in Plant Science. Cambridge University Press, Cambridge.

73.   HERMAN, E.M. 1988. Immunocytochemical localization of macromolecules with the electron microscope. Annu. Rev. Plant Physiol. Plant Mol. Biol. 39:139-155.

74.   JEFFREE, C.E., YEOMAN, M.M., KILPATRICK, D.C. 1982. Immunofluorescence studies on plant cells. Int. Rev. Cytol. 80:231-265.

75.   PERROT-RECHENMANN, G., GADAL, P. 1986. Enzyme immunocytochemistry. In: Immunology in Plant Science. (T.L. Wang, ed.) Cambridge University Press, Cambridge, pp. 59-88.

76.   IBRAHIM, R.K. 1990. Immunocytochemical localization of plant secondary metabolites and the enzymes involved in their biosynthesis. Phytochem. Anal. 1:49-59.

77.   COLLINS, F.W., DE LUCA, V., IBRAHIM, R.K., VOIRIN, B., JAY, M. 1981. Polymethylated flavonol glucosides of *Chrysosplenium americanum*. Identification and enzymatic synthesis. Z. Naturforsch. 36c:730-736.

78.   DE LUCA, V., IBRAHIM, R.K. 1982. Characterization of three distinct O-methyltransferases from *Chrysosplenium americanum*. Phytochemistry 21:1537-1540.

79.   BAJAJ, K.L., DE LUCA, V., KHOURI, H.E., IBRAHIM, R.K. 1983. Partial purification and properties of flavonol ring B-O-glucosyltransferase from *Chrysosplenium americanum*. Plant Physiol. 72:891-896.

80. DE LUCA, V., IBRAHIM, R.K. 1985. Enzymatic synthesis of poly-methylated flavonols in *Chrysosplenium americanum*. I. Partial purification and some properties of *S*-adenosyl-*L*-methionine: flavonol 3-, 6-, 7-, and 4'-*O*-methyl-transferases. Arch. Biochem. Biophys. 238:596-605.

81. KHOURI, H.E., ISHIKURA, N., IBRAHIM, R.K. 1986. FPLC purification and some properties of a partially methylated flavonol glucoside 2'-/5'-*O*-methyl-transferase from *Chrysosplenium americanum*. Phytochemistry 25:2475-2479.

82. DE LUCA, V., IBRAHIM, R.K. 1985. Enzymatic synthesis of polymethylated flavonols in *Chrysosplenium americanum*. II. Substrate interaction and product inhibition studies of flavonol 3-, 6- and 4'-*O*-methyltransferases. Arch. Biochem. Biophys. 238:606-618.

83. KHOURI, H.E., DE LUCA, V., IBRAHIM, R.K. 1988. Enzymatic synthesis of polymethylated flavonols in *Chrysosplenium americanum*. III. Purification and kinetic analysis of *S*-adenosyl-*L*-methionine: 3-methylquercetin 7-*O*-methyltransferase. Arch. Biochem. Biophys. 265:1-7.

84. LATCHINIAN, L., KHOURI, H.E., IBRAHIM, R.K. 1987. Fast protein affinity chromatography of two flavonoid glucosyltransferases. J. Chromatogr. 388:235-242.

85. BOSS, B. 1986. An improved *in vitro* immunization procedure for the production of monoclonal antibodies. Methods Enzymol. 121:27-33.

86. KHOURI, H.E., IBRAHIM, R.K. 1984. Kinetic mechanism of a flavonol ring B *O*-glucosyltransferase from *Chrysosplenium americanum*. Eur. J. Biochem. 142:559-605.

87. BRISSON, L., VACHA, W., IBRAHIM, R.K. 1986. Localization of partially methylated flavonol glucosides in *Chrysospleñium americanum*. II. Immunofluorescence. Plant Sci. 44:175-181.

88. MARCHAND, L., CHAREST, P.M., IBRAHIM, R.K. 1987. Localization of partially methylated flavonol glucosides in *Chrysosplenium americanum*. III. Immunogold labeling. J. Plant Physiol. 131:339-348.

89. HRAZDINA, G., ZOBEL, A.M., HOCH, H.C. 1987. Biochemical, immunological and immunocytochemical evidence for the association of chalcone synthase with endoplasmic reticulum membranes. Proc. Natl. Acad. Sci. USA 84:8966-8970.

90. BEERHUES, L., ROBENECK, H., WIERMANN, R. 1988. Chalcone synthase from spinach (*Spinacia oleracea* L.). II. Immunofluorescence

and immuno-gold labeling. Planta 173:544-553.

91. LEMBACH, A., BEERHUES, L., WIERMANN, R. 1989. *In situ*
     localization of chalcone synthase in *Larix* needles by indirect
     immunofluorescence. Protoplasma 153:58-61.

92. SCHMELZER, E., JAHNEN, W., HAHLBROCK, K. 1988. *In situ*
     localization of light-induced chalcone synthase mRNA, chalcone
     synthase and flavonoid end products in epidermal cells of parsley
     leaves. Proc. Natl. Acad. Sci. USA 85:2989-2993.

93. KEHREL, B., WIERMANN, R. 1985. Immunochemical localization of
     phenylalanine ammonia-lyase and chalcone synthase in anthers.
     Planta 163:183-190.

94. BURMEISTER,G., HÖSEL, W. 1981. Immunohistochemical localiza-
     tion of β-glucosidases in lignin and isoflavone metabolism in *Cicer
     arietinum* L. seedlings. Planta 152:578-586.

95. SCHMID, G., HAMMER, D.K., RITTERBUSCH, A., GRISEBACH,
     H. 1982. Appearance and immunohistochemical localization of UDP-
     glucose: coniferyl alcohol glucosyltransferase in spruce (*Picea abies*)
     seedlings. Planta 156:207-212.

96. MARCINOWSKI, S., FALK, H., HAMMER, D.K., HOYER, B.,
     GRISE-BACH, H. 1979. Appearance and localization a β-glucosidase
     hydrolyzing coniferin in spruce *(Picea abies)* seedlings. Planta
     144:161-165.

97. LAMOUREUX, S., VACHA, W., IBRAHIM, R.K. 1986. Localization
     of partially methylated flavonol glucosides in *Chrysosplenium
     americanum*. I. Preparation and some properties of an anti-
     trimethylflavonol glucoside antibody. Plant Sci. 44:169-173.

98. PETERS, A., SCHNEIDER-POETSCH, H.A.W., SCHWARZ, H.,
     WEISSENBÖCK, G. 1988. Biochemical and immunological
     characterization of chalcone synthase from rye leaves. J. Plant
     Physiol. 133:178-182.

99. BENHAMOU, N., NOEL, S., GRENIER, J., ASSELIN, A. 1991.
     Microwave energy fixation of plant tissue: an alternative approach
     that provides excellent preservation of ultrastructure and antigenicity.
     J. Electon Microsc. Technique 17:81-94.

100. LATCHINIAN-SADEK, L., IBRAHIM, R.K. 1991. Flavonol ring B-
      specific *O*-glucosyltransferases: Purification, production of polyclonal
      antibodies and immunolocalization. Arch. Biochem. Biophys.
      289:230-236.

101. MOOREHEAD, G.B., PLAXTON, W.C. 1988. Binding of glycolytic

101. MOOREHEAD, G.B., PLAXTON, W.C. 1988. Binding of glycolytic enzymes to a particulate fraction in carrot and sugar beet storage roots. Plant Physiol. 86:348-351.
102. GONTERO, B., CARDENAS, M.L., RICARD, J. 1988. A funcional five-enzyme complex of chloroplasts involved in the Calvin cycle. Eur. J. Biochem. 173:437-443.
103. CHRISPEELS, M.J. 1980. The endoplasmic reticulum. In: The Biochemistry of Plants, Vol. 1. (N.E. Tolbert, ed.) Academic Press, New York, pp. 389-412.
104. IBRAHIM, R.K., DE LUCA, V., KHOURI, H.E., LATCHINIAN, L., BRISSON, L., CHAREST, P.M. 1987. Enzymology and compartmentation of polymethylated flavonol glucosides in *Chrysosplenium americanum*. Phytochemistry 26:1234-1245.
105. SMALL, C.J., PECKET, R.G. 1982. The ultrastructure of anthocyanoplasts in red cabbage. Planta 154:97-99.
106. BRISSON, L., CHAREST, P.M., DE LUCA, V., IBRAHIM, R.K. 1992. Immunocytochemical localization of vindoline in mesophyll protoplasts of *Catharanthus roseus*. Phytochemistry 31:465-470.
107. TOKUYASU, K.T. 1980. Immunocytochemistry of ultrathin frozen sections. Histochem. J. 12:381-403.
108. TOKUYASU, K.T. 1986. The application of cryoultramicrotomy to immunocytochemistry. J. Microsc. 143:139-149.
109. HARRIS, N. 1986. Organization of the endomembrane system. Annu. Rev. Plant Physiol. 37:73-92.

Chapter Three

GENERAL PHENYLPROPANOID METABOLISM: REGULATION BY ENVIRONMENTAL AND DEVELOPMENTAL SIGNALS

Carl J. Douglas, Mary Ellard, Karl D. Hauffe, Elizabeth Molitor, Mário Moniz de Sá, Susanne Reinold, Rajgopal Subramaniam and Frank Williams.

Department of Botany, University of British Columbia Vancouver, B.C. V6T 1Z4, Canada

INTRODUCTION

A tremendous array of plant natural products is synthesized from L-phenylalanine via a series of biosynthetic pathways involving phenylpropanoid metabolism. Phenylpropanoid metabolism can be divided into a central, general

Phenolic Metabolism in Plants, Edited by H.A. Stafford and R.K. Ibrahim, Plenum Press, New York, 1992

pathway ("general phenylpropanoid metabolism"), required for the synthesis of all phenylpropanoid metabolites, and specific branch pathways emanating from the general pathway which may require in addition other pathways. These lead to the synthesis of specific phenolic end products (Fig. 1). Phenylalanine ammonia-lyase (PAL) catalyzes the first step of general phenylpropanoid metabolism, the deamination of L-phenylalanine to produce cinnamic acid. Cinnamic acid is hydroxylated by cinnamate 4-hydroxylase (C4H) to yield 4-coumaric acid. The activities of hydroxylases and O-methyl transferases upon 4-coumaric acid can yield derivatives of 4-coumaric acid (e.g. ferulic acid, 3' methoxylated, and sinapic acid, 3',5' methoxylated). The enzyme 4-coumarate: CoA ligase (4CL) catalyzes the third and final step in the general pathway, the formation of activated CoA esters of hydroxycinnamic acids (4-coumaric acid or its methoxylated derivatives).

CoA esters of hydroxycinnamic acids serve as the substrates for specific branch pathways. Among the best studied of these pathways are those leading to

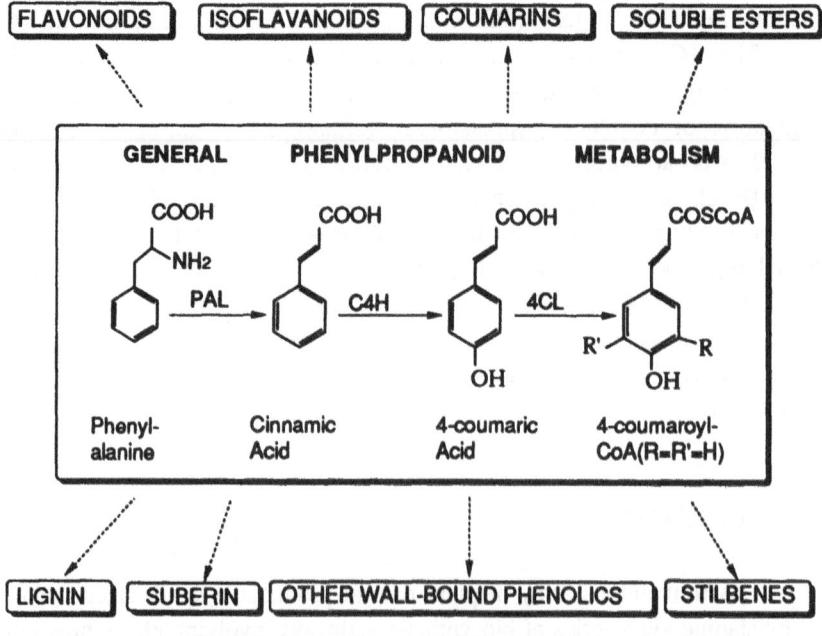

Fig. 1. The reactions of general phenylpropanoid metabolism. Dashed arrows indicate branch pathways emanating from the general pathway. PAL, phenylalanine ammonia-lyase; C4H, cinnamate 4-hydroxylase; 4CL, 4-coumarate:CoA ligase.

the synthesis of flavonoids [key enzymes include chalcone synthase (CHS) and chalcone isomerase (CHI)] and lignin [key enzymes include cinnamoyl-CoA reductase and cinnamyl alcohol dehydrogenase, (CAD)]. Activated 4-coumaric, ferulic, and sinapic acids are precursors for synthesis of p-hydroxyphenol, guaiacyl, and syringyl lignin, respectively.[1] Several chapters in this volume detail the properties of these and other phenylpropanoid metabolites as well as the regulation of their biosynthesis.

The chemistry, enzymology, genetics, and, molecular biology of phenylpropanoid metabolism has attracted the attention of large number of plant biologists over several decades. The reasons for this intense study lie in the important roles played by phenylpropanoid-derived compounds in many differentiated plant cells, tissues, and organs, and the roles many such compounds play in defensive or symbiotic interactions with microoganisms. As well, some phenylpropanoid derived compounds such as anthocyanin pigments are easily visible in pigmented plant organs. This has facilitated the isolation and characterization of mutants defective in the synthesis of these compounds. Such mutants have provided material for genetic and molecular genetic studies, which have revealed fundamental mechanisms by which plant genes are regulated[2,3] (see also Chapter 6 by Gerats).

Since plant secondary products play specific roles in differentiated cell types, it follows that the biosynthesis of these compounds is integrated into plant developmental programs. Wiermann, 1981, presented a review of the pertinent literature in this field as of 1981;[4] only a few examples illustrating the developmental regulation of phenylpropanoid metabolism are discussed here. An obvious and important example is the developmentally regulated biosynthesis and deposition of lignin in vessels and tracheids during xylem differentiation and during the differentiation of other highly lignified cell types. As mentioned above, another striking case is the synthesis of flavonoid pigments in plant organs such as petals and seeds,where pigment deposition often occurs in a cell type-specific manner (e.g. in epidermal cells of petals and the aleurone layer of maize kernels). Flavonoids are also synthesized in roots (where they function in legumes as signal in the symbiotic interaction with Rhizobia[5]) and in the epidermal cells of leaves.[6] In the latter case, flavonoid accumulation is UV-light and /or blue light-inducible, leading to the suggestion that UV light-absorbing flavonoids play a role as protectants against the harmful effects of UV irradiation. In oats, it was shown that the synthesis of flavonoids, lignin, and wall-bound hydroxycinnamic acid esters is regulated temporally and in a cell type-specific manner during leaf maturation.[7] In that case, the appearance of the biosynthetic enzymes was shown to be similarly regulated. Finally, the accumulation of compounds derived from phenylpropanoid metabolism and the activity of the corresponding enzymes, is temporally regulated during microsporogenesis in *Tulipa* ,[4] and in *Petunia* the flavonoid pigment 4,2',4',6'-tetrahydroxychalcone accumulates in mature pollen grains.[8]

Layered atop the developmentally regulated accumulation of phenylpropanoid metabolites is the stress-induced accumulation of many such compounds. In some plants, phenylpropanoid phytoalexins are synthesized in response to pathogen infection (e.g. isoflavonoids in legumes and furanocoumarins in parsley[9,10]). Deposition of cell wall-bound phenolic compounds at these sites also appears to be an important defense mechanism,[9] and the activation of general phenylpropanoid metabolism at sites of attempted pathogen infection seems to be ubiquitous in plants. Wounding also stimulates phenylpropanoid metabolism; here it is probably required for the synthesis and deposition of lignin and suberin at the sites of wound healing. In contrast to the developmentally regulated synthesis of phenylpropanoid metabolites, these stresses seem to induce the accumulation of phenylpropanoid defense compounds in a non-cell type-specific manner.[11]

Plant cell cultures have proved useful in analyzing the regulation of stress-activated phenylpropanoid metabolism. Parsley and French bean suspension cultures are the two best characterized systems,[9,10] but cell cultures have proven useful in several other plants, such as alfalfa (see Chapter 4 by Dixon). When parsley cells are treated with elicitors (fungal pathogen or plant cell wall-derived molecules which induce defense responses[11]), the biosynthesis and secretion of the furanocoumarin phytoalexins is induced but the flavonoid branch pathway remains inactive. Conversely, in UV light-treated parsley cells, vacuolar accumulation of flavonoids is induced but the furanocoumarin pathway remains inactive. In the bean system, both light and elicitor treatments lead to the activation of flavonoid synthesis. In parsley, immunohistochemistry, microspectrophotometry, and *in situ* hybridization have shown that the reactions observed in pathogen-infected or light-treated leaves of whole parsley plants closely parallel the induction of furanocoumarin and flavonoid accumulation in cell cultures.[9]

Genes encoding the enzymes of the general phenylpropanoid metabolism, flavonoid metabolism, and a few other branch pathways have been cloned from a number of plant species;[12] the structure and regulation of these genes have been the subject of several recent reviews.[9,10,12,13] Analysis of the stress-activated expression of genes encoding PAL, 4CL, and CHS in parsley suspension-cultured cells has shown that they are transcriptionally activated by treatment with elicitors or light.[9] Transcriptional activation is followed by increases in mRNA amount, enzyme activity, and accumulation of specific phenylpropanoid end products. The stimulus-specific activation of different branch pathways in parsley is regulated by gene transcription: PAL and 4CL transcription is activated by both stimuli, while CHS transcription is activated by light but not elicitor.[9] Genes encoding furanocoumarin-specific O -methyl-transferases are activated only by elicitor.[9] Similarly, phenylpropanoid genes are transcriptionally activated in suspension-cultured bean cells. In this case, however, CHS transcription is both elicitor and light-activated.

The mRNAs specific to PAL and 4CL accumulate in a cell type-independent manner at the sites of attempted pathogen infection or wounding of parsley leaves, and in an organ and cell type-specific manner during plant development.[11] As well, studies in several other plants (for example, *Antirrhinum*, bean, maize, pea, *Petunia*) have shown that genes encoding enzymes of general phenylpropanoid metabolism and flavonoid metabolism are expressed in a tissue and organ-specific fashion.[3,14,15,16] These, and an accumulation of genetic and molecular data in maize and *Antirrhinum*, [2,3,17] suggest that phenylpropanoid metabolism is regulated primarily at the level of the transcription of genes encoding individual enzymes. Genetic and molecular studies in *Antirrhinum*,[3] bean,[18] and maize[17] have defined trans-acting factors, likely to be transcriptional regulators, which interact with the promoters of a number of phenylpropanoid genes. A currently favored model proposes that the presence or activity of these transcription factors is modulated by developmental and environmental signals, and that these factors in turn regulate the expression of sets of phenylpropanoid genes and thus the activity of phenylpropanoid metabolic pathways.

In this chapter, we will focus on our recent work on the developmental and stress-regulated expression of the parsley 4CL-1 gene. We also discuss our current work in poplar, which should open the door towards a better understanding of the regulation of phenylpropanoid metabolism in woody plants.

4-COUMARATE:CoA LIGASE (4CL)

The 4CL Enzyme

Although it plays a key role as the last enzyme in general phenylpropanoid metabolism, supplying precursors for subsequent branch pathways, 4CL enzymes have been less studied than PAL. 4CL has been purified, or partially purified, from *Forsythia*,[19] soybean,[20] *Petunia*,[21] parsley,[22] pea, [23] poplar,[24] and maize.[25] The enzyme is monomeric and molecular weights of between 55,000 (soybean) and 67,000 (parsley) were estimated. More accurate estimations of the size of the parsley and potato enzymes from the deduced amino acid sequences of full-length cDNA clones[26,27] indicates that, in these two species, the protein has a molecular weight of just under 60,000. A search of the protein sequence data base[27] indicated that a seven amino acid motif clustered around a conserved cysteine residue within the potato and parsley enzymes is conserved in several apparently unrelated bacterial and insect enzymes. Each of these enzymes, like 4CL, binds and hydrolyzes ATP in order to activate relatively unrelated molecules (e.g. 4-coumaric acid, luciferase, phenylalanine).[27] The conserved cysteine could thus play an important role in the catalytic activity of 4CL.

As mentioned above, 4CL enzymes can utilize both 4-coumarate and its methoxylated derivatives as substrates. There is evidence from several plants that catalytically distinct isoforms of 4CL exist which preferentially utilize different hydroxycinnamic acids as substrates.[20,21,23-25] Potentially, the differential expression of 4CL gene family members encoding such isoforms could contribute to the control of metabolic flux into subsequent phenylpropanoid branch pathways by preferentially synthesizing branch pathway-specific hydroxycinnamic acid-CoA esters. However, no molecular or genetic evidence yet exists which demonstrates that different genes encode distinct isoforms in these plants.

Parsley 4CL (Pc4CL) Genes

The mechanisms underlying the regulation of the expression of genes encoding enzymes of general phenylpropanoid metabolism are of particular interest since expression of these genes is necessary in diverse cell types which require a large variety of phenylpropanoid metabolites. Furthermore, expression of the genes must be inducible under various stress conditions when such metabolites are also required. As a first step in elucidating the mechanisms regulating expression of parsley 4CL genes, 4CL clones were isolated from a genomic library, and the structure of the genes elucidated.[28] The enzyme is encoded by two highly homologous genes, Pc4CL-1 and -2, which show no evidence of differential regulation by light and elicitor in parsley suspension cells. The sequences of the genes are over 95% identical in coding regions and introns, and display a high degree of sequence identity for several hundred base pairs upstream of the transcription start sites.[26,28] The high degree of sequence identity between the two genes suggests that one of them arose by a recent gene duplication. The two genes encode slightly different, and physically separable, 4CL isoenzymes, but these have nearly identical abilities to accept a variety of hydroxycinnamic acids as substrates.[26]

Thus, in parsley, a single type of 4CL gene is capable of supplying substrates for all phenylpropanoid branch pathways, and both 4CL genes appear to be similarly regulated. The lack of differential expression of 4CL gene family members implies that each gene must be responsive to a wide array of environmentally and developmentally generated signals which activate expression at various times and places. An array of regulatory *cis*-acting DNA sequences capable of interacting with multiple signals would thus be predicted to be associated with each gene.

In contrast to Pc4CL, PAL is encoded by gene families in all angiosperms investigated so far, including parsley,[9,12,29] and differential expression of gene family members encoding distinct isoforms has been demonstrated.[14] In such cases, different PAL genes are responsive to different sets of developmentally or environmentally generated signals and each gene need not be responsive to the same array of signals as Pc4CL genes. Interestingly, the organization and expression of 4CL genes in potato (the only other plant from which 4CL genes have been characterized) is comparable to that in parsley: there are two highly similar genes which are not differentially regulated, and which probably encode catalytically similar enzymes.[27] In striking contrast, PAL is encoded by a large (> 20 members) gene family in potato.[9]

Thus, while multiple, divergent 4CL genes would be predicted in those plants having distinct 4CL isoforms, 4CL gene organization in parsley and potato is quite simple in comparison to that of genes encoding PAL. 4CL genes will need to be characterized from more plants, particularly those in which multiple isoforms have been described, to determine if differential expression of gene family members is a strategy adopted by some plants to regulate 4CL activity, as it has been for PAL and other phenylpropanoid genes.

REGULATION OF Pc4CL-1 EXPRESSION

Stress Activation of Pc4CL-1

To investigate the mechanisms by which parsley 4CL genes are regulated in cell cultures by light and elicitor, we chose to study the 4CL-1 gene. We first asked which 4CL-1 DNA sequences are required for inducible expression in these cells. 4CL gene constructions consisting of the entire structural gene and 1500 base pairs (bp), 600 bp, or 174 bp of flanking 5' DNA were created (Fig. 2) and used to generate transgenic parsley cell lines by direct gene transfer into protoplasts and subsequent selection for stable transformants. Introduction of a 15 bp in-frame deletion within the first intron of the 4CL-1 gene in these constructions (Fig. 2) allowed discrimination between transcripts originating from endogenous 4CL genes and introduced 4CL genes by the use of S1 nuclease protection assays. Using this assay, transcripts from the endogenous 4CL genes protect 475 bp and 415 bp fragments of a labelled probe from S1 digestion, while transcripts from the introduced 4CL genes protect a 213 bp fragment (Figs. 2 and 3). The amount of RNA specific to the introduced and endogenous genes in control, elicitor-treated, and light-treated transgenic cell cultures was determined. Figure 3 shows the results from an experiment in which cell cultures harboring introduced genes with either 600 bp or 174 bp of 5' DNA were elicitor-stimulated. Expression of the endogenous genes was strongly induced, and either 600 bp or 174 bp of 5' DNA, in conjunction with the

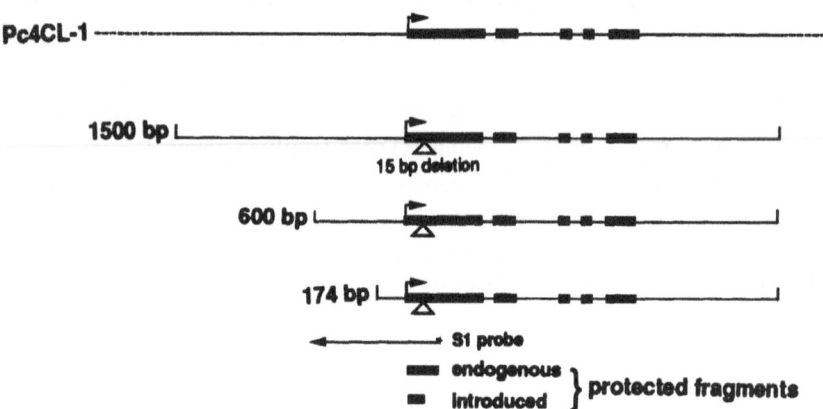

Fig. 2. Pc4CL-1 constructions used to generate stably transformed parsley cell lines. The structure of Pc4CL-1 is given above (filled boxes represent exons; the transcription start site is shown by an arrow). Constructions with 1500, 600 or 174 bp of 5' flanking sequences used to transform parsley cells are shown below. The location of a 15-bp deletion in the first exon in these constructions is shown. At the bottom, the location of a hybridization probe used for S1 nuclease protection assays and the fragments predicted to be protected by RNA transcribed from the endogenous and introduced are given.

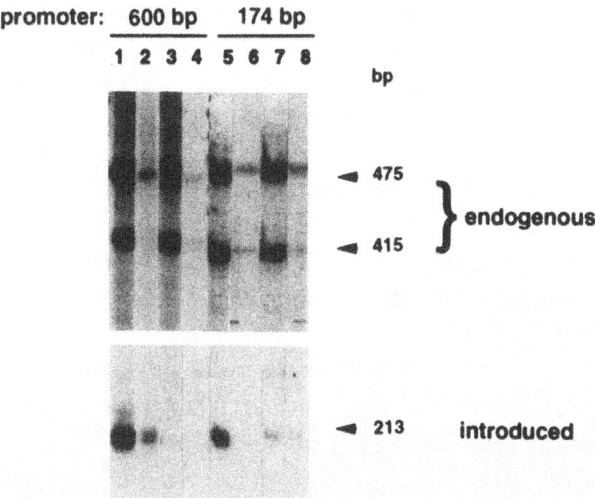

Fig. 3. Detection of RNA specific to endogenous and introduced genes in elicitor-treated and control parsley cells. RNA was extracted from two independent cell lines stably transformed with a construction containing a 600-bp promoter (lanes 1,2 and 3,4, respectively) or from two independent cell lines transformed with a construction containing a 174-bp promoter (lanes 5,6 and 7,8, respectively), hybridized to a S1 probe (see Fig.2) and digested with S1 nuclease. Lanes 1,3,5,7, elicitor-treated cells; lanes 2,4,6,8, untreated control cells. The protected fragments specific to the endogenous and introduced genes are indicated. The intensity of each signal is proportional to the amount of RNA in the sample.

structural gene, was sufficient for elicitor responsiveness. Expression of the 174-bp construct was relatively weak, and the levels of expression varied between individual transformed cell lines. Similar results were obtained with the 1500-bp promoter and expression of introduced genes was also light-inducible in cell lines harboring all three constructions (not shown).

These results showed that 174 bp or less of the 4CL-1 promoter are sufficient to confer light- and elicitor-inducible expression upon the 4CL structural gene. To test whether these promoter sequences alone were sufficient to confer inducibility upon a heterologous gene, the same promoter fragments were fused to the β-glucuronidase (GUS) reporter gene, and transgenic cell lines generated. Surprisingly, and in contrast to the above experiments, expression of these introduced genes was not inducible by light or elicitor when either GUS enzyme activities or mRNA levels were measured.[30] These and a large number of other gene constructions were also not elicitor or light activated in transient assays in parsley protoplasts, although parsley CHS and PR-1 genes are light[31] and elicitor-inducible[32] in such assays. Further experiments demonstrated that only 4CL-1 coding sequences, and not 3' flanking DNA or introns, are required along with 4CL-1 promoter sequences to activate 4CL-1 expression in response to elicitor and light.[30]

To confirm and extend these observations, two lines of transgenic tobacco plants were generated. One contained the entire 4CL-1 gene with 1500 bp of promoter and the other a 4CL-1 promoter-GUS gene fusion. The fact that the parsley 4CL-1 gene does not cross-hybridize to tobacco 4CL facilitated the detection of transcripts from the introduced parsley gene; expression of the 4CL-1-GUS gene was monitored by GUS enzyme activity and mRNA amount. The results were similar to those obtained in parsley cells: the complete 4CL-1 gene was light and elicitor responsive, but 4CL-1 promoter-GUS fusions were non-responsive to these stimuli.[30] Interestingly, expression of both the entire 4CL-1 gene and 4CL-1-GUS fusions were wound inducible in a manner similar to that of 4CL genes in parsley plant.[30] As controls, we monitored the expression of endogenous tobacco 4CL genes in these experiments, taking advantage of the fact that a potato 4CL cDNA clone[27] cross-hybridizes to tobacco 4CL but does not cross-hybridize to the Pc4CL-1 gene. In all cases, expression of tobacco 4CL was elicitor, light, and wound inducible in a manner very similar to that of the introduced 4CL-1 gene.

A final set of experiments was performed in transgenic tobacco to determine if 4CL-1 promoter coding sequences alone are sufficient to confer elicitor and light inducibility upon a heterologous promoter. We chose the Cauliflower Mosaic Virus (CaMV) 35S promoter for these experiments, since it is not responsive to these stimuli. Figure 4 shows that expression of 35S-4CL gene fusions was not responsive to light and elicitor in transgenic tobacco, although expression of the endogenous 4CL gene(s) was clearly induced. A fusion of the 4CL cDNA to a truncated version of the 35S gene was more weakly expressed, but this expression was also not inducible (not shown). These results are in contrast to those reported for the pea ferridoxin I gene, which also required transcribed sequences for light-inducible expression.[33] In that case, the intronless structural gene rendered the CaMV 35S promoter light-responsive in transgenic tobacco. From these experiments we conclude that a combination

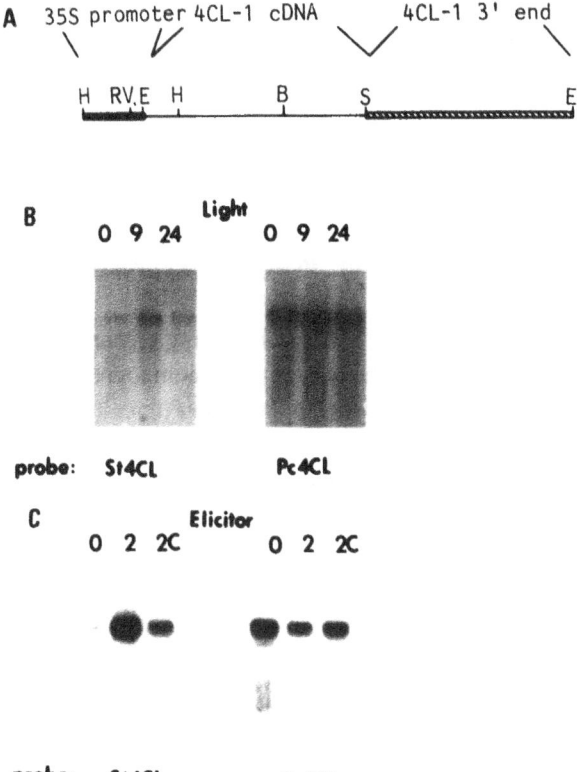

Fig. 4. Expression of 35S-4CL-1 fusions in light- or elicitor-treated transgenic tobacco plants. (A) Structure of plasmid BIN-300, used to generate transgenic tobacco plants. 35S promoter, 4CL-1 cDNA and 4CL-1 3' end regions are shown. Restriction sites: B, BamHI; E, EcoRI; H, HindIII; RV, Eco RV; S, SstI. (B) and (C) Expression of 35S-4CL-1 fusions was analyzed by RNA blots. Endogenous tobacco 4CL transcripts were detected by hybridization to a potato 4CL cDNA (St4CL) and transcripts specific to the introduced 35S-4CL-1 gene were detected by hybridization to a parsley 4CL cDNA (Pc4CL). (B) RNA samples (10μg/lane) from dark-grown plants irradiated with u.v.-containing white light for 0, 9, and 24 h. (C) RNA samples (10μg/lane) from leaves 0 and 2 h after treatment with Pmg elicitor solution and 2 h after treatment with water alone (2C).

of promoter sequences and sequences within the coding portion of the 4CL-1 gene is required for light and elicitor-activated expression. The location of the gene-internal sequences and their mode of action is currently unknown but is under investigation. This arrangement of *cis*-linked controlling sequences is novel for phenylpropanoid genes analyzed to date. In each of these cases,[16,34,35,36,37] promoter sequences were sufficient to regulate environmentally triggered reporter gene expression in heterologous or homologous transgenic plants.

Developmental Regulation of Pc4CL-1 Expression

Based on the key role of 4CL in supplying precursors for the synthesis of developmentally required phenylpropanoid compounds, expression of the parsley 4CL genes was expected to be under developmental regulation. Schmelzer et al [11] used *in situ* hybridization to show that 4CL transcripts accumulate preferentially in the epidermal cells, xylem, and oil duct cells in parsley leaves.

We used transgenic tobacco plants harboring both 4CL-GUS fusions and the complete Pc4CL-1 gene to further investigate the developmental regulation of the gene. The histochemical assay for GUS gene expression allowed us to ask whether or not the 4CL-1 promoter is sufficient to drive cell type-specific expression in the patterns predicted from expression in parsley leaves and from similar experiments carried out using PAL-GUS fusions in tobacco.[34,35,37] Analysis of plants containing the 600 bp promoter fragment fused to GUS demonstrated that, although it was insufficient to specify elicitor- or light-inducible expression in tobacco (see above), this promoter has the ability to direct developmental expression of GUS in a large number of vegetative and floral organs[38] (Table 1).

The expression patterns were generally consistent with *in situ* hybridization patterns in parsley, with patterns of expression directed by PAL promoters, and with the requirement for phenylpropanoid derivatives in certain cell types. For example, high expression was observed in the xylem of all organs, where lignified tracheary elements are differentiating; this expression was also observed in parsley[11] and in tobacco plants expressing PAL-GUS fusions.[34,35] Expression patterns in secondary xylem, roots and flower petals were very similar to those directed by the bean PAL promoter.[34,35] Expression was observed in floral nectaries, stigmata and ovules. It is likely that 4CL expression here is required for the elaboration of specific phenylpropanoid compounds in these cells and tissues.

Table 1. Expression of a 600-bp 4CL-1 promoter-GUS fusion in transgenic tobacco plants.

Organ	Tissue/cell expressed[a]
stem	primary xylem
	secondary xylem
leaf	xylem
root	xylem
	root hair
	subapical cells
flower	xylem
	petal epidermis (pigmented)
	stigma epidermis
	ovule epidermis
	nectary
	pollen

[a] Expression was assayed histologically by staining with X-GLUC.

Activity was not observed in epidermal cells of leaves where, in parsley, light is critical in triggering flavonoid biosynthesis.[11] Expression of PAL-GUS fusions was observed in leaf epidermal cells in transgenic tobacco,[34,35] and 4CL transcripts accumulate in parsley leaf epidermal cells.[11] These results are consistent with those suggesting that 4CL-1 gene-internal sequences are required for light-activated 4CL-1 expression. The fact that 4CL-1 promoter sequences are sufficient to direct tissue-specific expression, but not elicitor or light-induced expression, indicates that there is at least a partial separation of the cis-acting elements directing developmental expression and stress-induced expression of the gene.

Since expression of additional phenylpropanoid genes in floral tissues other than petals and anthers had not been described, we wished to establish whether the endogenous tobacco 4CL genes and Pc4CL-1 were regulated in parallel in the floral organs where GUS activity was observed. To do so, we capitalized on the fact that we could discriminate between Pc4CL transcripts and tobacco 4CL transcripts by using parsley and potato probes (see above). We hybridized these probes to sectioned flowers from transgenic tobacco plants containing the complete Pc4CL-1 gene. Figure 5 shows that RNA from the endogenous tobacco 4CL gene(s) and 4CL-1 accumulated in stigma epidermal/ hair cells and ovule epidermal cells in an identical manner. This expression mirrors the GUS expression directed by the 4CL-1 promoter, and indicates that this expression is an accurate reflection of 4CL gene activity in tobacco.

Flowers of different developmental stages were analyzed for accumulation of 4CL RNA by *in situ* hybridization or for GUS enzyme activity by histochemical staining. These experiments revealed that 4CL expression in tobacco is temporally as well as spatially regulated during flower development. A summary of the results is presented in Figure 6. In each of several organs, 4CL was expressed in a temporally distinct manner, implying that the gene responds to a variety of developmental signals which are generated during the differentiation of cells in different floral organs.

cis-acting Elements Regulating Developmental Expression

Our results indicate that expression of the parsley 4CL-1 gene is temporally and spatially regulated in a complex array of cells and tissues. It is thus likely that this expression is regulated by the interaction of 4CL-1 *cis*-acting elements with *trans*-acting transcriptional regulators, whose presence or activity during the differentiation of floral and vegetative organs is in turn regulated. The 4CL-1 promoter may thus have a complex arrangement of *cis*-acting elements capable of directing expression in many cell types.

As a first step in mapping 4CL-1 *cis*-acting elements and assessing their functions in specifying expression in various cells, we transferred a number of 4CL-GUS fusions with promoter deletions into tobacco plants, and assayed their developmental expression using the histochemical stain for GUS activity.[38] Table 2 shows that a promoter 210 bp in length specifies expression in all cell types described above. Further deletion of the 4CL-1 promoter to -174 bp abolishes part of the cell type-specific expression, while no expression was observed in plants expressing fusions with 120-bp or 78-bp promoters. Thus, the region downstream of -210 appears to be critical for regulating developmental expression.

In a further set of experiments, the ability of a series of internally deleted 4CL-1 promoters to drive tissue-specific expression of the GUS gene in tobacco was tested (K.D. Hauffe and C. Douglas, unpublished). When upstream 4CL-1 promoter sequences (from -244 to -597) were fused to the 78-bp promoter, overall GUS expression increased over ten-fold. In these plants,

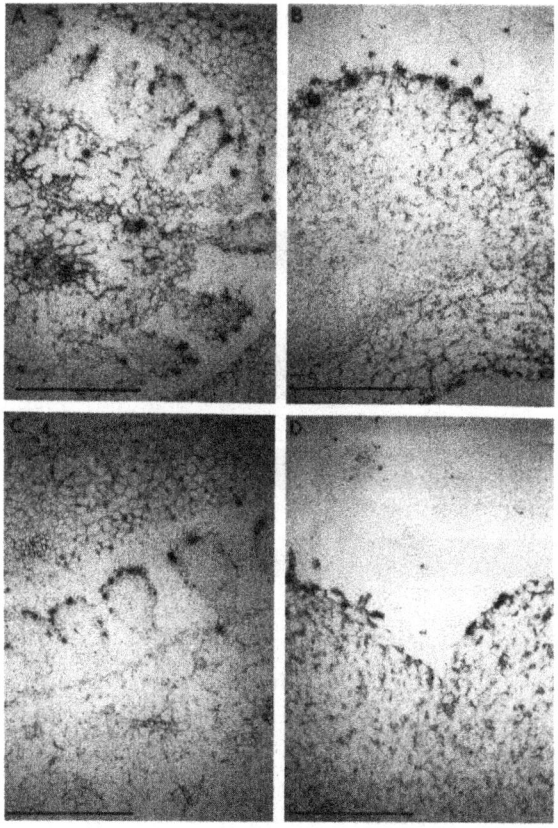

Fig. 5. *In situ* localization of Pc4CL-1 and tobacco 4CL RNA in developing flowers of a tobacco plant transgenic for the entire Pc4CL-1 gene. (A) and (B) were hybridized with ^{35}S antisense RNA from a Pc4CL cDNA clone; (C) and (D) were hybridized with ^{35}S antisense RNA from a potato 4CL cDNA clone. (A) and (C), cross-sections through a developing ovary, showing hybridization to the epidermal cells of ovules; (C) and (D), cross-sections through a developing stigma, showing hybridization to cells at the stigmatic surface. Bars equal 50 μm.

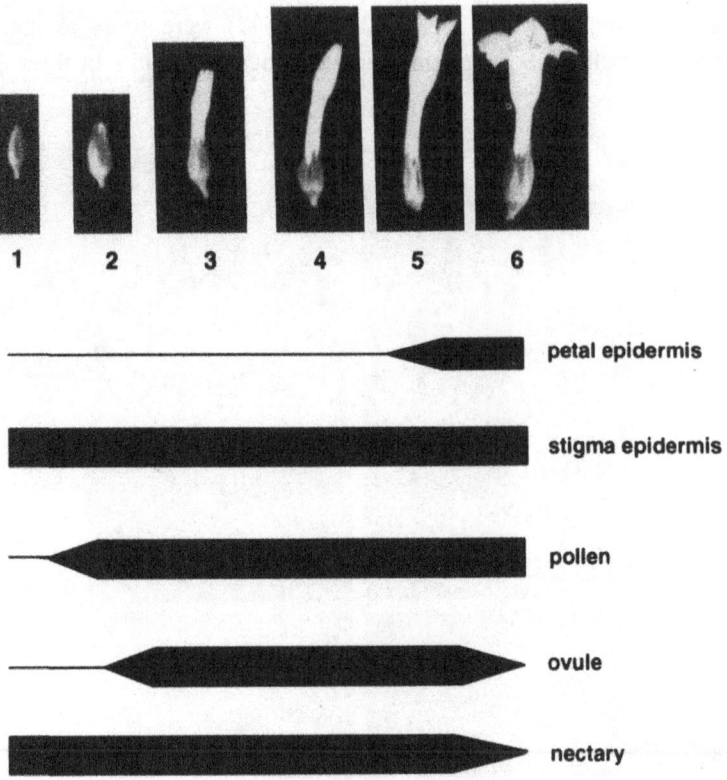

Fig. 6. Temporal regulation of Pc4CL-1 expression during tobacco
flower development. Flowers at different developmental stages (1-6,
top) from plants transgenic for the entire Pc4CL-1 gene or 4CL-1-
GUS fusions were examined by *in situ* hybridization or for histochemically detectable GUS expression. The expression observed in
various organs is shown below (width of the bars indicates amount of
expression).

histochemically detectable GUS expression was observed only in pollen. Fusion
of these same upstream sequences to the 120-bp promoter led to partial
restoration of tissue-specific expression: in addition to pollen, GUS activity was
evident in leaf and stem xylem and petal epidermal cells. Finally, fusion of the
upstream sequences to the 174-bp promoter was required to restore the complete

Table 2. Summary of histochemically localized GUS expression in transgenic tobacco plants.[a]

Construction	5'endpoint (bp)	GUS positive/ total plants[b]	Histochemically-detectable expression						
			vascular[c]	nectary	stigma	pollen	petal	ovule/ seed coat	root tip
99-G1-801	-600	6/8	+	+	+	+	+	+	+
808	-339	4/5	+	+	+	+	+	+	+
809	-252	5/7	+	+	+	+	+	+	+
810	-210	6/11	+	+	+	+	+	+	+
803	-174	0/14[d]	-	+	+	+	-	-	-
813	-120	0/8	-	-	-	-	-	-	-
812	-78	0/7	-	-	-	-	-	-	-

[a] © 1991 American Society of Plant Physiologists, reproduced with permission

[b] Individual plants harboring each construction were histochemically screened for vascular expression.

[c] Vascular expression, when present, was observed in vascular tissue of leaves, cotyledons, roots, and in primary and secondary vascular tissue of stems.

[d] In one out of the five plants in which floral organs were examined, weak but detectable expression was observed in nectaries and stigmata; in two out of five, expression was detected in pollen. None of the five plants had detectable expression in petals or ovules.

array of cell- and tissue-specific expression. Therefore, sequences between -174 and -210 do not appear to be required for developmentally regulated expression and probably act to increase the quantity of expression. Sequences between -174 bp and -78 bp are required for developmental regulation of 4CL-1 expression in all cells except pollen. Provided that they are linked to upstream sequences to boost the quantity of transcription, sequences downstream of -78 bp are sufficient for pollen expression, and sequences downstream of -120 bp are sufficient for xylem and petal expression but not expression in the remaining floral organs. Thus, different sets of *cis*-acting elements contained on these different promoter fragments appear to specify 4CL-1 expression in different cells and tissues. This suggests that the 4CL-1 gene is responsive to different signals in different cell types. Such signals could regulate the presence or activity of different sets of transcription factors in different cell types, which interact with different combinations of *cis*-acting elements.

Interestingly, *in vivo* footprinting of the 4CL-1 and -2 promoters in parsley tissue culture cells revealed that the regions between -200 bp and -78 bp are the sites of at least six distinct protein-DNA interactions.[38] The observed footprints are identical in elicitor-treated and untreated cells, and only minor changes were observed after light treatment.[29] One of the footprinted regions corresponds to a *cis*-acting element of unknown function conserved in all PAL genes examined so far[37] and in several other phenylpropanoid and stress-inducible genes.[29] Several of the footprints contain clusters of C residues and are spaced about 10 bp apart so that they are on the same face of the DNA double helix. When the promoter sequence of the potato 4CL genes was compared to those of the parsley genes, a surprisingly high degree of sequence conservation, limited to the first 170 bp of Pc4CL genes, was observed.[27] The conserved sequences include some of the footprinted Pc4CL sequences and three of the corresponding potato 4CL elements were also constitutively footprinted in cultured potato cells,[27] consistent with their potential importance in regulating 4CL expression.

The potential role that the *cis*-elements identified by *invivo* footprinting may play in the developmental regulation of 4CL-1 expression is also supported by our recent finding that internal deletion of DNA containing some of these potential regulatory sequences correlates with loss of tissue-specific expression in tobacco. For example, three major footprints are located between -78 bp and -120 bp; deletion of these abolishes xylem and petal expression, but not pollen expression (K.D. Hauffe and C. Douglas, unpublished). We are currently making more refined mutations in these elements to test their roles in specifying expression of 4CL in certain cells and tissues.

Taken together, our data show that the parsley 4CL-1 gene is regulated in a complex manner, integrating both developmental and environmental signals to insure expression at the proper times and places. 4CL-1 promoter and gene-internal sequences together are required for light- and elicitor-responsiveness. A group of *cis*-acting elements within 174 bp of the transcription start site controls a complex pattern of tissue-specific expression. When coordinately expressed with other phenylpropanoid genes, 4CL expression results in the elaboration phenylpropanoid products at the correct times and places. Part of the complexity of 4CL-1 regulation may be related to the fact that, as an essentially single copy gene encoding a critical enzymatic activity, it must be responsive to all signals regulating phenylpropanoid product synthesis.

PHENYLPROPANOID METABOLISM IN POPLAR

PAL and 4CL Genes

Although progress has been made in understanding the regulation of general phenylpropanoid metabolism in certain plants, little is known about such processes in woody plants. Phenylpropanoid metabolism plays an important role in the growth and development of woody plants, since an important component of wood is lignin. Furthermore, while it is likely that phenylpropanoid metabolites are important in defense against pathogen infection in trees, there is a paucity of information about such potential defense responses in these plants. We have chosen hybrid poplar (*Populus deltoides* X *P. trichocarpa*; DT) as a model woody plant to begin studies on the regulation of genes encoding phenylpropanoid genes, and present here some of our initial findings.

Poplar has attributes which make it a useful tree for these studies. It has a relatively small genome size and is easily propagated vegetatively. Three generations of certain DT hybrids are available[39] allowing genetic analysis of potential gene family members. These poplar hybrids grow much more rapidly than either parent in many sites in northwestern North America,[39] which has led to interest in establishing poplar plantations to supply woody biomass as an alternative to wild or managed softwood forests.

In initial studies, we found that PAL, 4CL and CAD enzyme activities were highest in very young leaves of apical buds, and in developing secondary xylem isolated by scraping stem tissue from 2-4 year-old trees. Since activities were highest in apical buds, cDNA libraries were constructed from young leaf poly(A^+) RNA isolated from DT hybrids H11-11[39] and 53-246. Preliminary data indicated that potato PAL and parsley 4CL cDNA probes gave the best hybridization signals on Southern blots of poplar genomic DNA washed at low stringency. These probes were used to screen the cDNA libraries, and several putative poplar PAL and 4CL cDNA clones were isolated. Sequence analysis of these clones indicates that they are homologous to PAL and 4CL from other plants. The longest PAL clone (2.6 kb in length) appears to be nearly full length, containing the entire coding region, while the longest 4CL clone (1.5 kb) is about 500 bp short of being full length.

In order to obtain an estimate of the number of poplar genes encoding PAL and 4CL in poplar, poplar PAL and 4CL clones were hybridized to Southern blots of genomic DNA from DT hybrids H11-11 and 53-246 as well as the male and female parents of 53-246 (*P. deltoides* clone ILL 129 and *P. trichocarpa* clone 93-968, respectively). Genomic DNA was cut with four different restriction enzymes; the inheritance of each hybridizing 53-246 fragment could be traced to one of either parents, allowing hybrid alleles to be distinguished from different genes. These preliminary data are consistent with the presence of at least 4 gene copies of both PAL and 4CL in the hybrids and parents. Further Southern blot analyses of F_2 and backcross generations and hybridization of blots with PAL and 4CL 5' and 3' probes should help to better estimate the size of the two gene families.

Elicitor Activation

We set up a poplar cell culture system in order to investigate possible poplar defense responses associated with the activation of phenylpropanoid metabolism. Treatment of suspension-cultured H11-11 cells with any of three non-specific elicitors (*Erwinia carotovora* polygalacturonic acid lyase or cell wall elicitors derived from *Phytophthora megasperma* f.sp.*glycinea* or *Fusarium oxysporum*) resulted in activation of general phenypropanoid metabolism (Fig. 7). PAL and 4CL enzyme activities increased coordinately and transiently, with maximum activities at about 7 h after elicitation. These increases in enzyme activity were preceded by large, rapid, and transient increases in PAL and 4CL mRNA amounts, detected by Northern blots hybridized to either homologous or heterologous PAL and 4CL probes.

We also assayed the activities of CHS and CAD, specific to flavonoid and lignin branch pathways, respectively. Although both enzyme activities were detected, they were not inducible above background levels. Thus, while the activation of general phenylpropanoid metabolism provides the potential for the

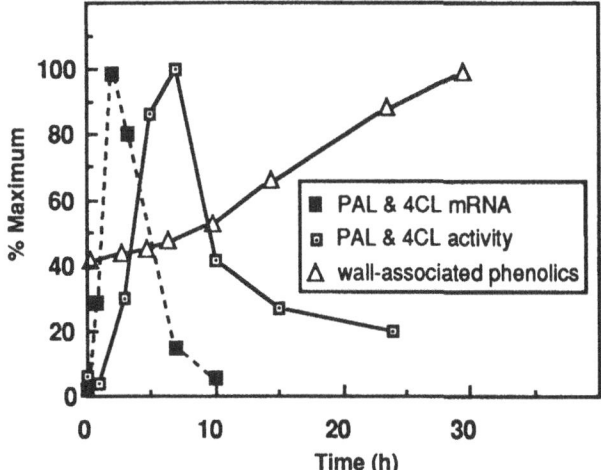

Fig. 7. Summary of elicitor-activated phenylpropanoid metabolism in poplar suspension-cultured cells. Relative PAL and 4CL mRNA amounts, PAL and 4CL enzyme activities, and amounts of wall-associated phenolic compounds at various times after elicitor treatment are shown.

increased synthesis of phenylpropanoid products in response to elicitor stimulation, the nature of such products is unclear. Phenolic compounds do accumulate in the cell walls of elicitor-treated poplar cells. This was determined by phloroglucinol-HCl staining and thiolglycolic acid extraction of isolated cell walls. The identity of the wall-associated phenolic compounds (e.g. as lignin) is under investigation. HPLC chromatography of cell extracts and culture filtrates indicates that several soluble compounds, potentially phenolic in nature, also accumulate in response to elicitation (M. Moniz de Sá and C. Douglas, unpublished). The identity of these compounds, and their possible roles as poplar defense compounds, is also being investigated.

As discussed in the introduction, earlier work suggested the existence of three 4CL isoforms, with varying specificities for differently methoxylated 4-coumarate derivatives, in poplar.[24] We are interested in the possibility that the multiple genes which appear to encode 4CL in the DT poplar hybrids may, in fact, encode catalytically distinct 4CL isoforms. To begin testing this possibility, we assayed the activity 4CL in control and elicitor-treated suspension-cultured cells against substrates other that 4-coumarate. Table 3 shows that, in control cells, 4CL is active against 4-coumarate, caffeate, sinapate, and, to small degree, ferulate. In elicitor-treated cells, however, only the activity against 4-coumarate and ferulate was elicitor inducible; activity against caffeate and sinapate remained unchanged. This suggests that there is a differential activation in elicitor-treated cells of one or more 4CL isoforms which preferentially utilize 4-coumarate and ferulate as substrates. These observations are consistent with the findings of Grand et al[24] who described an isoform of 4CL in poplar which is active against 4-coumarate and ferulate, but inactive against caffeate and sinapate. We now intend to determine if there is differential activation of different 4CL genes in elicitor-treated cells and if different 4CL genes encode catalytically distinct isoenzymes. These studies will be aided by the finding that 4CL activity is present in extracts of *E. coli* strains expressing 4CL cDNAs.[26]

Developmental Regulation

As in other plants, the expression of phenylpropanoid metabolic pathways is likely to be compartmentalized in those poplar cells and tissues where specific products are required. We used Northern blots and *in situ* hybridization to quantify and localize PAL and 4CL expression in DT hybrid H11-11. PAL and 4CL expression patterns were identical in Northern blots, which showed that RNA accumulation was highest in young leaves and

Table 3. Substrate specificities of 4CL in untreated and elicitor-treated H11-11 poplar cells.

substrate	% 4-coumarate activity in elicitor-treated cells	
	untreated	elicitor-treated
4-coumarate	19.4	100.0
caffeate	21.7	26.4
ferulate	2.0	32.9
sinapate	51.2	51.0

developing xylem scraped off woody stems, lower in cambium scraped off peeled bark and in young, non-woody stem tissue (to which xylem contributed a relatively small part), and lowest in fully expanded leaves. *In situ* hybridization was performed on the soft tissues of leaves, petioles and young, non-woody stem tissue. In the young leaves of the apical bud, PAL and 4CL were highly expressed in epidermal cells, and at lower levels in differentiating xylem. In young stems, expression was limited mainly to developing xylem.

The genus *Populus* is characterized by its richness in phenolic compounds.[40] HPLC profiles of methanol extracts of young leaves from hybrid H11-11 showed a complex mixture of phenolic compounds, and many of these disappeared from extracts of older leaves (F. Williams and C. Douglas, unpublished). We believe that the high expression of PAL and 4CL in young leaf epidermal cells is associated with the very active synthesis of phenylpropanoid derivatives which accumulate in these leaves. High PAL and 4CL expression in secondary xylem is probably associated with lignin biosynthesis in this tissue. We intend now to determine whether there is differential expression of poplar PAL and 4CL genes different tissues, and how these developmentally regulated expression patterns compare to expression in response to elicitors or pathogen infection. It will be of particular interest to determine if differential expression of 4CL gene family members, encoding distinct 4CL isoforms, plays a role in channeling phenylpropanoid metabolism in different cells and tissues in poplar.

CONCLUSION

The regulation of genes encoding enzymes of general phenylpropanoid metabolism is complex, reflecting the diversity of phenylpropanoid compounds required at numerous times and places during plant growth and development. Thus, the complex arrangement of regulatory *cis*-acting elements in the parsley 4CL-1 promoter and structural gene is likely a function of the requirement for intricately regulated temporal and spatial patterns of expression. Since the synthesis of most phenylpropanoid compounds requires the coordinate action of both PAL and 4CL enzyme activities, there has probably been strong selection for coordinate expression of the corresponding genes in plants. The mechanisms by which this coordinate expression is achieved remain to be elucidated. Some *cis*-acting elements in PAL and 4CL genes are conserved. This, coupled with the observation that regulatory genes encoding probable transcription factors regulate the spatially and temporally coordinated expression of maize genes encoding enzymes of flavonoid metabolism,[2,17] suggests that the interaction of common, developmentally and/or environmentally regulated transcription factors with PAL and 4CL *cis*-acting elements governs part of their coordinate expression. Significant differences in PAL and 4CL gene organization and regulation, however, suggest that at least partially independent regulatory modes may be operative. Further information on the mechanisms regulating the expression PAL and 4CL genes should help shed light on these problems.

ACKNOWLEDGEMENTS

We thank Jeff Dangl, Maria-Ester Ites-Morales, and Paul Schulze-Lefert, and Uta Paszkowski, at the Max-Planck-Institute für Züchtungsforschung, Köln, Germany, for their collaboration on part of the work described here. This work was supported by Operating and Strategic Grants from the Natural Sciences and Engineering Research Council of Canada to C.J.D.

REFERENCES

1.   LEWIS, N.G.,YAMAMOTO, E. 1990. Lignin: Occurance, biogenesis, and degradation. Ann. Rev. Plant Physiol. Plant Mol. Biol. 41: 455-496.

2.   LUDWIG, S.R., WESSLER, S.R. 1990. Maize *R*-gene family: tissue-specific helix-loop-helix proteins. Cell 62: 849-851.

3.   ALMEIDA, J., CARPENTER, R., ROBBINS, T.P., MARTIN, C., COEN, E.S. 1989. Genetic interactions underlying flower color patterns in *Antirrhinum majus*. Genes Dev. 3: 1758-1767.

4.   WIERMANN, R. 1981. Secondary plant products and cell and tissue differentiation. In: The Biochemistry of Plants, vol. 7, (P.K. Stumpf and E.E. Conn, eds.), Academic Press, New York, pp. 85-116.

5.   LYNN, D.G., CHANG, M. 1990. Phenolic signals in cohabitation: Implications for plant development. Annu. Rev. Plant Physiol. Plant Mol. Biol. 41: 497-526.

6.   SCHMELZER, E., JAHNEN, W., HAHLBROCK, K. 1988. *In situ* localization of light-induced chalcone synthase mRNA, chalcone synthase, and flavonoid end products in epidermal cells of parsley leaves. Proc. Natl. Acad. Sci. U.S.A. 85: 2989-2993.

7.   KNOGGE, W., WEISSENBÖCK, G. 1986. Tissue-distribution of secondary phenolic biosynthesis in developing primary leaves of

Avena sativa L. Planta 167:196-205.

8. de VLAMING, P., KHO, K.F.F. 1976. 4,2',4',6'-Tetrahydroxychalcone in pollen of *Petunia hybrida*. Phytochemistry 15:348-349.

9. HAHLBROCK, K., SCHEEL, D. 1989. Physiology and molecular biology of phenylpropanoid metabolism. Annu. Rev. Plant Physiol. Plant Mol. Biol. 40:347-369.

10. DIXON, R.A., LAMB, C.J. 1990. Molecular communication in interactions between plants and microbial pathogens. Annu. Rev. Plant Physiol. Plant Mol. Biol. 41:339-367.

11. SCHMELZER, E., KROEGER-LEBUS, S., HAHLBROCK, K. 1989. Temporal and spatial patterns of gene expression around sites of attempted fungal infection in parsley leaves. Plant Cell, 1:993-1001.

12. DIXON, R.A., HARRISON, M.J. 1990. Activation, structure, and organization of genes involved in microbial defense in plants. Adv. Genet. 28:165-234.

13. DANGL, J.L. 1991. Regulatory elements controlling developmental and stress induced expression of phenylpropanoid genes. In: Plant Gene Research, Vol. 8, (T. Boller and F. Meins, eds.), Springer Veralag, Vienna, in press.

14. LIANG, X., DRON, M., CRAMER, C., DIXON, R.A., LAMB, C.J. 1989. Differential regulation of phenylalanine ammonia-lyase genes during development and by environmental cues. J. Biol. Chem. 264:14486-14492.

15. HARKER, C.L., ELLIS, T.H.N., COEN, E.S. 1990. Identification and genetic regulation of the chalcone synthase multigene family in pea. Plant Cell 2:185-194.

16. KOES, R.E, van BLOKLAND, R., QUATTROCHIO, F., van TUNEN, A.J., MOL, J.N.M. 1990. Chalcone synthase promoters in *Petunia* are active in pigmented and unpigmented cell types. Plant Cell 2:379-392.

17. LUDWIG, S.R., HABERA, L.F., DELLAPORTA, S.L., WESSLER, S.R. 1989. LC, a member of the maize R-gene family responsible for tissue-specific anthocyanin production, encodes a protein similar to transcriptional activators and contains the myc homology region. Proc. Natl. Acad. Sci. U.S.A. 86:7092-7096.

18. HARRISON, M.J., LAWTON, M.A., LAMB, C.J., DIXON, R.A. 1991. Characterization of a nuclear protein that binds to three elements within the silencer region of a bean chalcone synthase gene promoter. Proc. Natl. Acad. Sci. U.S.A. 88:2515-2519.

19.   GROSS, G.G., ZENK, M.H. 1974. Isolation and properties of hydroxy-
      cinnamate:CoA ligase from lignifying tissue of *Forsythia*. Eur. J.
      Biochem. 42:453-459.

20.   KNOBLOCH, K.-H., HAHLBROCK, K. 1975. Isoenzymes of p-
      coumarate:CoA ligase from cell suspension cultures of Glycine max.
      Eur. J. Biochem. 52:311-320.

21.   RANJEVA, R., BOUDET, A.M., FAGGION, R. 1976. Phenolic
      metabolism in *Petunia* tissues. IV. Properties of *p*-coumarate:
      coenzyme A ligase isoenzymes. Biochimie 58:1255-1262.

22.   KNOBLOCH, K. H., HAHLBROCK, K. 1977. 4-Coumarate:CoA ligase
      from cell suspension cultures of *Petroselinum hortense* Hoffm. Arch.
      Biochem. Biophys. 184:237-248.

23.   WALLIS, P.J., RHODES, M.J.C. 1977. Multiple forms of hydroxy-
      cinnamte: CoA ligase in etiloated pea seedlings. Phytochemistry
      16:1891-1894.

24.   GRAND, C., BOUDET, A., BOUDET, A.M. 1983. Isoenzymes of
      hydroxycinnamate:CoA ligase from poplar stems and tissue distribu-
      tion. Planta 158:225-229.

25.   VINCENT, J.R., NICHOLSON, R.L. 1987. Evidence for isoenozymes of
      hydroxycinnamate:CoA ligase in maize mesocotyls and their response
      to infection by *Heminthosporium maydis* race 0. Physiol. Mol. Plant
      Pathol. 30:121-129.

26.   LOZOYA, E., HOFFMANN, H., DOUGLAS, C.J., SCHULZ, W.,
      SCHEEL, W., HAHLBROCK, K. 1988. Primary structures and
      catalytic properties of isoenzymes encoded by the two 4-coumarate:
      CoA ligase genes in parsley. Eur. J. Biochem. 176:661-667.

27.   BECKER-ANDRÉ, M., SCHULZE-LEFERT, P., HAHLBROCK, K.
      1991. Structural comparison, modes of expression, and putative *cis*-
      acting elements of the two 4-coumarate:CoA ligase genes in potato.
      J. Biol. Chem. 266:8551-8559.

28.   DOUGLAS, C.J., HOFFMANN, H., SCHULZ, W., HAHLBROCK, K.
      1987. Structure and elicitor or u.v.-light stimulated expression of two
      4-coumarate: CoA ligase genes in parsley. EMBO J. 6:1189-1195.

29.   LOIS, R., DIETRICH, A., HAHLBROCK, K., SCHULZ, W. 1989. A
      phenylalanine ammonia-lyase gene from parsley: structure, regulation
      and identification of elicitor and light responsive elements. EMBO J.
      8:1641-1648.

30.   DOUGLAS, C.J., HAUFFE, K.D., ITES-MORALES, M.-E., ELLARD,
      M., PASZKOW-SKI, U., HAHLBROCK, K., DANGL, J.L. 1991.

Exonic sequences are required for elicitor and light activation of a plant defense gene, but promoter sequences are sufficient for tissue specific expression. EMBO J., 10:1767-1775.

31. SCHULZE-LEFERT, P., DANGL, J.L., BECKER-ANDRE, M., HAHLBROCK, K., SCHULZ,W. 1989. Inducible *in vivo* footprints define sequences necessary UV-light activation of the parsley chalcone synthase gene. EMBO J. 8:651-656.

32. van de LÖCHT, U., MEIER, I., HAHLBROCK, K., SOMSSICH, I.E. 1990. A 125 bp promoter fragment is sufficient to for strong elicitor-mediated gene activation in parsley. EMBO J. 9:945-2950.

33. ELLIOTT, R.C., DICKEY, L.F., WHITE, M.J., THOMPSON, W.F. 1989. *cis*-acting elements for light regulation of pea ferredoxin I gene expression are located within transcribed sequences. Plant Cell 1:691-698.

34. BEVAN, M., SHUFFLEBOTTOM, D., EDWARDS, K., JEFFERSON, R., SCHUCH,W. 1989. Tissue- and cell-specific activity of a phenylalanine ammonia-lyase promoter in transgenic plants. EMBO J. 8:1899-1906.

35. LIANG, X., DRON, M., SCHMID, J., DIXON, R.A., LAMB, C.J. 1989. Developmental and environmental regulation of a phenyl-alanine ammonia-lyase beta-glucuronidase gene fusion in transgenic tobacco plants. Proc. Natl. Acad Sci. U.S.A. 86:9284-9288.

36. SCHMID, J., DOERNER, P.W., CLOUSE, S.D., DIXON, R.A., LAMB, C.J. 1990. Developmental and environmental regulation of a bean chalcone synthase promoter. Plant Cell 2:619-631.

37. OHL, S., HEDRICK, S.A., CHORY, J., LAMB, C.J. 1990. Functional properties of the phenylalanine ammonia-lyase promoter from *Arabidopsis*. Plant Cell 2:837-848.

38. HAUFFE, K.D., PASZKOWSKI, U., SCHULZE-LEFERT, P., HAHL-BROCK, K., DANGL, J.L., DOUGLAS, C.J. 1991. A parsley 4CL-1 promoter fragment specifies complex expression patterns in transgenic tobacco. Plant Cell 3:435-443.

39. HEILMAN, P.E., STETTLER, R.F. 1985. Genetic variation and productivity of *Populus trichocarpa* and its hybrids. II. biomass production in a 4-year plantation. Can. J. For. Res. 15:384-388.

40. WOLLENWEBER, E. 1973. Flavonoidmuster als systematisches Merkmal in der Gattung Populus. Biochem. Systematics Ecol. 3:35-45.

Chapter Four

MOLECULAR BIOLOGY OF STRESS-INDUCED PHENYLPROPANOID AND ISOFLAVONOID BIOSYNTHESIS IN ALFALFA

Richard A. Dixon, Arvind D. Choudhary,[*] Karen Dalkin, Robert Edwards,[*] Theo Fahrendorf, Ganesan Gowri, Maria J. Harrison, Christopher J. Lamb,[#] Gary J. Loake, Carl A. Maxwell, John Orr and Nancy L. Paiva

Plant Biology Division, The Samuel Roberts Noble Foundation, P.O. Box 2180, Ardmore, Oklahoma 73402, U.S.A.

[*]Department of Botany, Nagpur University Campus, Nagpur 440010, India (A.D.C.); Department of Biological Science, University of Durham, South Road, Durham DH1 3HP, United Kingdom (R.E.)

[#]Plant Biology Laboratory, The Salk Institute for Biological Studies, 10010 North Torrey Pines Road, La Jolla, California 92037, U.S.A.

Phenolic Metabolism in Plants, Edited by H.A. Stafford
and R.K. Ibrahim, Plenum Press, New York, 1992

INTRODUCTION

Alfalfa (*Medicago sativa* L.) is probably the world's most important forage crop. As a nitrogen-fixing legume, it is of value as a source of reduced nitrogen, and is becoming increasingly popular as a soil-conserving perennial in low-input sustainable agricultural systems. It is a cross-pollinated autotetraploid which, being highly heterozygous, exhibits substantial genetic diversity between individual plants of a particular cultivar. The complexity of alfalfa genetics is a hindrance to any form of molecular investigation which involves genetic crosses; however, alfalfa is amenable to stable genetic transformation with *Agrobacterium tumefaciens* or *A. rhizogenes*,[1-6] and is readily, if slowly, regenerable from culture via somatic embryogenesis.[7-9] It is therefore a suitable target for improvement by recombinant DNA technology.

A large part of the literature on alfalfa biochemistry and molecular biology is concerned with nodulation in response to *Rhizobium meliloti* [10-13] and subsequent nitrogen metabolism. In 1988, we initiated a program aimed at understanding the molecular biology of alfalfa defense responses as a basis for manipulating disease resistance in transgenic plants. Previous studies had implicated the isoflavonoid phytoalexin medicarpin as an important factor in defense against a number of fungal pathogens of alfalfa,[14-17] and the basic pathways for medicarpin biosynthesis had been elucidated.[18-21] The early enzymatic stages of the phenylpropanoid pathway leading to isoflavonoids are common to the biosynthesis of lignin and to flavone and chalcone inducers of rhizobial nodulation genes.[22] The synthesis of functionally different phenylpropanoid and flavonoid end products is under differential environmental and developmental control. These pathways thus provide an excellent subject for investigation of the molecular mechanisms underlying the activation of specific plant genes by microbial attack, fungal elicitation and endogenous developmental signals.[23,24] Additional goals in alfalfa include (1) understanding the roles of isoflavonoids and substituted chalcones in defense and symbiosis, respectively; (2) manipulating the phytoalexin pathway via gene transfer in order to improve resistance both qualitatively and quantitatively; (3) manipulating lignin content/ composition in order to improve forage quality.

The present chapter summarizes our current understanding of the enzymology and molecular biology of stress-induced phenylpropanoid and flavonoid biosynthesis in alfalfa. We describe preliminary studies on nuclear factors which regulate transcription of phenylpropanoid biosynthetic genes in this species, and outline strategies involving genetic manipulation of the phenylpropanoid and isoflavonoid pathways for alfalfa improvement.

ACCUMULATION OF ISOFLAVONOIDS IN INFECTED OR ELICITED ALFALFA CELLS

Alfalfa is host to a wide range of potential fungal pathogens.[25] These include *Pseudopeziza*, *Leptosphaerulina* and *Stemphylium* leaf spot diseases, anthracnose *(Colletotrichum trifolii)*, *Fusarium* and *Verticillium* wilts and *Phytophthora* root rot. The genetic diversity in the crop has greatly facilitated the development of resistance to such diseases. Of a range of isoflavonoids with antimicrobial activity characterized in *Medicago* species,[26-30] the pterocarpan medicarpin (Fig. 1) has been implicated in the resistance of alfalfa to *C. trifolii*, *Verticillium albo-atrum* and several leaf spot diseases.[14-17] This proposed role is based primarily on indirect evidence such as correlations between the timing of phytoalexin accumulation in resistant and susceptible interactions, and the potency of medicarpin as an antimicrobial agent *in vitro*. The cloning of genes for medicarpin biosynthesis may now facilitate the direct testing of the role of the phytoalexin in resistance, if phenotypic mutants in phytoalexin synthesis can be engineered by expression of antisense or ribozyme constructs (see below).

Alfalfa callus and cell suspension cultures maintain the capacity to produce isoflavonoid phytoalexins in response to a suitable stimulus. Callus cultures have been selected which maintain resistance or susceptibility to infection with spores of *Verticillium albo-atrum*.[17] In such cases, resistance is associated with greater and more rapid production of medicarpin, with parallel effects on the timing and extent of increase in the activities of enzymes of medicarpin biosynthesis.[31] Alfalfa cell suspension cultures can be induced to synthesize medicarpin on exposure to crude elicitor preparations from the cell walls of *C. lindemuthianum* or *C. trifolii*, from yeast, or from autoclaved alfalfa leaf extracts.[32] In contrast to the situation reported in other legumes such as

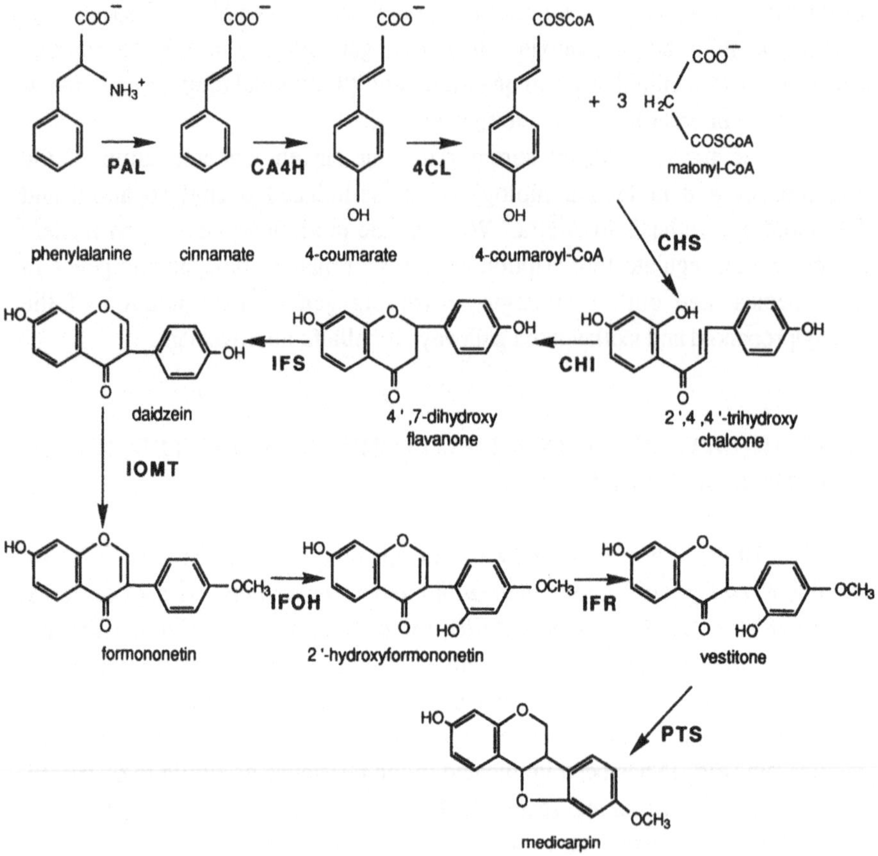

Fig. 1. Biosynthesis of the isoflavonoid phytoalexin medicarpin in alfalfa. The enzymes are: PAL, L-phenylalanine ammonia-lyase; CA4H, cinnamic acid 4-hydroxylase; 4CL, 4-coumarate:CoA ligase; CHS, chalcone synthase; CHI, chalcone isomerase; IFS, isoflavone synthase; IFMT, isoflavone 4'-*O*-methyltransferase; IFOH, isoflavone 2'-hydroxylase; IFR, isoflavone reductase; PTS, pterocarpan synthase.

bean or soybean, pectic fragments do not elicit medicarpin production in alfalfa cell suspension cultures.[32] Likewise, the tripeptide glutathione, which is a strong inducer of the phytoalexin response in bean,[33] pea,[34] soybean[35] and even the cactus *Cephalocereus senilis*,[36] is inactive as an elicitor in alfalfa cell suspensions.[33] This may possibly be the result of a greater insolubilization of the tripeptide during uptake into the cells than is observed in a glutathione-inducible species such as bean.[33]

Unelicited cell suspension cultures of alfalfa cultivar Calwest 475 contain low levels (approximately 40 nmol/g fresh wt) of medicarpin 3-O-glucoside-6"-O-malonate (MGM), but undetectable levels of the free aglycone.[27] Exposure of the cells to fungal elicitor results in the appearance of high concentrations of both medicarpin and MGM in the cells (Fig. 2). Medicarpin, but not the glycoside, also accumulates in the culture medium.[37] The major constitutive isoflavonoids present in these alfalfa cell cultures are the 7-O-glucoside and 7-O-glucoside-6"-O-malonate of afrormosin (4',6-dimethoxy-7-hydroxy isoflavone).[27] Their levels do not change significantly during elicitation, although these compounds do incorporate significant label from ^{14}C-phenylalanine, suggesting that they may be undergoing turnover. In labeling experiments in which elicited cells were treated with L-α-aminooxy-β-phenylpropionic acid, a potent and specific inhibitor of the first enzyme of the phenylpropanoid pathway, the initial accumulation of medicarpin of low specific activity suggests that some conversion of MGM to medicarpin may occur in the early stages of elicitation.[27] MGM is also present in low levels in roots, but not leaves, of healthy alfalfa seedlings[37] (N.L. Paiva, unpublished results). If its function is to act as a preformed source of phytoalexin, its levels can only make a minor contribution to the overall phytoalexin response.

BIOCHEMISTRY AND MOLECULAR BIOLOGY OF THE PHENYLPROPANOID AND ISOFLAVONOID PATHWAYS IN ALFALFA

Coordinated Induction of Enzymes for the Synthesis of Medicarpin from L-Phenylalanine

Figure 1 shows the proposed biosynthetic pathway to medicarpin. The later stages of this pathway beyond the chalcone synthase (CHS) reaction were elucidated by radiolabelled precursor feeding studies in alfalfa seedlings elicited by exposure to $CuCl_2$,[18-20] and by direct isolation of the intermediates from infected white clover.[21] Much of this pathway has now been confirmed by the characterization of the specific enzymes involved, primarily in soybean, chickpea and pea,[38-45] which make related isoflavonoid phytoalexins, albeit with different ring substitution patterns. There is still, however, a major question over the stage at which the 4'-methoxy substituent is added to the isoflavone nucleus.

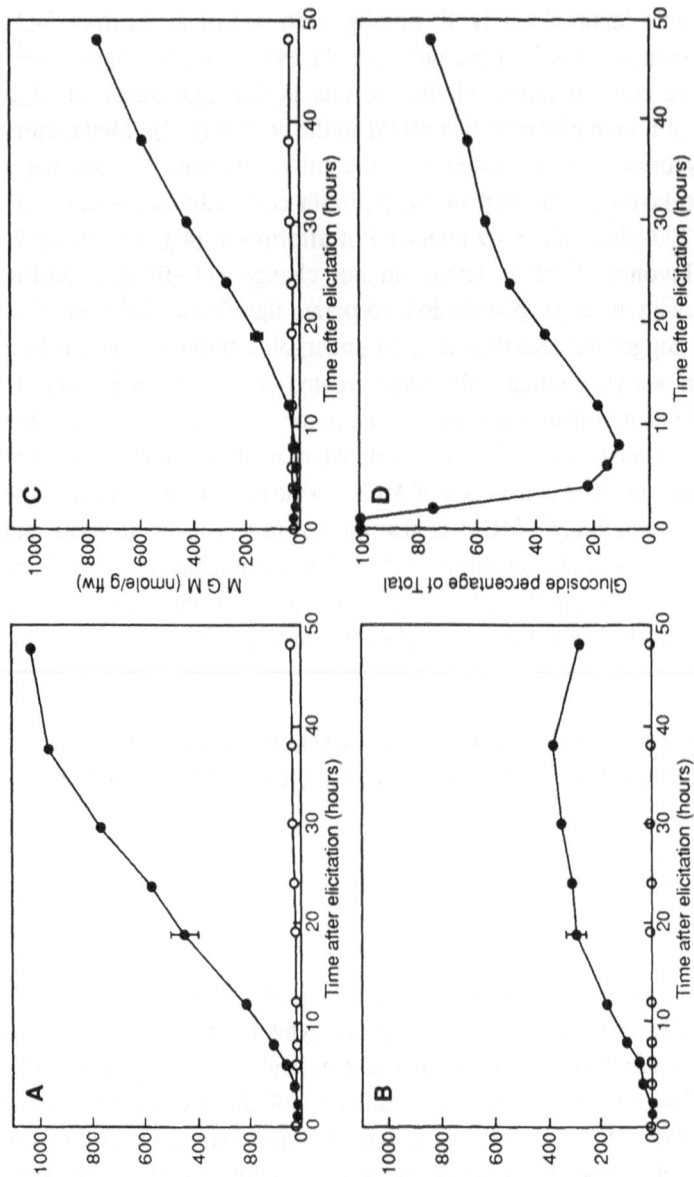

Fig. 2. Appearance of medicarpin and medicarpin malonyl glycoside (MGM) in alfalfa cell suspension cultures treated with elicitor from the cell walls of *Colletotrichum lindemuthianum* (●) or water-treated controls (o). **A:** Total medicarpin (medicarpin plus MGM). **B:** Medicarpin. **C:** MGM. **D:** MGM as a proportion of total medicarpin.

The isoflavone O-methyltransferase activity in Figure 1 is shown as methylating the first isoflavonoid intermediate in the pathway, daidzein, to yield formononetin. However, the labelling experiments in $CuCl_2$-treated alfalfa seedlings suggested that daidzein was not a precursor of formononetin, although 2',4,4'-trihydroxychalcone was efficiently incorporated.[19] This led to the proposal that methylation of the 4'-hydroxyl group may be an integral part of the isoflavone synthase reaction. Subsequent characterization of isoflavone synthase enzymatic activity from soybean[41] and alfalfa[37] does not appear to substantiate this proposal. This stage of the pathway clearly requires further investigation (see below).

The elicitor-induced accumulation of medicarpin in alfalfa cell suspension cultures is preceded by increases in the extractable activities of all the enzymes shown in Figure 1.[32,37] With the exception of 4-coumarate:CoA ligase (4CL), which is present at a relatively high basal level and only induced approximately 2-fold by elicitor, the activities are all strongly induced from low basal levels. The times at which the induced activities attain their maximum levels post-elicitation, and their approximate fold inductions, are summarized in Table 1. Although the kinetics of appearance of the ten individual enzymes in Figure 1 are not strictly co-ordinated through 48 h post-elicitation, they all exhibit their initial increases within 1 to 2 h of exposure to elicitor, suggesting that their induction is probably in response to a signal common to the activation of each of their genes. This is further supported by studies at the RNA transcript level (Table 1). In contrast, expression of defense related hydroxyproline-rich glycoprotein genes in bean cells is delayed several hours following elicitation, suggesting the operation of a signal pathway different from, but possibly dependent on, that involved in activation of phytoalexin biosynthesis.[46] Six of the enzymes in Figure 1 are transiently induced by elicitor, whereas the activities of chalcone isomerase (CHI), isoflavone O-methyltransferase (IOMT), isoflavone reductase (IFR) and pterocarpan synthase (PTS) remain at elevated levels up to 48 h post-elicitation.

Changes in mRNA Populations in Elicited Alfalfa Cells

In order to obtain a general picture of the relative changes in mRNA species occurring in response to fungal elicitor treatment, polysomal mRNA was isolated from alfalfa cells at various times up to 48 h post-elicitation, and translated *in vitro* in a message-dependent rabbit reticulocyte lysate. ^{35}S-labelled translation products were then analyzed by 1- and 2-D gel electrophoresis. In these experiments, at least 15 distinct *in vitro* translation products resolved on 1-D gels were strongly up-regulated by elicitor; these include phenylalanine

Table 1. Induction of enzymes of isoflavonoid biosynthesis and related pathways in alfalfa cell suspension cultures exposed to elicitor from the cell walls of *Colletotrichum lindemuthianum*.

	Time of attainment of maximum extractable enzyme activity (h)	Approximate fold induction	Induction of transcripts by elicitor	Ref. #
L-Phenylalanine ammonia-lyase	10	14	+	32,47,54
Cinnamic acid 4-hydroxylase	15	5	nd	32,37
4-Coumarate:CoA ligase	24	2	+	32,47
Chalcone synthase	12	30	+	32,37,47
Chalcone isomerase	24	5	nd	32
Isoflavone synthase	12	10	nd	37
Isoflavone 4'-*O*-methyltransferase	48	2-3	nd	67
Isoflavone 7-*O*-methyltransferase	12	200	nd	32,67
Isoflavone 2'-hydroxylase	15	20	nd	37
Isoflavone reductase	30	10	+	70
Pterocarpan synthase	30	2	nd	70
Chalcone 2'-*O*-methyltransferase[a]	24	5	nd	_b
Caffeic acid *O*-methyltransferase	42-48	1.5-2.0	+	65,69
Coniferyl alcohol dehydrogenase	24	2	nd	32

nd = not determined.

[a] cells were induced by yeast elicitor.

[b] C. Maxwell, unpublished.

ammonia-lyase (PAL) and CHS subunits, plus major unidentified products of M_r 41,000, 37,000, 34,000, 25,000 and 24,000.[47] Elicitor-induced increases in translatable mRNA activity are generally seen within 1 h of elicitation; however, the Mr 25,000 and 24,000 products appear after a lag of approximately 3 h. The unidentified Mr 34,000 polypeptide, which is second to CHS in relation to the amount of label incorporated from ^{35}S-methionine *in vitro*, is resolved into approximately 4 putative subunit isoforms on 2-D gel analysis.[47] As the major increases in translation products occur at early times post-elicitation, RNA from cells exposed to elicitor for 2, 3 and 4 h was pooled and used for construction of a cDNA library in λZAPII for isolation of alfalfa cDNAs encoding the enzymes of phenylpropanoid and isoflavonoid biosynthesis described below.

Phenylalanine ammonia-lyase (PAL, EC.4.3.1.5)

PAL has been purified from a number of species, including several legumes.[22,48,49] Previous studies on bean PAL showed that the enzyme existed in cell cultures in four distinct forms separable by chromatofocussing.[49] These forms were differentially induced by fungal elicitor, with preferential appearance of the form with the lowest K_m value for L-phenylalanine. Multiple active forms, and subunits separable by 2-D gel electrophoresis, were shown to arise due to the expression of different members of a family of four bean PAL genes.[50,51]

PAL from alfalfa cell suspension cultures can be resolved into three distinct forms separable on the basis of chromatofocussing or hydrophobic interaction chromatography.[52] These forms have identical subunit M_r (79,000), although polypeptides of M_r 71,000, 67,000 and 57,000 are observed in all preparations. These probably arise by degradation of the intact PAL subunit, a situation observed with the enzyme from bean.[53] The active alfalfa PAL enzyme is a tetramer of M_r 311,000; it is not yet known whether the multiple forms observed during purification are homotetramers or heterotetramers with respect to the different subunit isoforms which can be resolved on 2-D gel analysis of purified preparations or of *in vitro* translated subunits after specific immunoprecipitation. As with bean PAL, the individual forms of the alfalfa PAL tetramer exhibit classical Michaelis-Menten kinetics, whereas mixtures of the forms, as found in partially purified preparations, appear to exhibit negative rate cooperativity with respect to L-phenylalanine.[52] Furthermore, the different forms are differentially induced as a function of time in response to fungal elicitors, with the least hydrophobic isoform 1 (lowest K_m for phenylalanine) accumulating to highest activity. Differences are also observed in the extent of induction of the highest K_m form (isoform 3) by two different elicitors, from the cell walls of *C. lindemuthianum* or from yeast. The activity of isoform 3 as a percentage of total PAL activity remains constant at approximately 15% in cells elicited by yeast extract, whereas it declines to only 3% 10 h after exposure to *Colletotrichum* elicitor.[52]

A polyclonal antiserum has been raised against purified alfalfa PAL, and used to screen the λZAP cDNA expression library. A single, full length alfalfa PAL cDNA (APAL1) was isolated consisting of a 2,175 bp open reading frame,

96 bp 5'-untranslated leader and 128 bp 3' non-coding region.[54] The deduced amino acid sequence was 86.5% similar to that of the PAL2 gene from bean. Potato PAL has recently been shown to be glycosylated.[55] The APAL1 deduced amino acid sequence contains seven potential N-glycosylation sites,[54] although there is no evidence to suggest that this enzyme is processed through the endoplasmic reticulum.

Re-screening the cDNA library with the APAL1 coding sequence led to the isolation of a further 60 putative PAL clones, five of which were sequenced

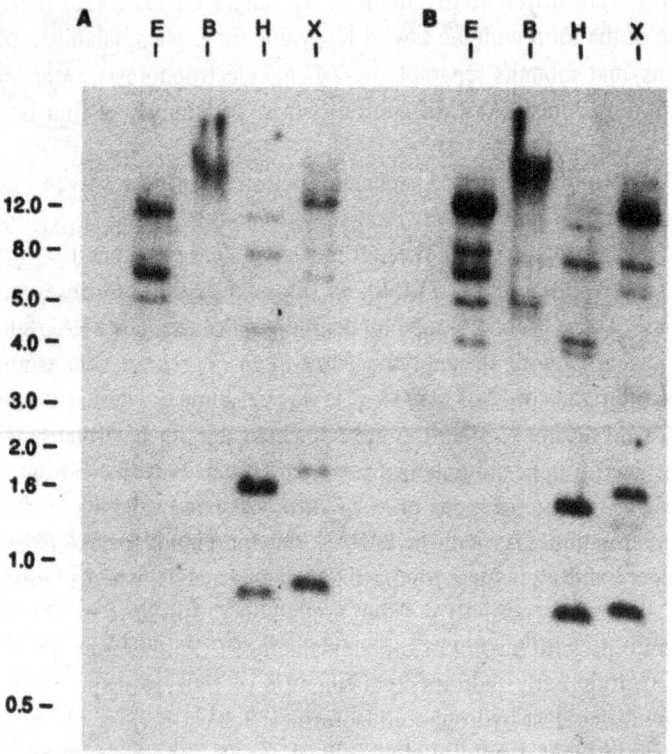

Fig. 3. Southern blot analysis of alfalfa genomic DNA for L-phenylalanine ammonia-lyase sequences. DNA from whole seedlings was digested with *Eco*RI (E), *Bam*HI (B); *Hind*III (H) or *Xba*I (X). After electrophoresis and blotting, membranes were probed with ^{32}P-labelled sequences, generated by polymerase chain reaction, complementary to the 3'-untranslated regions of the APAL1 and APAL3 cDNAs.

1 2 3 4 5 6   7 8 9 10 11 12 13 14

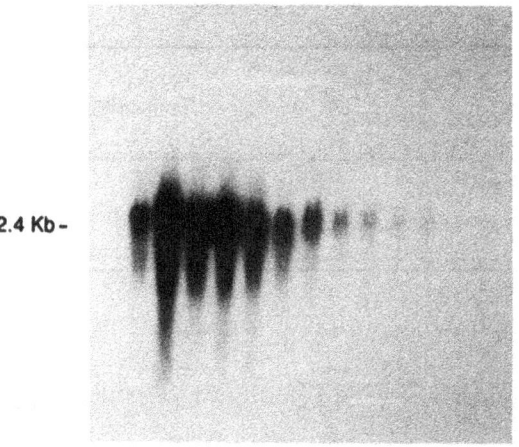

2.4 Kb -

Fig. 4. Northern blot analysis of total PAL transcripts in elicitor-treated alfalfa cell suspension cultures. Cells were exposed to elicitor from the cell walls of *Colletotrichum lindemuthianum* for 0,1,2,4,6, 8,12,14,24,30,38, and 48 h respectively (lanes 1-12) or to an equal volume of water for 6 or 48 h (lanes 13, 14). Total RNA was extracted, electrophoresed and, after blotting, membranes were hybridized with a ^{32}P-labelled 471 bp internal coding fragment of APAL1.

from their 3' ends. The 3' end of a clone designated APAL3 was 86% identical to the 3'-untranslated region of APAL1.[54] Southern blot hybridization of genomic DNA using the 2.4 kb APAL1 sequence as probe revealed a multigene family of up to six PAL genes, or possible allelic variants, in alfalfa.[54] Hybridization with probes specific for the 3'-ends of APAL1 and APAL3 also revealed multiple bands in alfalfa genomic DNA, suggesting that several of the individual PAL genes may have similar 3'-sequences (Fig. 3).

Alfalfa PAL transcripts are rapidly and strikingly induced in fungal elicitor-treated cell cultures, reaching maximum levels approximately 2 h post-elicitation, then declining to near basal levels by 24 h (Fig. 4). Northern blot analysis of organ samples from alfalfa seedlings at various stages of development reveals maximum levels of PAL transcripts in roots, stems, and petioles (Table 2), consistent with the requirement for PAL activity in lignifying tissues.

Table 2.   Organ-specific expression of phenylpropanoid and isoflavonoid pathway gene transcripts in young alfalfa seedlings.

Organ	Transcript				
	PAL	CHS	IFR	COMT	AdoMet Synthetase
Roots	++++	++++	++++	+++	++
Root nodules	+	+++	++++	+	++
Stems	++++	++	+	++++	++++
Growing points	+	+	nd	++	+
Leaves	+	+	+	+	+
Petioles	+++	+	+	++	++
Buds	+	++	nd	++	+
Flowers	+	++	nd	+++	+

nd = not detected.

Cinnamic Acid 4-Hydroxylase (CA4H, EC 1.14.13.11) and Other Cytochrome P-450s

Three reactions in the biosynthesis of medicarpin are catalyzed by cytochrome P450 enzymes. These are CA4H, isoflavone synthase (IFS), and isoflavone 2'-hydroxylase (IFOH) (Fig. 1). To date, none of the genes encoding these enzymes has been cloned. The purification of these P-450s presents serious technical difficulties in view of their being integral membrane proteins of low abundance which require reconstitution with cytochrome P-450 reductase in order to assay after solubilization. Their primary characterization has been reported from several systems,[41,56,57] but only their induction kinetics have been determined in alfalfa (Table 1).[37] All three P-450 activities are induced in elicitor-treated alfalfa cell suspension cultures, but the IFS and IFOH activities in particular are extremely labile. Several groups, including our own, are employing indirect molecular approaches to try to isolate plant cytochrome P-450s. These include polymerase chain reaction amplification of cDNA or

genomic sequences using series of synthetic oligonucleotide primers based on
sequences with the best levels of conservation among members of the vertebrate,
bacterial and fungal P-450 families, and the use of a heterologous antiserum
raised against an abundant P-450 from ripening avocado fruit.[58] The avocado
antibody detects a major polypeptide of M_r 52,000 and minor species of slightly
lower M_r when used at low dilutions to probe Western blots of alfalfa cell
culture proteins (T. Fahrendorf, R.A. Dixon, unpublished results). These polypeptides are not, however, significantly induced by fungal elicitor. Knowing
their amino acid sequences may, however, aid the design of oligonucleotide
primers more specific for alfalfa P-450s. Putative P-450 cDNAs obtained by the
above strategies could be analyzed functionally by two criteria: first, elicitor
inducibility of their transcripts, and second, ability of antisense transcripts to
inhibit the elicitor-induced enzymatic activity of a specific P-450 species on
electroporation into alfalfa protoplasts, in which CA4H, IFS and IFOH are all
inducible by *Colletotrichum* elicitor.[37] The feasibility of this latter approach
has been shown by successful inhibition of elicitor-induced CHS activity in
alfalfa protoplasts by electroporation of a plasmid expressing bean CHS in the
antisense orientation.[59]

Chalcone Synthase (CHS, EC 2.3.1.74)

CHS is the first committed step channelling phenylpropanoid and
acetate metabolites into the flavonoid/isoflavonoid branch pathway. In most
species, the enzyme is a dimer of M_r 43,000 subunits. The details of its
complex catalytic reaction, involving the head-to-tail condensation of three
molecules of malonyl coenzyme A with one molecule of 4-coumaroyl coenzyme
A, have been reviewed elsewhere.[22] CHS catalyzes the chalcone condensation
reaction; in the case of the 5-deoxy flavonoid/isoflavonoid derivatives, such as
occur in legumes, a separate reductase co-acts with CHS to catalyze the removal
of the hydroxyl group which would become the 5-hydroxyl group on the A-ring.
This reaction appears to occur prior to release of the chalcone from the
condensing enzyme. The reductase has recently been characterized from soybean,
where it is co-induced with CHS.[45] It has yet to be investigated in alfalfa (see
also Chapter 5 by Barz).

The CHS condensing enzyme has been purified from a number of
sources, including buckwheat,[60] parsley,[61] bean[62] and spinach,[63] and has been
cloned from an even wider range of species.[23] In most legume species studied to
date it is encoded by a small multigene family. This apparent amplification of
CHS genes may be related to the special involvement of flavonoids in microbial

resistance and symbiosis in the *Leguminosae*. In alfalfa, 2-D gel analysis of immunoprecipitated *in vitro* translation products indicates the presence of a number of CHS subunits isoforms differing in isoelectric point, very similar to the pattern of bean or soybean CHS subunits.[47] Increases in transcripts encoding each of these subunit isoforms are observed on elicitation of cell cultures. This parallels the observed induction of all members of the bean CHS gene family, albeit with somewhat different kinetics, in response to fungal elicitor. Southern blot analysis confirms that CHS is encoded by a multigene family in alfalfa as in other legumes, the pattern of hybridizing fragments being somewhat more complex than observed for bean CHS sequences (Fig. 5).

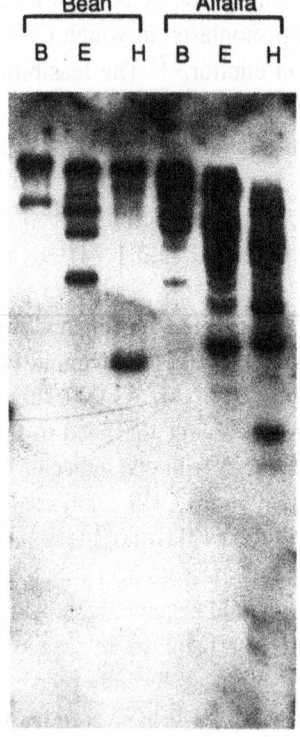

Fig. 5. Southern blot analysis of alfalfa and bean genomic DNA for chalcone synthase sequences. DNA was digested with *Bam*HI (B), *Eco*RI (E) or *Hind*III (H). After electrophoresis and blotting, membranes were probed with a ^{32}P-labelled 1.4 kb fragment from the coding sequence of the bean CHS1 gene.

cDNAs encoding several different alfalfa CHS transcripts have been isolated and sequenced (K. Dalkin and R.A. Dixon, unpublished results). They show strong sequence similarity in their open reading frames to CHS from other legumes. Alfalfa CHS transcripts are strongly induced by elicitor, and are expressed preferentially in root tissues during seedling development (Table 2).

Flavonoid derivatives synthesized via the CHS reaction can function as flower pigments, UV-protectants, phytoalexins, nodulation gene inducers[10,12] and possibly as regulators of auxin transport.[64] There is considerable interest as to how these various functions are integrated into the developmental program of the plant. For this reason, the CHS gene promoter has become a model for studying the factors which interact with plant genes to determine their environmentally and developmentally regulated expression patterns. Studies of cis-elements and trans-factors conditioning expression of a bean CHS promoter in alfalfa cells are outlined in a later section of this chapter.

Isoflavone 7-and 4'-O-methyltransferases

Fractionation studies indicate that alfalfa cell cultures contain at least three distinct O-methyltransferase activities which can methylate products of the phenylpropanoid pathway (Fig. 6). The major activity catalyzes the methylation of caffeic and 5-hydroxyferulic acids,[65] and is therefore involved in the lignin branch pathway. A distinct activity, of considerably lower abundance, catalyzes the 2'-O-methylation of 2',4,4'-trihydroxychalcone (C.A. Maxwell, R. Edwards, R.A. Dixon, unpublished results), and is thus believed to be involved in the formation of signals for induction of *Rhizobium meliloti* nodulation genes. As discussed above, the origin of the methyl group on the B-ring of the isoflavonoid phytoalexins of alfalfa is still a matter of debate. Early studies on the enzymatic O-methylation of daidzein did not conclusively demonstrate the position of methylation.[66] Isoflavone O-methyltransferase activity is induced approximately 200-fold on exposure of alfalfa cells to fungal elicitor.[67] However, this activity catalyzes the conversion of daidzein to the A-ring methoxy derivative isoformononetin, which is unlikely to be a precursor for medicarpin biosynthesis. This isoflavonoid-specific O-methyltransferase activity can be separated into two forms, IOMT I and IOMT II, of which IOMT II has been purified to near homogeneity.[67] IOMT I and IOMT II can methylate a range of isoflavones, and IOMT I can also methylate pterocarpans. Neither IOMT I nor IOMT II effectively methylates daidzein in the 4'-position; indeed, the best substrate for both IOMT I and IOMT II is 6,7,4'-trihydroxyisoflavone. K_m values for the methylation of daidzein to isoformononetin by IOMT II are 20 μM

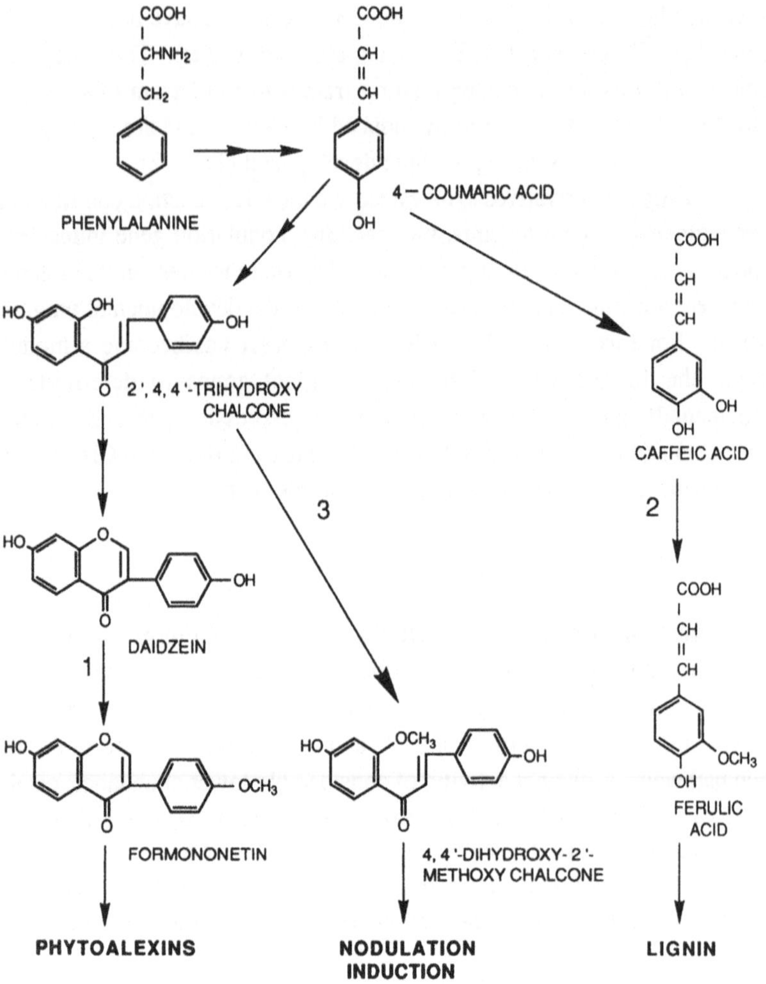

Fig. 6. The involvement of distinct O-methyltransferase activities in the biosynthesis of phytoalexins, lignin and nodulation gene-inducing chalcones. The enzymes are: 1, isoflavone 4'-O-methyl-transferase; 2, caffeic acid O-methyltransferase; 3, chalcone 2'-O-methyltransferase.

for daidzein and 150 μM for the methyl group donor S-adenosyl-L-methionine (SAM). Both isoflavonoid O-methyltransferases have a subunit M_r of 41,000, can be photoaffinity labelled at the active site by tritiated SAM, and are

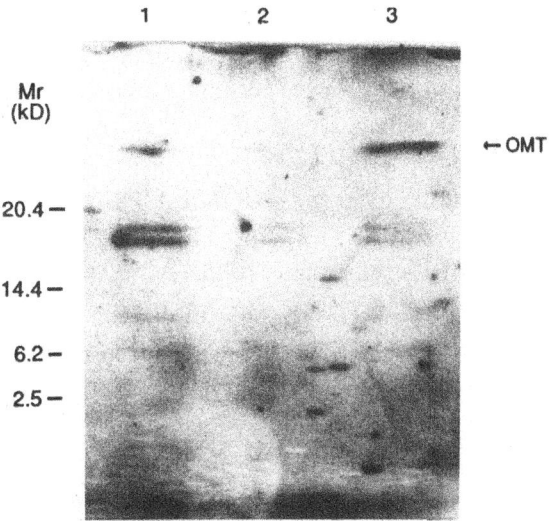

Fig. 7. Peptide maps of purified caffeic acid O-methyltransferase and isoflavone 7-O-methyltransferase from alfalfa cell suspension cultures. The silver-stained gel shows a separation of Cleveland peptide digests of COMT I (lane 1), COMT II (lane 2) and IOMT II (lane 3). OMT = undigested O-methyltransferase.

catalytically active as the monomers.[67] IOMT II appears to be structurally related to the caffeic acid O-methyltransferase on the basis of sharing an identical peptide map pattern (Fig. 7). However, the two enzymes do not appear to be closely related antigenically.[67]

The role of the elicitor-induced isoflavone 7-O-methyltransferase remains to be established. This activity could be involved in the formation of afrormosin (if it also possessed 6-O-methyltransferase activity) or its isomer alfalone (6-hydroxy-7,4'-dimethyoxy isoflavone), both of which have been shown to be present in alfalfa cells. However, as described above, afromosin and its glucoside conjugates do not accumulate in elicitor-treated alfalfa cultures, although their rate of synthesis, as determined by incorporation of label from ^{14}C-phenylalanine, is stimulated on elicitation.[27] It seems unlikely, however, that a 200-fold induction of a biosynthetic enzyme would occur if the fate of its products was simply to be a rapid turnover. A second major question concerns the nature of the 4'-O-methyltransferase activity. Is daidzein the natural substrate for this activity, or does methylation occur at a later stage? A very low level of 4'-O-methyltransferase activity is observed in alfalfa cultures, but this is only very slowly and weakly induced by fungal elicitor.[67] Perhaps the 4'-O-methyltransferase requires more stringent conditions for measurement of activity than the other alfalfa O-methyltransferases.

Caffeic Acid O-methyltransferase (COMT, EC 2.1.1.42)

Elicitation of alfalfa cell cultures results in the accumulation of wall-bound phenolic material. Although these phenolics have not been characterized, it is likely that they consist of methoxylated cinnamic acids or their related alcohols, as their accumulation is preceded by increases in the extractable activities of the lignin biosynthetic enzymes COMT and coniferyl alcohol dehydrogenase (Table 1). COMT (Fig. 6) has been purified to near homogeneity from elicited alfalfa cultures.[65] It exists as two isoforms (COMT I and II) with identical Mrs of 41,000, whose peptide maps appear identical to that of IOMT II (Fig.7). The two forms have very similar substrate specificities, being most active against the lignin precursor caffeic acid. K_m values are in the range of 55 μM for caffeic acid and 10 μM for SAM. Fractionation of extracts using Mono Q FPLC indicates that COMT I is preferentially induced in response to fungal elicitor, although this effect is much less striking than observed with the PAL isoforms, and the extent of induction of COMT activity by elicitor is very low compared to that of PAL and the isoflavonoid biosynthetic enzymes (Table 1).

Using an antiserum against the bifunctional caffeic/5-hydroxyferulic acid O-methyltranferase from aspen,[68] we isolated a COMT cDNA from the alfalfa cDNA expression library.[69] The deduced amino acid sequence of the alfalfa COMT was 86% identical to that of the aspen enzyme. The identity of the clone was confirmed by expression in *E. coli*; lysates from bacterial cells harboring the COMT sequence convert caffeic and 5-hydroxyferulic acids to ferulic and sinapic acids, respectively, in the presence of SAM.[69] Daidzein is not methylated by the *E. coli* extracts. Southern blot analysis of genomic DNA indicates the presence of a maximum of two COMT genes in alfalfa. As one of these could possibly be an allelic variant, and as all the cDNAs obtained on re-screening the library with the COMT sequence appear to correspond to the same transcript, it is not yet clear whether the two isoforms of COMT observed in alfalfa cultures arise from the expression of different genes or via post-translational modification of the product of a single transcript.

In spite of the small and slow increase in COMT activity in elicited alfalfa cultures (Table 1), COMT transcripts are rapidly and strongly induced, reaching maximum levels around 19 h post-elicitation.[69] This suggests that COMT may be regulated post-translationally as well as transcriptionally. Transcripts encoding ATP: L-methionine-S-adenosyl transferase (AdoMet synthase, EC 2.5.1.6), are co-induced, although more transiently than COMT, in response to elicitor.[69] This suggests that the increased demand for methylation reactions during the synthesis of isoflavonoids, lignin precursors and methoxy-chalcone requires increased production of the methyl group donor SAM. With the exception of increased levels in flowers, the organ-specific pattern of COMT and AdoMet synthetase transcripts in developing alfalfa seedlings is similar to that of PAL (Table 2), presumably reflecting a requirement for COMT in lignifying tissues.

Chalcone 2'-O-methyltransferase

Alfalfa produces a number of flavonoid derivatives which can induce expression of the nodulation (*nod*) genes on the megaplasmid of *Rhizobium meliloti*. The *nod*-gene inducers present in alfalfa seed extracts are luteolin (5,7,3',4'-tetrahydroxyflavone) and chrysoeriol (3'-methoxy-5,7,4',-trihydroxy-flavone).[10] A different group of *nod*-gene inducing flavonoid derivatives is found in alfalfa root exudates; 7,4'-dihydroxyflavone, 7,4'-dihyroxyflavanone and 2'-methoxy-4,4'-dihydroxychalcone.[12] The methoxychalcone is of particular interest as methylation at the 2'-position (Fig. 6) will prevent the chalcone isomerase (CHI)-catalyzed cyclization of chalcone to flavanone, therefore blocking the formation of any potential antimicrobial flavonoid derivatives.

Whether CHI and chalcone 2'-O-methyltransferase compete for substrate *in vivo* is not yet clear; the O-methyltransferase activity is co-induced with CHI in elicitor treated alfalfa cell cultures, where the isomerase activity is a factor of approximately 1,000 greater than that of the O-methyltransferase. However, the K_m of the O-methyltransferase for 4,2',4'-trihydroxychalcone is somewhat lower than that of CHI (C.A. Maxwell, R. Edwards, R.A. Dixon, unpublished results). Chalcone O-methyltransferase activity is predominantly restricted to the roots of alfalfa seedlings, where its activity increases during the early stages of development. We do not yet know whether chalcone O-methyltransferase and CHI activities occur in the same cells in alfalfa roots; if a particular cell type is involved in producing *nod*-gene inducers, it would appear to be necessary for both enzymes to be present in order to make the observed chalcone, flavone and flavanone inducers.

Chalcone O-methyltransferase has been purified to near homogeneity from elicited alfalfa cell suspension cultures (C.A. Maxwell, R. Edwards, R. A.

Dixon, unpublished results). It has a subunit M_r of 43,000, similar to that of COMT and IOMT. It is, however, a distinct enzyme on the basis of its separation from the other O-methyltransferases during purification, and its different activity patterns in developing and elicited alfalfa roots. Presumably, this enzyme could be a potential regulator of the balance between production of flavonoid derivatives with distinctly opposite functions; for example, establishment of a symbiotic association and antimicrobial defense. In view of this, the cloning of the chalcone O-methyltransferase gene, and gaining an understanding of its expression in relation to other flavonoid pathway genes, are important goals for future work.

Isoflavone Reductase and Pterocarpan Synthase

The pterocarpan nucleus of medicarpin contains two chiral centers. Medicarpin in alfalfa occurs in the 6aR, 11aR configuration ((-)-medicarpin), whereas in pea the related pterocarpan maackiain exists predominantly as the 6aS, 11aS ((+)isomer) (Fig. 8). The first of the chiral centers in pterocarpans is introduced by the enzyme isoflavone reductase (IFR), the penultimate enzyme in medicarpin biosynthesis. The pterocarpan is then formed from the 2'-hydroxy-isoflavanone by ring-closure, catalyzed by pterocarpan synthase (PTS) (Fig. 8).

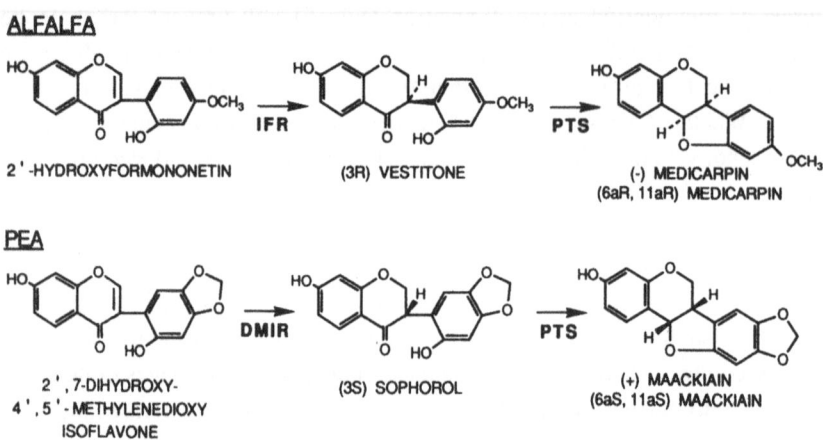

Fig. 8. Reactions catalyzed by isoflavone reductase (IFR) and pterocarpan synthase (PTS) in alfalfa and pea. Note the opposite stereochemistries of the isoflavone products vestitone and sophorol. (See Chapter 5 by Barz re numbering of methylene dioxy moiety in products such as maackiain).

IFR from alfalfa forms the isoflavanone $(3R)$-vestitone from 2'-hydroxyformononetin. Its activity is strongly induced in *Colletotrichum* elicitor treated cell suspension cultures, reaching maximum levels at approximately 12 h post-elicitation (Table 1).[70] This activity remains elevated through 48 h post-elicitation. Induction of PTS activity follows similar kinetics.[70] An isoflavone reductase activity from peas (DMIR, forming the isoflavanone of opposite stereochemistry to that found in alfalfa) (Fig. 8) has been purified from $CuCl_2$-treated seedlings and a polyclonal antiserum raised against it.[43]

This antiserum strongly recognizes three polypeptides on Western blots of SDS-gel separations of protein extracts from elicited alfalfa cultures; a constitutively expressed polypeptide of M_r 36,500, and polypeptides of M_r 37,500 and 35,000 whose induction kinetics parallel those of IFR activity.[70] In view of its similarity to the M_r of the pea DMIR, we assume that the elictor-inducible M_r 37,500 polypeptide corresponds to alfalfa IFR, and that the other proteins may be related reductases, although it can not be ruled out that the M_r 35,500 polypeptide could be IFR. The anti-(pea DMIR) serum was used to screen the alfalfa cDNA expression library, and 45 positive clones were isolated out of a total of 25,000 screened.[70] In view of the lack of absolute specificity of the antibody for alfalfa antigens, antibody positive clones were screened for expression of IFR activity. Lysates from *E. coli* cells harboring one clone exhibited high levels of IFR activity, specific for 2'-hydroxyisoflavonoids as substrate (Fig. 9). Circular dichroism spectroscopy indicates that the vestitone formed from the cloned IFR is of the $3R$-configuration, as expected for the alfalfa enzyme (Fig. 10).

The alfalfa IFR cDNA contained a 954 bp open reading frame encoding a 318 amino acid polypeptide with a deduced M_r of 35,410.[70] The amino acid sequence showed homologies to regions from a number of NADPH-utilizing reductases and ATP-binding proteins; a particularly strong similarity was observed in the N-terminal 195 amino acids to the dihydroflavonol reductase of *Antirrhinum majus*.[71]

A 1.4 kb transcript recognized by the IFR cDNA was highly induced on elicitation of alfalfa cell suspension cultures; its induction kinetics were consistent with the appearance of the IFR activity, reaching maximum levels between 2 and 8 h post-elicitation.[70] Although IFR activity remained at maximum induced levels up to 48 h post-elicitation, IFR transcripts began to decline after 12 h post-elicitation, and reached virtually undetectable levels by 38 h. This suggests that IFR enzyme activity has a relatively long half life in alfalfa cultures. IFR transcripts were expressed at relatively high levels in alfalfa roots and root nodules, with only trace amounts being detected in leaves, stems and

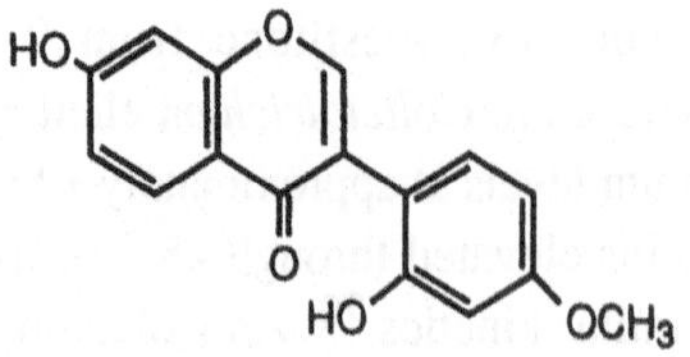

2'-HYDROXYFORMONONETIN

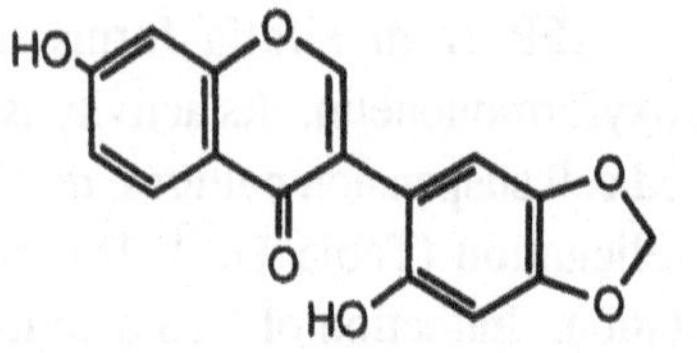

2'-HYDROXYPSEUDOBAPTIGENIN

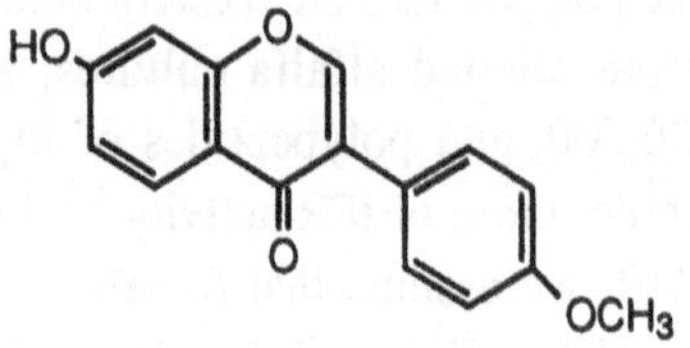

FORMONONETIN

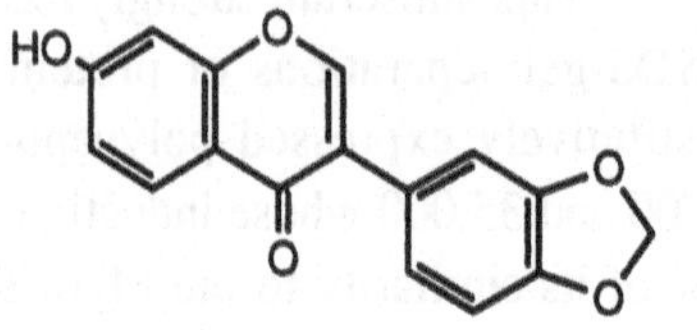

PSEUDOBAPTIGENIN

Fig. 9. Substrates for the cloned alfalfa isoflavone reductase. Extracts from *E. coli* cells harboring the IFR cDNA in an expression plasmid catalyzed the reduction to the corresponding isoflavanones of 2'-hydroxy-formononetin and 2'-hydroxypseudobaptigenin, whereas formononetin and pseudobaptigenin were not reduced.

petioles (Table 2). This may be related to the presence of low levels of MGM in root tissues. We do not yet know whether similar accumulation of IFR transcripts is observed in non-nodulated roots.

ANALYSIS OF DEFENSE GENE PROMOTER: REPORTER GENE FUSIONS IN ALFALFA PROTOPLASTS AND TRANSGENIC PLANTS

In all cases studied to date, elicitor-induced increases in phytoalexin biosynthetic enzyme activities result from increased transcription of the gene or genes encoding the enzyme.[23] Thus, in alfalfa, it would be predicted that the signal(s) generated by interaction of elicitor with its putative receptor somehow lead to changes in a transcription factor or factors which are recognized by regulatory regions on each of the genes encoding the enzymes of the biosynthetic

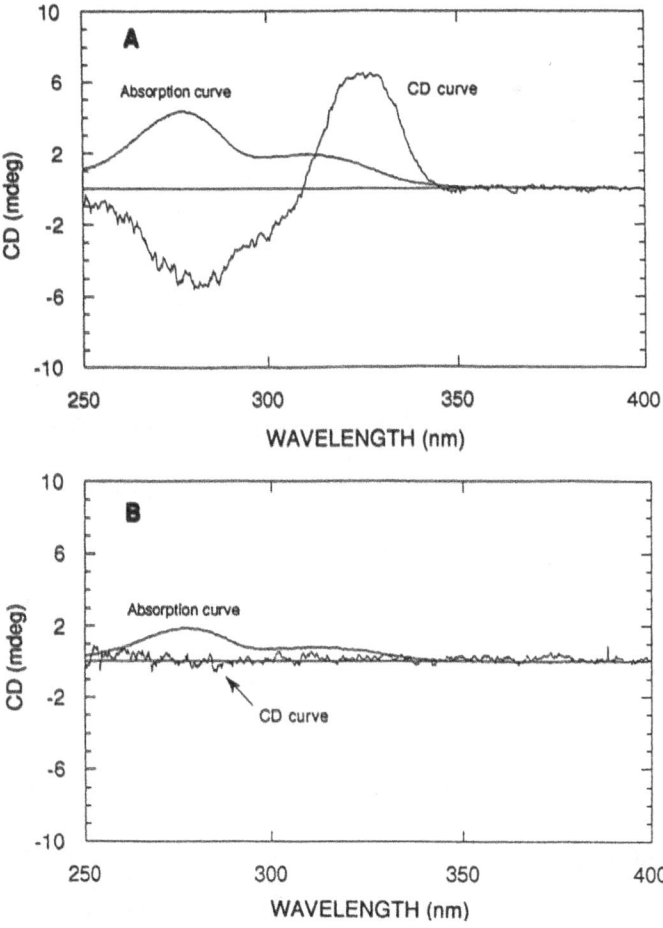

Fig. 10. A: Circular dichroism and UV absorption spectra of (3R)-vestitone produced from 2'-hydroxyformononetin in extracts from *E. coli* harboring an expression plasmid containing the cloned alfalfa isoflavone reductase cDNA. B: Spectra for a sample of chemically synthesized vestitone (racemic mixture).

pathway from L-phenylalanine to medicarpin. At the same time, it must be remembered that some of these genes, particularly those encoding enzymes of the early stages in the pathway which are common to the synthesis of other phenylpropanoid metabolites, are under tissue- or organ-specific developmental regulation.[51,72-74] Such regulation may be overridden by stress signals; for example, PAL and CHS are preferentially expressed in roots (Table 2), but wounding, infection or elicitation induce increased transcript levels in most organs of the plant challenged by the stress.[51,73,75]

Current strategies for defining the regulatory DNA sequences which control expression of a particular gene in a specific tissue or in response to a particular stimulus involve isolation of the promoter region of the gene, and production of constructs in which the promoter can drive expression of a readily assayable reporter gene. Such constructs may be introduced into homologous or heterologous plant species via *Agrobacterium tumefaciens*-mediated transformation. Deletions or mutations in the promoter can then be examined for their effects on tissue specific and developmental expression. Such experiments can be extremely time consuming, and subject to much variability due to so called "position effects" resulting from integration of the transgene into apparently different regions of the genome in different independent transformants. This approach, or variants involving alternative methods of DNA delivery, e.g. particle bombardment, is, however, the only one applicable to studies of qualitative expression patterns. Gene promoter elements affecting quantitative expression, including elicitor inducibility, can be studied by a more rapid approach involving transient expression of the promoter-reporter fusion in electroporated protoplasts.[35,76] Unlike many legumes, alfalfa is amenable to both stable transformation[2] and transient expression assays in protoplasts.[59]

Techniques for the isolation of alfalfa protoplasts and the introduction into them of promoter-reporter constructs by electroporation in the presence of polyethylene glycol have been described.[59] To date, we have concentrated our studies on constructs containing sequences of the bean (*Phaseolus vulgaris*) CHS15 promoter[77] linked to the bacterial chloramphenicol acetyltransferase (CAT) reporter gene and *nos* 3'-terminator in the plasmid pUC19. As shown in Figure 11, the bean CHS15 promoter (to position -326 or -173 from the transcription start site) drives CAT expression when electroporated into alfalfa protoplasts, and this expression is inducible by *Colletotrichum* elicitor or glutathione, although to a lesser extent than observed for induction of alfalfa or bean CHS transcripts by elicitor in their respective cell suspension cultures.

The responsiveness of the CHS15 promoter to glutathione in alfalfa protoplasts decreases on repeated subculture of the cell suspensions from which the protoplasts are isolated.[78] The absolute basal level of expression of pCHC1 (containing the CHS15 promoter to position -326) varies between different batches of protoplasts. This basal expression is probably caused by elicitation during preparation of the protoplasts, as cell wall degrading enzymes are known to possess elicitor activity. An alternative explanation is that the alfalfa cell suspension cultures are, on the basis of their phenylpropanoid secondary metabolite profile, physiologically similar to root cells,[37] and the CHS15 promoter is expressed at relatively high levels in unelicited roots of transgenic alfalfa plants (K. Dalkin, R.A. Dixon, unpublished results). Whatever the reason, the basal expression increases as a function of time (Fig. 11b). This is in contrast to the behavior of other protoplast systems, e.g. soybean, in which initially high basal expression then declines,[35] or parsley, in which protoplasts remain unelicited during preparation.[79] Although the response of alfalfa cell suspension protoplasts to fungal elicitor is not high, it is highly reproducible. The variation of absolute expression levels between different batches of protoplasts requires, however, that an internal control construct be included in each experiment so that data can be normalized relative to its expression level.[80]

The data in Figure 11 indicate that alfalfa cells contain the necessary transcription factors for expression of an elicitor-inducible phenylpropanoid gene promoter. Deletion of the promoter to -173 significantly reduces overall expression, but does not remove elicitor responsiveness. More detailed deletion and mutational analyses of the CHS15 promoter in the alfalfa protoplast system have revealed a number of putative *cis*-elements which affect quantitatively the level of expression (G.J. Loake, O. Faktor, C.J. Lamb, R.A. Dixon, unpublished results). The region from -326 to -141 contains both activator and silencer elements.[81] As deletion to -173 results in decreased expression, the deleted sequences appear to function overall as an activator. This is in contrast to the situation in electroporated soybean protoplasts, where the -173 deletion is expressed at higher levels than the -326 promoter: CAT construct.[35] The region from -326 to -214 contains 3 elements, boxes I - III (Fig. 12d), each of which is characterized by a variant of the consensus sequence GGTTAA, the binding sequence for factor GT-1 involved in the light-mediated expression of the pea rbcS-3A gene.[82] A protein factor has been isolated and purified from bean cell culture nuclei which binds to each of the GT-1 sites in the CHS15 promoter. [83,84] The factor has been named SBF-1 (*silencer box factor*) because of the negative effects on CHS promoter activity of its cognate *cis*-element in soybean protoplasts. That a factor related or identical to SBF-1 may act as a positive regulator of CHS expression in alfalfa cells of root origin (either root suspension protoplasts or intact roots) is suggested by (a) the deletion analysis in protoplasts described above; (b) the decrease in CHS15 promoter expression on co-electroporation of the SBF-1 binding sequence (as a tetrameric synthetic oligonucleotide in plasmid pUC19, see Fig. 12b and discussion below) along

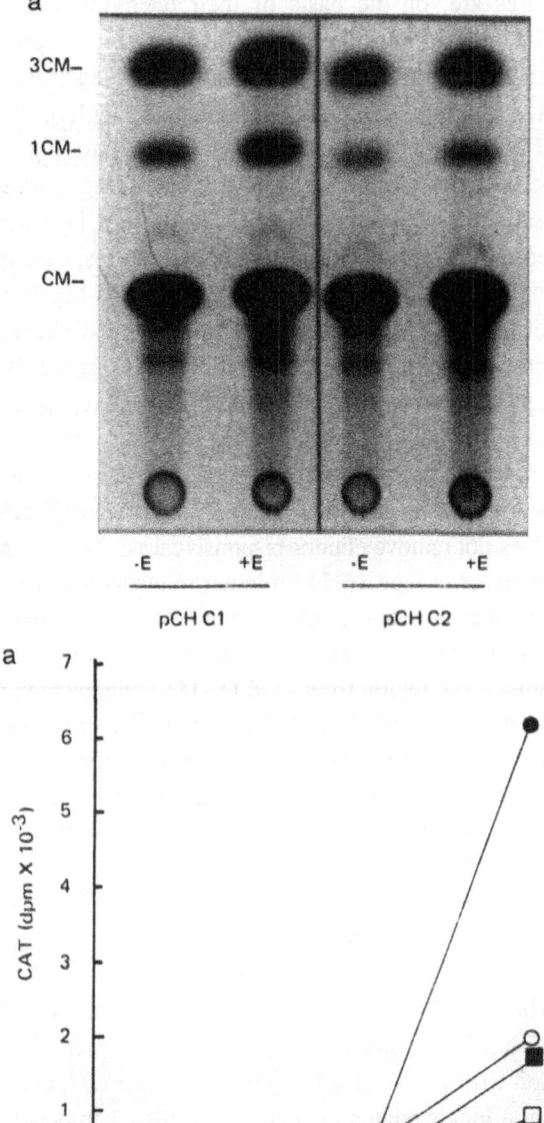

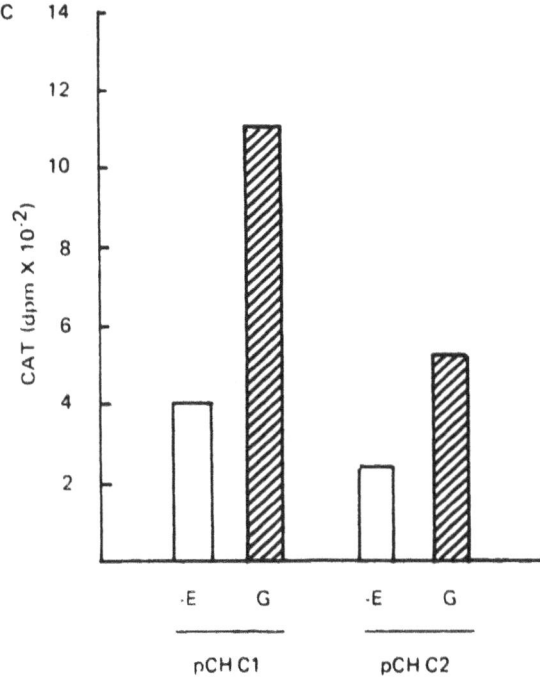

Fig. 11. (11a: facing page, 11b: above) Transient expression of bean chalcone synthase promoter constructs in electroporated alfalfa protoplasts. pCHC1 contains a transcriptional fusion of the bean CHS15 promoter to position -326 to the bacterial chloramphenicol acetyltransferase (CAT) reporter gene and nopaline synthase 3'-terminator in plasmid pUC19. pCHC2 is an identical construct except that the CHS promoter is 5'-deleted to position -173. A: pCHC1 or pCHC2 (40 µg) were electroporated into alfalfa proto-plasts which were incubated for 8 h in the absence (-E) or presence (+E) of fungal (*Colletotrichum*) elicitor. Protoplasts were then extracted and assayed for CAT activity. Chloramphenicol substrate (CM) and acetylated products (1CM, 3CM) were separated by thin layer chromatography. B: Time course of CAT expression in alfalfa protoplasts electroporated with pCHC1 in the presence (●) or absence (○) of fungal elicitor. A parallel experiment is shown using the construct pCHC2 (■, □). C: CAT expression from pCHC1 and pCHC2 in electroporated alfalfa protoplasts incubated for 8 h in the absence of elicitor (-E) or in the presence of 1 mM glutathione (G).

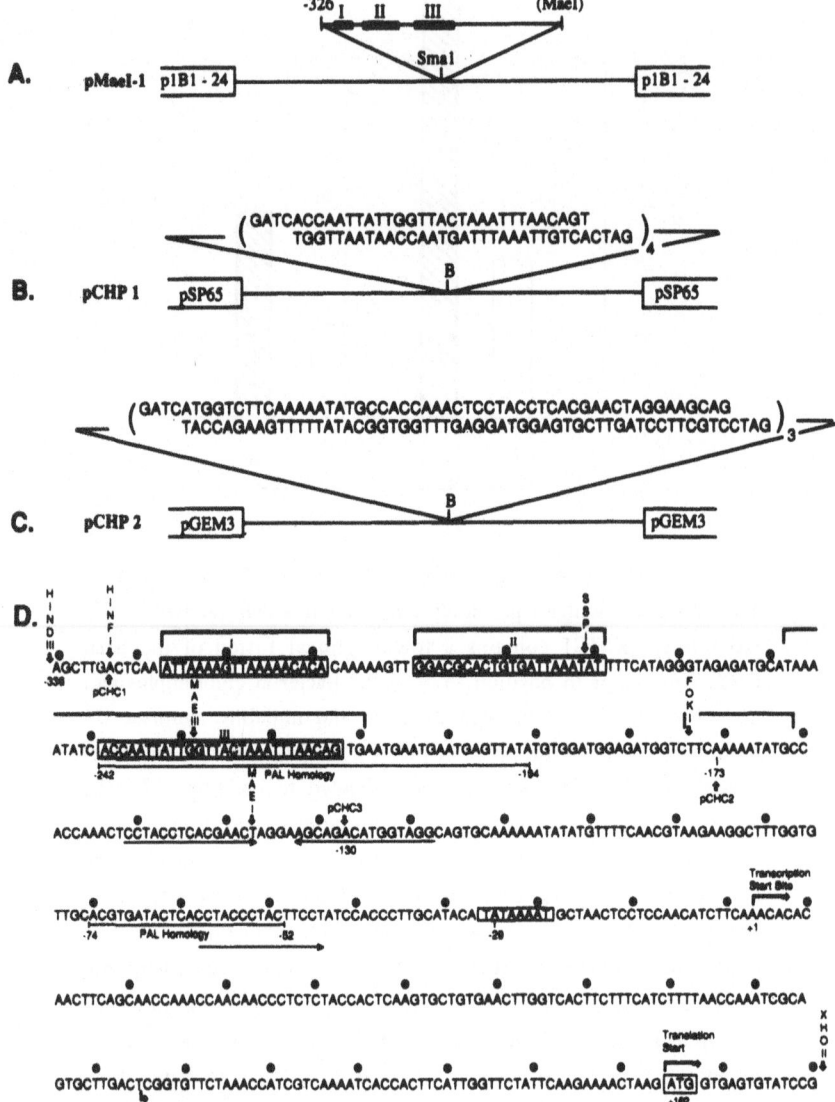

with the intact promoter: CAT construct into suspension cell protoplasts;[81] (c) the loss in root expression of a CHS15 promoter: β–glucuronidase reporter gene construct in transgenic alfalfa plants when the promoter is 5'-deleted with endpoints between -299 to -130 (K. Dalkin, R.A. Dixon, unpublished results).

Table 3 summarizes the results of deletion and mutational analyses aimed at defining cis-elements in the CHS15 promoter functional in alfalfa cells. The GT-1 elements have also been defined on the basis of *in vitro* footprinting with bean[84] and alfalfa[81] nuclear extracts (see below). Likewise, the CCTACC-(N_7)CT element has been shown, by *in vivo* genomic footprinting, to be important in the elicitation response of a parsley PAL gene.[85] Elicitor-responsiveness of the bean CHS15 promoter is lost on 5'-deletion beyond position -72.[35] It is likely that sequences between positions -74 and -52 are important in conferring elicitor inducibility to the CHS15 promoter. This region of the promoter shows extensive sequence homology to a region of the similarly regulated bean PAL2 promoter,[50] and contains G-box,[86] GATA box and CCTACC(N_7)CT elements. One potential role of the G-box sequence in alfalfa cells is discussed below.

TRANS-ACTING FACTORS FOR REGULATION OF DEFENSE GENE EXPRESSION IN ALFALFA

If a particular *cis*-element is functional in alfalfa cells (i.e. its removal or deletion significantly effects expression of the remainder of the promoter), it is likely that the element interacts with a cognate *trans*-acting factor. It would therefore be predicted that, if the factor were not in excess, introducing extra copies of the element, unlinked to a reporter gene, into protoplasts containing

---

Fig. 12. (facing page) A - C: Constructs containing CHS15 promoter sequences used in the co-electroporation experiments described in the text. D: Binding sites on the bean CHS15 promoter for bean and alfalfa nuclear proteins, and *cis*-acting elements functional in alfalfa protoplasts or transgenic plants. The TATA box and three SBF-1 binding sites determined by footprinting with bean nuclear extracts (I, II and III) are boxed. The areas overlined are regions footprinted by alfalfa nuclear extracts. The long horizontal arrows show the three CCTACC(N_7)CT elements. Regions with strong homology to the bean PAL2 promoter are underlined.

Table 3. *Cis*-acting elements in the bean CHS15 promoter functional in alfalfa cell suspension protoplasts. Unless stated, assays determined the level of expression of constructs containing 5'-deleted or internally deleted CHS15 promoters driving expression of the CAT reporter gene in elicited protoplasts. Refer to Figure 11 for the full sequence of the CHS15 promoter.

Position upstream from transcription start site[a]	Core Sequence	Name of motif	Effect on expression of deletion of element
-313	AGTTAA	SBF-1 (Box I)	Loss of root-expression and UV-induction in transgenic alfalfa
-282	GATTAA	SBF-1 (Box II)	Down-regulation
-230	GGTTAC	SBF-1 (Box III)	Up-regulation (x 1.5 - 2.0)
-161	GCCACCAAA	AC-rich element	Up-regulation (x 3.5)
-152	CCTACC		Up-regulation (x 3.5); reduces coumarate stimulation by ~60%
		H-Box [b]	
-145	(N7)CT		Up-regulation (x 11)
-129	CCTACC(N7)CT	H-Box, lower strand	Not known
-72	CACGTG	G-box	Down-regulation; abolishes coumarate stimulation
-59	CCTACC		Down-regulation (x 2); reduces coumarate stimulation by 80%
		H-box	
	(N7)CT		Not known

[a] Approximate center of element. Transcription start site = +1.
[b] H-box sequence is CCTACC(N7)CT. Deletion of each half can have different effects.

the intact promoter: CAT construct would result in competition between the element construct and the intact promoter for binding of the factor. The outcome of such a co-electroporation experiment would depend on the nature of the factor being competed; if it were an activator, co-electroporation of excess copies of its cognate *cis*-element sequence would result in decreased promoter expression, and the opposite would be true if the factor were a negative regulator.

The involvement of putative *trans*-acting factors in the expression of the CHS15 promoter in alfalfa protoplasts has been demonstrated by this indirect functional co-electroporation strategy. Co-electroporation of pMaeI-1 (Fig. 12a), containing the three GT-1 consensus sequences, into elicitor-treated protoplasts reduces by a factor of 3 to 8 the expression of the -326 promoter: CAT construct; this is consistent with the overall function of this region of the promoter as an activator in fungal elicitor-treated alfalfa cells.[78,81] Co-electroporation of pCHP1, containing four copies of the most 3' of the GT-1 elements and its flanking sequences (Fig.12b), results in only a very slight decrease in promoter expression, suggesting that more than one of the GT-1 elements is necessary for strong binding *in vivo*, or perhaps that correct spacing between the elements is important.[78] The construct pCHP2 (Fig. 12c) contains three copies of the -183 to -130 sequence of the CHS15 promoter; within this sequence is a region which is footprinted by alfalfa nuclear extracts *in vitro*.[81] Co-electroporation of pCHP2 reduces promoter expression in a similar manner to pMaeI-1, suggesting the presence of a binding site for a positively acting factor between -183 and -130.[81]

In vitro DNAase I footprinting experiments with alfalfa cell suspension nuclear extracts reveal three footprints on a -326 to -141 fragment of the CHS15 promoter, indicating binding of factor(s) to putative *cis*-elements in the promoter.[81] The elements footprinted are the three GT-1-like sequences and their flanking regions (Fig. 12d). The same regions have been identified as binding sites for factor SBF-1 from bean cell suspension nuclear extracts.[83,84] Gel-retardation analysis using the -326 to -141 fragment as probe reveals a binding factor in alfalfa nuclear extracts which retards the probe to an identical position as observed using bean extracts (Fig. 13). Unlabelled MaeI-1 and the 28 nucleotide box III sequence (Fig. 12d) compete for the binding of alfalfa factor, but the box III sequence in which the two G residues of the GGTTAA consensus are mutated to C residues is not a competitor.[81] This indicates that the binding in the alfalfa extracts is sequence-specific, and reflects the presence in alfalfa nuclei of a GT-1-like factor, probably the equivalent of bean SBF-1. A single weak footprint is also observed on a probe spanning the -183 to -130 region of the CHS15 promoter, mapping to positions -177 to -164 (Fig. 12d).

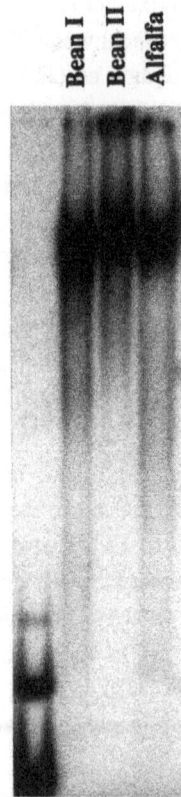

Fig. 13. Gel retardation analysis of GT-1-like DNA binding
activities in nuclear extracts from bean and alfalfa cell suspension
cultures. The left hand lane shows electrophoresis of ^{32}P-labelled
probe alone (the -326 to -141 region of the bean CHS15 promoter).
In the other lanes, probe was pre-incubated with 2 μg of bean (two
duplicate preparations) or alfalfa nuclear protein extracts prior to
electrophoresis.

A sequence-specific alfalfa factor recognizing this sequence has been identified by
gel retardation analysis.[81] Unlabelled -183 to -130 oligonucleotides compete for
its binding, but a mutant oligonucleotide in which the C at position -174 is altered to a G is not a competitor. The function of this element is not yet known.

REGULATION OF DEFENSE GENE EXPRESSION BY PHENYLPROPANOID COMPOUNDS

Previous experiments on pea epicotyl sections[87] and bean cell suspension cultures[88-90] have demonstrated that cinnamic acid (CA), the product of the PAL reaction, inhibits the synthesis of PAL and brings about a decrease in PAL enzyme activity. Although the concentrations of exogenously applied CA that cause such effects are relatively high (usually greater than 10^{-4}M), a possible physiological relevance for this effect is supported by the super-induction of PAL enzyme activity and PAL and CHS transcript levels in elicited cells when CA is prevented from being formed as a result of inhibition of PAL activity by the potent inhibitor L-α-aminooxy-β-phenylpropionic acid (AOPP). [88,90,91] Treatment of elicited bean cells with AOPP also increases the levels of CHS transcripts,[88] including transcripts from the CHS15 gene (M. Mavandad, R.A. Dixon, unpublished results). Alfalfa cell suspension cultures respond to exogenous additions of CA and AOPP in a similar manner to bean cells (J. Orr, A.D. Choudhary, R.A. Dixon, unpublished results). This facilitates analysis of putative *cis*-acting elements involved in the response to phenylpropanoid compounds by transient assay of CHS15 promoter constructs in electroporated cell suspension protoplasts.

Expression of the CHS15 promoter and a constitutively expressed cauliflower mosaic virus 35S promoter: CAT construct in alfalfa protoplasts, as a function of the concentration of exogenously added CA, is shown in Figure 14a. Expression from both promoters is weakly up-regulated by low CA concentrations, but then severely depressed at higher concentrations. The CHS promoter is, however, more sensitive to the higher CA concentrations than is the 35S promoter. The high CA concentrations do not appear to affect protoplast viability (Fig. 14b). Expression of bean HRGP (hydroxyproline-rich glycoprotein) and rice chitinase promoter: CAT constructs is relatively unaffected over the whole concentration range of CA used in the experiments in Figure 14, indicating that the effect of CA is promoter-specific.[80] In contrast to these results, treating protoplasts with 4-coumarate (4CA, the product of the cinnamic acid 4-hydroxylase reaction; Fig. 1) has the opposite effect on expression of the CHS15 promoter, which is stimulated approximately 4.5-fold at 5×10^{-4} M 4CA.[80] These data suggest that alfalfa cells contain the components of a mechanism for regulating CHS expression as a function of CA or 4CA concentration. Such a mechanism might underlie the very rapid, transient induction of CHS transcripts observed in elicitor treated alfalfa or bean cells, with 4CA regulating an initial accelerated increase in transcription, and

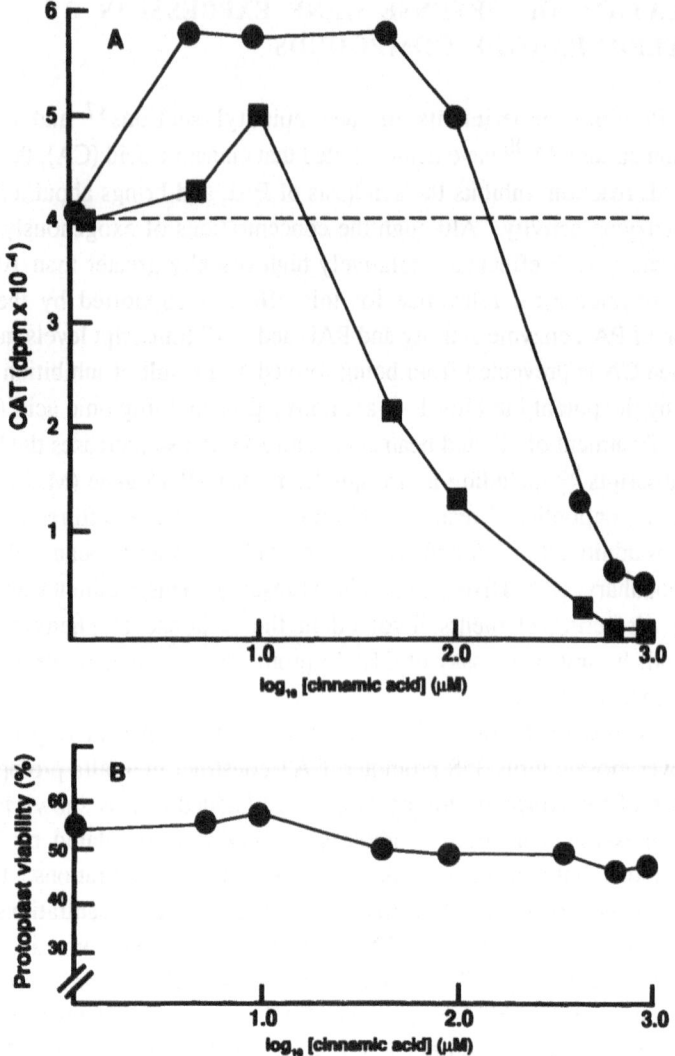

Fig. 14. Dose response curves for CA effects on expression of
promoter-CAT constructs in alfalfa protoplasts, and on viability. A:
Alfalfa protoplasts were electroporated in the presence of 40 μg of
pCHC1 (bean CHS15 promoter) (■), or cauliflower mosaic virus 35S
promoter: CAT construct (●), treated with elicitor from *Colletotrichum lindemuthianum* (final concentration 50 μg glucose equiva-
lents ml^{-1}) plus a range of concentrations of CA, cultured for a
further 10 h prior to extraction and assay of CAT activity. B:
Viability, determined by Evans blue staining, of protoplasts analyzed in A at time of harvest (10 h after addition of elicitor plus CA).

then being responsible for the following rapid decline in transcription rate. A similar hypothesis has been presented earlier in relation to the effects of CA on PAL expression in bean cells.[92]

Several predictions can be made if the above hypothesis is to be valid for gene regulation under physiological conditions *in vivo*. First, genes regulated by CA or 4CA would be expected to contain specific *cis*-elements involved in these responses. Second, levels of endogenous CA and 4CA should change in elicitor treated cells consistent with a regulatory role and third, artificial manipulation of CA and 4CA levels *in vivo* should give predictable results consistent with the model. These predictions are now being tested in alfalfa cell cultures and electroporated protoplasts.

A specific element conditioning down-regulation of the CHS15 promoter by exogenously applied CA has not yet been defined by deletion analysis. However, the small but reproducible up-regulation by low CA concentrations is abolished by deletion of the promoter to -130. This deletion also has a strongly negative effect on up-regulation by 4CA, abolishing it at low 4CA concentrations and inhibiting it by approximately 60% at high concentrations ($5 \times 10^{-4}M$).[80] The involvement of a putative *trans*-acting factor for CA or 4CA-mediated up-regulation of CHS expression in alfalfa cells is suggested by the abolition of this up-regulation by co-electroporation of a plasmid containing multiple copies of the -130 to -183 sequence of the CHS15 promoter.[80] Analysis of a set of CHS15 promoter internal deletion constructs has indicated that elements essential for the coumarate response map to two regions of the promoter known to be of importance for plant gene regulation; the CCTACC(N7)CT or H-box element located between positions -61 and -56, and the G-box at positions -75 to -69 (Fig. 12d) (G. Loake, O. Faktor, C.J. Lamb, R.A. Dixon, unpublished results). The H-box has been implicated in elicitor-induction of a parsley PAL gene,[85] and the G-box is strongly conserved in a number of plant genes including *rbcS*, *Chs* and *Adh*. It appears to be a site for the binding of a general transcription factor which is found in yeast as well as plant cells.[86] It has been suggested, on the basis of fine-structure mapping of light-responsive elements in the parsley CHS promoter, that the G-box may gain its functional capacity and specificity by combinatorial diversity involving flanking protein-binding elements.[93]

Further strong support for the involvement of the above sequences in coumarate regulation of CHS expression in alfalfa is provided by the observation that four copies of the CHS15 sequence containing the G-box and downstream CCTACC(N7)CT element confer 4CA-responsiveness when fused to a 35S minimal promoter: CAT construct, which is itself unresponsive to 4CA.

Furthermore, co-electroporation of a cloned oligonucleotide containing multiple copies of the CCTACC(N7)CT element prevents 4CA-mediated up-regulation of the CHS15 promoter: CAT construct, whereas this effect is not observed if the first two Cs in each copy of the element are changed to Gs (G. Loake, O. Faktor, C.J. Lamb, R.A. Dixon, unpublished results).

Elicitation of alfalfa cell suspension cultures leads to changes in endogenous CA and 4CA levels, as determined by capillary GLC analysis of derivatized cell samples (J. Orr, R. Edwards and R.A. Dixon, unpublished results). The changes are over the total cellular concentration ranges 2.5 to 25 μM (CA) and 0 to 55 μM (4CA); these are within the values for stimulatory effects on transcription, although it is necessary to postulate the presence of endogenous pools of higher concentration for the down-regulatory effects of CA. Treatment of elicited alfalfa cell suspension cultures with AOPP decreases the levels of CA and 4CA, with a corresponding increase in PAL transcripts. Conversely, treatment with the cytochrome P-450 inhibitor tetcyclasis[94] causes an increase in CA and a decrease in 4CA levels (presumably as a result of inhibition of cinnamic acid 4-hydroxylase), with a corresponding decrease in PAL transcripts (J. Orr, R.A. Dixon, unpublished results). Experiments are now in progress to determine the effects of these inhibitors on CHS and a number of control transcripts in the alfalfa system. Results to date remain consistent with the hypothesis that a gene regulatory mechanism involving sensing of the levels of pathway intermediates operates in alfalfa cells to control the rate of flux through the phenylpropanoid and flavonoid pathways.

PROSPECTS FOR MANIPULATING PHENYLPROPANOID AND FLAVONOID METABOLISM IN ALFALFA BY GENE TRANSFER STRATEGIES.

In terms of practical application, genetic manipulation of phenyl-propanoid and flavonoid metabolism in alfalfa could be of value with respect to improving disease resistance and nodulation, or modifying lignin content or composition. Although reasonably efficient transformation strategies have been developed for alfalfa, regeneration from culture via somatic embryogenesis is strongly genotype-dependent.[8,9] The best regeneration response is observed in cultivars containing a strong genetic contribution from the two landrace germplasm sources *M. falcata* and Ladak; these cultivars are creeping-rooted types. Recurrent selection has led to the development of cultivars such as Regen-S in which two-thirds of the plants are capable of regeneration from callus culture.[95] Introduction of specific genes into such model cultivars would then be followed by introduction of the transgene(s) into elite cultivars by conventional crossing. Recently, considerable attention has been paid to the development of diploid alfalfa lines[96] or relatives[97] which will provide simplified genetic systems for fundamental and applied studies.

The nature of disease resistance genes, as opposed to the defense response genes described in this chapter, is not understood. These genes, which often act as single, dominant determinants of resistance to a particular pathogen race, are believed to act as receptors/initial transducers for the signal molecule(s) determining avirulence in the pathogen.[24] As such, they may be key components in controlling expression of infection-induced phenylpropanoid and flavonoid metabolism. Strategies for the isolation of such genes, in the absence of any knowledge of their products, include transposon tagging and RFLP mapping/chromosome walking. Diploid alfalfa[96] or *Medicago truncatula*[97] could now be developed for such approaches. Indeed, DNA reassociation analysis has shown that the haploid alfalfa genome contains 1.0×10^9 bp or 1.1 pg DNA, with 36% unique sequences, 42% medium repetitive sequences and 22% highly repetitive sequences,[98] making RFLP mapping/walking strategies at least feasible in this species. Furthermore, recent studies of unstable anthocyanin pigmentation indicate that alfalfa contains endogenous transposable elements.[99] The development of genetic approaches could now begin to make an impact on our understanding of environmental regulation of phenylpropanoid biosynthesis in alfalfa.

A number of strategies have been proposed for manipulating defense responses in transgenic plants.[100] These include transfer of single dominant resistance genes *per se*, altering the levels of expression of endogenous defense response genes such as chitinase or glucanase, or transferring novel defense response genes from other sources. The isoflavonoid phytoalexin pathway comprises a complex series of biosynthetic reactions with the involvement of many individual genes; modification of the terminal stages only may present a feasible strategy for engineering increased, or wider spectrum, resistance. For example, the work of VanEtten and colleagues has drawn attention to the role of phytoalexin detoxification as a determinant of virulence in several plant-pathogen interactions.[101] Where phytoalexins can exist as isomeric pairs, (e.g., (+) and (-)pterocarpans), it is often seen that the pathogen is more sensitive to the isomer that it is not exposed to in nature than to the natural phytoalexin of the host.[101] This increased sensitivity may result from an inability of the pathogen to detoxify the "unusual" stereoisomer. Production of transgenic alfalfa expressing an IFR and pterocarpan synthase that form (+)-medicarpin might result in increased resistance towards pathogens which can detoxify (-)medicarpin. The potential of this strategy is currently under investigation. Initial experiments are aimed at assessing whether inhibition of IFR synthesis by expressing an antisense IFR transgene leads to increased fungal susceptibility. Such direct evidence of a key role for medicarpin in resistance would justify attempts to engineer altered phytoalexin composition in alfalfa.

As alfalfa stems mature, their lignin content increases, and this may result in impaired forage digestibility.[102] The cloning of alfalfa COMT and other enzymes of the lignin pathway opens up the possibility for genetic manipulation of lignin quantity and quality, again using antisense technology. Such studies are now in progress in our laboratories.

CONCLUSION

Our studies on phenylpropanoid and isoflavonoid biosynthesis in alfalfa, reviewed in this chapter, have concentrated on obtaining a basic understanding of enzymology, obtaining homologous cDNA probes for a number of key biosynthetic enzymes, and developing systems for stable and transient transgene expression. Future studies will utilize transgenic technology to study further the *cis*-elements, *trans*-acting factors and signal transduction components which together regulate the expression of phenylpropanoid and flavonoid biosynthetic genes within the developmental program of the plant and in response to external stimuli such as pathogen attack. Together with the ability to alter endogenous transcript levels by antisense or overexpression strategies, this knowledge will hopefully form a basis for genetically manipulating phenylpropanoid and flavonoid metabolism for alfalfa improvement.

ACKNOWLEDGEMENTS

We are grateful to our colleagues and collaborators who have contributed materials and/or ideas to the work described in this chapter. These include Drs. Wilbur Campbell, Michel Dron, Ian Dubery, Ouriel Faktor, Geza Hrazdina, Yuejin Sun and Hans VanEtten. We thank Allyson Wilkins for preparation of the manuscript. The work described from the authors' laboratories was supported by The Samuel Roberts Noble Foundation.

REFERENCES

1. BEACH, K.H., GRESSHOFF, P.M. 1988. Characterization and culture of *Agrobacterium rhizogenes* transformed roots of forage legumes. Plant Sci. 57:73-81.

2. CHABAUD, M., PASSIATORE, J.E., CANNON, F., BUCHANAN-WOLLASTON, V. 1988. Parameters affecting the frequency of kanamycin resistant alfalfa obtained by *Agrobacterium tumefaciens* mediated transformation. Plant Cell Rep. 7:512-516.

3. DEAK, M., KISS, G.B., KONCZ, C., DUDITS, D. 1985. Transformation of *Medicago* by *Agrobacterium* mediated gene transfer. Plant Cell Rep. 5:97-100.

4. REICH, T.J., IYER, V.N., MIKI, B.L. 1986. Efficient transformation of alfalfa protoplasts by the intranuclear microinjection of Ti plasmids. Bio/Technology 4:1001-1004.

5. SHAHIN, E.A., SPEILMANN, A., SUKHAPINDA, K., SIMPSON, R.B., YASHAR, M. 1986. Cell biology and molecular genetic transformation of cultivated alfalfa using disarmed *Agrobacterium tumefaciens*. Crop Sci. 26:1235-1239.

6. SUKHAPINDA, K., SPIVEY, R., SHAHIN, E.A. 1986. Ri-plasmid as a helper for introducing vector DNA into alfalfa plants. Plant Mol. Biol. 8:209-216.

7. ATANASSOV, A., BROWN, D.C.W. 1984. Plant regeneration from suspension culture and mesophyll protoplasts of *Medicago sativa* L. Plant Cell Tissue Organ Culture 3:149-162.

8. BROWN, D.C.W., ATANASSOV, A. 1984. Role of genetic background in somatic embyrogenesis in *Medicago*. Plant Cell Tissue Organ Culture 4:111-122.

9. KRIS, M.H.S., BINGHAM, E.T. 1988. Interactions of highly regenerative genotypes of alfalfa (*Medicago sativa*) and tissue culture protocols. In: Vitro Cell. and Devel. Bio. 24:1047-1052.

10. HARTWIG, U.A., MAXWELL, C.A., JOSEPH, C.M., PHILLIPS, D.A. 1990. Chrysoeriol and luteolin released from alfalfa seeds induce *nod* genes in *Rhizobium meliloti*. Plant Physiol. 92:116-122.

11. KISS, G.B., VINCZE, E., VEGH, Z., TOTH, G., SOOS, J. 1990. Identification and cDNA cloning of a new nodule-specific gene, Nms-25 (nodulin-25) of *Medicago sativa*. Plant Mol. Biol. 14:467-475.

12. MAXWELL, C.A., HARTWIG, U.A., JOSEPH, C.M., PHILLIPS, D.A. 1989. A chalcone and two related flavonoids released from

alfalfa roots induce *nod* genes of *Rhizobium meliloti*. Plant Physiol. 91:842-847.

13. EGLI, M.A., GRIFFITH, S.M., MILLER, S.S., ANDERSON, M.P., VANCE, C.P. 1989. Nitrogen assimilating enzyme activities and enzyme protein during development and senescence of effective and plant gene-controlled ineffective alfalfa nodules. Plant Physiol. 91: 898-904.

14. BAKER, C.J., O'NEILL, N.R., BAUCHAN, G.R. 1988. Production of phytoalexins in race-clone interactions of alfalfa and *Colletotrichum trifolii*. Phytopathology 78: 1590.

15. FLOOD, J., KHAN, F.Z., MILTON, J.M. 1978. The role of phyto-alexins in *Verticillium* wilt of lucerne (*Medicago sativa*). Ann. Appl. Biol. 89:329-332.

16. HIGGINS, V.J. 1972. Role of the phytoalexin medicarpin in three leaf spot diseases of alfalfa. Physiol. Plant Pathol. 2:289-300

17. LATUNDE-DADA, A.O., LUCAS, J.A. 1984. Involvement of the phytoalexin medicarpin in the differential response of callus lines of lucerne (*Medicago sativa*) to infection by *Verticillium albo-atrum*. Physiol. Plant Pathol. 26:31-42.

18. DEWICK, P.M., MARTIN, M. 1979. Biosynthesis of pterocarpan and isoflavan phytoalexins in *Medicago sativa*:: the biochemical inter-conversion of pterocarpans and 2'-hydroxyisoflavans. Phytochemistry 18:591-596.

19. DEWICK, P.M., MARTIN, M. 1979. Biosynthesis of pterocarpan, isoflavan and coumestan metabolites of *Medicago sativa*: chalcone, isoflavone and isoflavanone precursors. Phytochemistry 18:597-602.

20. MARTIN, M., DEWICK, P.M. 1980. Biosynthesis of pterocarpan, isoflavan and coumestan metabolites of *Medicago sativa:* the role of an isoflav-3-ene. Phytochemistry 19:2341-2346.

21. WOODWARD, M.D. 1981. Identification of the biosynthetic precursors of medicarpin in inoculation droplets on white clover. Physiol. Plant Pathol. 18:33-39.

22. DIXON, R.A., DEY, P.M., LAMB, C.J. 1983. Phytoalexins: enzymo-logy and molecular biology. Adv Enzymol. Related Areas Mol. Biol. 55:1-136.

23. DIXON, R.A., HARRISON, M.J. 1990. Activation, structure and organization of genes involved in microbial defense in plants. Adv. in Genetics 28:165-234.

24. DIXON, R.A., LAMB, C.J. 1990. Molecular communication in plant:

microbial pathogen interactions. Annu. Rev. Plant Physiol. and Plant Mol. Biol. 41:339-367.

25. STUTEVILLE, D.L., ERWIN, D.C., eds. 1990. Compendium of Alfalfa Diseases. Second Edition, APS Press. St. Paul, MN. pp. 84.

26. INGHAM, J.L. 1989. Isoflavonoid phytoalexins of the genus *Medicago*. Biochem. Systematics and Ecol. 7: 29-34.

27. KESSMANN, H., EDWARDS, R., GENO, P., DIXON, R.A. 1990. Stress responses in alfalfa (*Medicago sativa* L.) V. Constitutive and elicitor-induced accumulation of isoflavonoid conjugates in cell suspension cultures. Plant Physiol. 94: 227-232.

28. MILLER, R.W., SPENCER, G.F. 1989. (-)-5'-Methoxysativan, a new isoflavan from alfalfa. J. Nat. Prod. 52: 634-636.

29. OLAH, A.F., SHERWOOD, R.T. 1970. Flavones, isoflavones and coumestans in alfalfa infected by *Ascochyta imperfecta*. Phyto-pathology 61: 65-69.

30. SALEH, N.A.M., BOULOS, L., EL-NEGOUMY, S.I., ADBALLA, M.F. 1982. A comparative study of the flavonoids of *Medicago radiata* with other *Medicago* and related *Trigonella* species. Biochem. System. and Ecol. 10:33-36.

31. LATUNDE-DADA, A.O., DIXON, R.A., LUCAS, J.A. 1987. Induction of phytoalexin biosynthetic enzymes in resistant and susceptible lucerne callus lines infected with *Verticillium albo-atrum*. Physiol. Mol. Plant Pathol. 31:15-23.

32. DALKIN, K., EDWARDS, R., EDINGTON, B., DIXON, R.A. 1989. Stress responses in alfalfa (*Medicago sativa* L.) I. Elicitor-induction of phenylpropanoid biosynthesis and hydrolytic enzymes in cell suspension cultures. Plant Physiol. 92:440-446.

33. EDWARDS, R., BLOUNT, J.W., DIXON, R.A. 1991. Glutathione as an elicitor of the phytoalexin response in legume cell cultures. Planta 184:403-409.

34. YAMADA, T., HASHIMOTO, M., SHIRAISHI, T., OKU, H. 1989. Suppression of pisatin, phenylalanine ammonia-lyase mRNA and chalcone synthase mRNA accumulation by a putative pathogenicity factor from the fungus *Mycosphaerella pinodes*. Mol. Plant Microbe Interact 2:256-261.

35. DRON, M., CLOUSE, S.D., DIXON, R.A., LAWTON, M.A., LAMB, C.J. 1988. Glutathione and fungal elicitor regulation of a plant-defense gene promoter in electroporated protoplasts. Proc. Natl. Acad. Sci. USA 85:6738-6742.

36. PARE, P., MISCHKE, C.F., EDWARDS, R., DIXON, R.A., NORMAN, H., MABRY, T.J. 1991. Induction of phenylpropanoid pathway enzymes in elicitor-treated cultures of the cactus *Cephalocereus senilis*. Phytochemistry, submitted.

37. KESSMANN, H., CHOUDHARY, A.D., DIXON, R.A. 1990. Stress responses in alfalfa (*Medicago sativa* L.) III. Induction of medicarpin and cytochrome P450 enzyme activities in elicitor-treated cell suspension cultures and protoplasts. Plant Cell Rep. 9:38-41.

38. BLESS, W., BARZ, W. 1988. Isolation of pterocarpan synthase, the terminal enzyme of pterocarpan biosynthesis in cell suspension cultures of *Cicer arietinum*. FEBS Lett. 235:47-50.

39. DANIEL, S., TIEMANN, K., WITTKAMPF, U., BLESS, W., HINDERER, W., BARZ, W. 1990. Elicitor-induced metabolic changes in cell cultures of chickpea (*Cicer arietinum* L.) cultivars resistant and susceptible to *Aschochyta rabiei*. I. Investigations of enzyme activities involved in isoflavone and pterocarpan phytoalexin biosynthesis. Planta 182:270-278.

40. FISCHER, D., EBENAU-JEHLE, C., GRISEBACH, H. 1990. Phytoalexin synthesis in soybean: Purification and characterization of NADPH: 2'-hydroxydaidzein oxidoreductase from elicitor-challenged soybean cell cultures. Arch. Biochem. Biophys. 276:390-395.

41. HAGMANN, M., GRISEBACH, H. 1984. Enzymatic rearrangement of flavanone to isoflavone. FEBS Lett. 175:199-202.

42. PREISIG, C.L., BELL, J.N., SUN, Y., HRAZDINA, G., MATTHEWS, D.E., VANETTEN, H.D. 1990. Biosynthesis of the phytoalexin pisatin. Isoflavone reduction and further metabolism of the product sophorol by extract of *Pisum sativum*. Plant Physiol. 94:1444-1448.

43. SUN, Y., WU, Q., VANETTEN, H.D., HRAZDINA, G. 1991. Stereoisomerism in plant disease resistance: induction and isolation of the 7,2'-dihydroxy-4',5'-methylenedioxyisoflavone oxidoreductase, an enzyme introducing chirality during synthesis of isoflavonoid phytoalexins in pea (*Pisum sativum* L). Arch. Biochem. Biophys. 284:167-173.

44. TIEMANN, K., HINDERER, W., BARZ, W. 1987. Isolation of NADPH: isoflavone oxidoreductase, a new enzyme of pterocarpan phytoalexin biosynthesis in cell suspension cultures of *Cicer arietinum*. FEBS Lett. 213:324-328.

45. WELLE, R., GRISEBACH, H. 1989. Phytoalexin synthesis in soybean cells: Elicitor induction of reductase involved in biosynthesis of 6'-

deoxychalcone. Arch. Biochem. Biophys. 272:97-102.

46. LAWTON, M.A., LAMB, C.J. 1987. Transcriptional activation of plant defense genes by fungal elicitor, wounding and infection. Mol. Cell. Biol. 7: 335-341.

47. DALKIN, K., JORRIN, J., DIXON, R.A. 1990. Stress responses in alfalfa (*Medicago sativa* L.) VII. Induction of defense-related mRNAs in elicitor-treated cell suspension cultures. Physiol. Mol. Plant Pathol. 37:293-307.

48. GIVEN, N.K., VENIS, M.A., GRIERSON, D. 1988. Purification and properties of phenylalanine ammonia-lyase from strawberry fruit and its synthesis during ripening. J. Plant Physiol. 133:31-37.

49. BOLWELL, G.P., BELL, J.N., CRAMER, C.L., SCHUCH, W., LAMB, C.J., DIXON, R.A. 1985. L-Phenylalanine ammonia-lyase from *Phaseolus vulgaris*. Characterization and differential induction of multiple forms from elicitor-treated cell suspension cultures. Eur. J. Biochem. 149:411-419.

50. CRAMER, C.L., EDWARDS, K., DRON, M., LIANG, X., DILDINE, S.L., BOLWELL, G.P., DIXON, R.A., LAMB, C.J., SCHUCH, W. 1989. Phenylalanine ammonia-lyase gene organization and structure. Plant Mol. Biol. 12:367-383.

51. LIANG, X., DRON, M., CRAMER, C.L., DIXON, R.A., LAMB, C.J. 1989. Differential regulation of phenylalanine ammonia-lyase genes during plant development and by environmental cues. J. Biol. Chem. 264:14486-14492.

52. JORRIN, J., DIXON, R.A. 1989. Stress responses in alfalfa (*Medicago sativa* L.). II. Purification, characterization and induction of phenylalanine ammonia-lyase isoforms from elicitor treated cell suspension cultures. Plant Physiol. 92:447-455.

53. BOLWELL, G.P., SAP, J., CRAMER, C.L., SCHUCH, W., LAMB, C.J., DIXON, R.A. 1985. L-Phenylalanine ammonia-lyase from *Phaseolus vulgaris*: Partial degradation of enzyme subunits *in vitro* and *in vivo*. Biochim. Biophys. Acta 881:210-221.

54. GOWRI, G., PAIVA, N.L., DIXON, R.A. 1991. Stress responses in alfalfa (*Medicago sativa* L.) IX. Sequence analysis of phenylalanine ammonia-lyase (PAL) cDNA clones and appearance of PAL transcripts in elicitor-treated cell cultures and developing seedlings. Plant Mol. Biol., in press.

55. SHAW, N.M., BOLWELL, G.P., SMITH, C. 1990. Wound-induced phenylalanine ammonia-lyase in potato (*Solanum tuberosum*) tuber

discs. Significance of glycosylation and immunolocalization of enzyme subunits. Biochem. J. 267:163-170.

56. HINDERER, W., FLENTJE, U., BARZ, W. 1987. Microsomal isoflavone 2' and 3'-hydroxylases from chickpea (*Cicer arietinum* L.) cell suspension cultures induced for pterocarpan phytoalexin formation. FEBS Lett. 214:101-106.

57. KOCHS, G., GRISEBACH, H. 1989. Phytoalexin synthesis in soybean: purification and re-constitution of cytochrome P450 3,9-dihydroxypterocarpan 6a-hydroxylase and separation from cytochrome P450 cinnamate 4-hydroxylase. Arch. Biochem. Biophys. 273:543-553.

58. O'KEEFE, D.P., LETO, K.J. 1989. Cytochrome P450 from the mesocarp of avocado (*Persea americana*). Plant Physiol. 89:1141-1149.

59. CHOUDHARY, A.D., KESSMANN, H., LAMB, C.J., DIXON, R.A. 1990. Stress responses in alfalfa (*Medicago sativa* L.) IV. Expression of defense gene constructs in electroporated suspension cell protoplasts. Plant Cell Rep. 9:42-46.

60. HRAZDINA, G., LIFSON, E., WEEDEN, N.F. 1986. Isolation and characterization of buckwheat (*Fagopyrum esculentum* M.) chalcone synthase and its polyclonal antibodies. Arch. Biochem. Biophys. 247:414-419.

61. HELLER, W., HAHLBROCK, K. 1980. Highly purified "flavanone synthase" from parsley catalyzes the formation of naringenin chalcone. Arch. Biochem. Biophys. 200:617-619.

62. WHITEHEAD, I.M., DIXON, R.A. 1983. Chalcone synthase from cell suspension cultures of *Phaseolus vulgaris*. Biochim. Biophys. Acta 747:298-303.

63. BEERHUES, L., WIERMANN, R. 1988. Chalcone synthases from spinach (*Spinacia oleracea* L.). I. Purification, peptide patterns, and immunological properties of different forms. Planta 173:532-543.

64. JACOBS, M., RUBERY, P.H. 1988. Naturally occurring auxin transport regulators. Science 241:346-349.

65. EDWARDS, R., DIXON, R.A. 1991. Purification, characterization and photoaffinity labelling of S-adenosyl-L-methionine: caffeic acid 3-*O*-methyltransferase from suspension cultures of alfalfa (*Medicago sativa* L.). Arch. Biochem. Biophys. 287:372-379.

66. WENGENMAYER, H., EBEL, J., GRISEBACH, H. 1974. Purification and properties of a S-adenosyl methionine: isoflavone 4'-*O*-methyltransferase from cell suspension cultures of *Cicer arietinum* L. Eur. J. Biochem. 50:135-143.

67. EDWARDS, R., DIXON, R.A. 1991. Isoflavone O-methyltransferase activities in elicitor-treated cell suspension cultures of *Medicago sativa*. Phytochemistry, in press.

68. BUGOS, R.C., CHIANG, V.L., CAMPBELL, W.H. 1989. Isolation of O-methyltransferase associated with lignin biosynthesis in aspen. Wood Pulp Chem. TAAPI Proc. pp. 345-347.

69. GOWRI, G., BUGOS, R.C., CAMPBELL, W.H., MAXWELL, C.A., DIXON, R.A. 1991. Molecular cloning and expression of alfalfa S-adenosyl-L-methionine: caffeic acid 3-O-methyltransferase, a key enzyme of lignin biosynthesis. Plant Physiol., in press.

70. PAIVA, N.L., EDWARDS, R., SUN, Y., HRAZDINA, G., DIXON, R.A. 1991. Molecular cloning and expression of alfalfa isoflavone reductase, a key enzyme of isoflavonoid phytoalexin biosynthesis. Plant Mol. Biol., in press.

71. BELD, M., MARTIN, C., HUITS, M., STUITJE, A.R., GERATS, A.G.M. 1989. Flavonoid synthesis in *Petunia hybrida*: partial characterization of dihydroflavonol-4-reductase genes. Plant Mol. Biol. 13:491-502.

72. BEVAN, M., SHUFFLEBOTTOM, D., EDWARDS, K., JEFFERSON, R., SCHUCH, W. 1989. Tissue-and cell-specific activity of a phenylalanine ammonia-lyase promoter in transgenic plants. EMBO J. 8:1899-1906.

73. LIANG, X., DRON, M., SCHMID, J., DIXON, R.A., LAMB, C.J. 1989. Developmental and environmental regulation of a phenylalanine ammonia-lyase-β-glucuronidase gene fusion in transgenic tobacco plants. Proc. Natl. Acad. Sci. USA 86:9284-9288.

74. DANGL, J.L., HAHLBROCK, K., SCHELL, J. 1989. Regulation and structure of chalcone synthase genes: In: Cell Culture and Somatic Cell Genetics of Plants (I.K. Vasil, J. Schell, eds.), vol 6, Plant Nuclear Genes and Their Expression, Academic Press, New York, pp 155-174.

75. STERMER, B.A., SCHMID, J., LAMB, C.J., DIXON, R.A. 1990. Infection and stress activation of bean chalcone synthase promoters in transgenic tobacco. Mol. Plant Microbe Int. 3:381-388.

76. WINGENDER, R., ROHRIG, H., HORICKE, C., SCHELL, J. 1990. *cis*-Regulatory elements involved in ultravoilet light regulation and plant defense. Plant Cell 2:1019-1026.

77. RYDER, T.B., HEDRICK, S.A., BELL, J.N., LIANG, X., CLOUSE, S.D., LAMB, C.J. 1987. Organization and differential activation of a

gene family encoding the plant defense enzyme chalcone synthase in *Phaseolus vulgaris*. Mol. Gen. Genet. 210:219-233.

78.   CHOUDHARY, A.D., LAMB, C.J., DIXON, R.A. 1990. Stress responses in alfalfa (*Medicago sativa* L.) VI. Differential responsiveness of chalcone synthase induction to fungal elicitor or glutathione in electroporated protoplasts. Plant Physiol. 94:1802-1807.

79.   DANGL, J.L., HAUFFE, K.D., LIPPHARDT, S., HAHLBROCK. K., SCHEEL, D. 1987. Parsley protoplasts retain differential responsiveness to u.v. light and fungal elicitor. EMBO J. 6:2551-2556.

80.   LOAKE, G., CHOUDHARY, A.D., HARRISON, M.J., MAVANDAD, M., LAMB, C.J., DIXON, R.A. 1991. Phenylpropanoid pathway intermediates regulate transient expression of a chalcone synthase gene promoter in electroporated protoplasts. Plant Cell, in press.

81.   HARRISON, M.J., CHOUDHARY, A.D., DUBERY, I., LAMB, C.J., DIXON, R.A. 1991. *Cis*-elements and *trans*-acting factors for the quantitative expression of a bean chalcone synthase gene promoter in electroporated alfalfa protoplasts. Plant Mol. Biol., in press.

82.   GREEN. P.J., YONG, M-H., CUOZZO, M., KANO-MURAKAMI, Y., SILVERSTEIN, P. CHUA, N-H. 1988. Binding site requirements for pea nuclear protein factor GT-1 correlate with sequences required for light-dependent transcriptional activation of the rbcS-3A gene. EMBO J. 7:4035-4044.

83.   HARRISON, M.J., LAWTON, M.A., LAMB, C.J., DIXON, R.A. 1991. Characterization of a nuclear protein which binds to three elements within the silencer region of a bean chalcone synthase gene promoter. Proc. Natl. Acad. Sci. USA 88:2515-2519.

84.   LAWTON, M.A., DEAN, S.J., DRON, M., KOOTER, J., KRAGH, K., HARRISON, M.J., YU, L., TANGUAY, L. DIXON, R.A., LAMB, C.J. 1991. Silencer region of a chalcone synthase promoter contains multiple binding sites for a factor, SBF-1, closely related to GT-1. Plant Mol. Biol. 16:235-249.

85.   LOIS, R., DIETRICH, A., HAHLBROCK. K., SCHULZ, W. 1989. A phenylalanine ammonia-lyase gene from parsley: structure, regulation and identification of elicitor and light-responsive *cis*-acting elements. EMBO J. 8:1641-1648.

86.   DONALD, R.G.K., SCHINDLER, U., BATSCHAUER, A., CASHMORE, A.R. 1990. The plant G box promoter sequence activates transcription in *Saccharomyces cerevisiae*, and is bound *in vitro* by a yeast activity similar to GBF, the plant G box binding factor. EMBO

J. 9:1727-1735.

87. SHIELDS, S.E., WINGATE, V.P.M., LAMB, C.J. 1982. Dual control of phenylalanine ammonia-lyase production and removal by its product cinnamic acid. Eur. J. Biochem. 123:389-395.

88. BOLWELL, G.P., MAVANDAD, M., MILLAR, D.J., EDWARDS, K.H., SCHUCH, W., DIXON, R.A. 1988. Inhibition of mRNA levels and activities by *trans*-cinnamic acid in elicitor-induced bean cells. Phytochemistry 27:2109-2117.

89. BOLWELL, G.P., CRAMER, C.L., LAMB, C.J., SCHUCH, W., DIXON, R.A. 1986. L-phenylalanine ammonia-lyase from *Phaseolus vulgaris*. Modulation of the levels of active enzyme by *trans*-cinnamic acid. Planta 169:97-107.

90. MAVANDAD, M., EDWARDS, R., LIANG, X., LAMB, C.J., DIXON, R.A. 1990. Effects of *trans*-cinnamic acid on expression of the bean phenylalanine ammonia-lyase gene family. Plant Physiol. 94:671-680.

91. AMRHEIN, N., GERHART, J. 1979. Superinduction of phenylalanine ammonia-lyase in gherkin hypocotyls caused by the inhibitor, L-α-aminooxy-β-phenylpropionic acid. Biochim. Biophys. Acta 53:434-442.

92. DIXON, R.A., BOLWELL, G.P. 1986. Modulation of the phenylpropanoid pathway in bean (*Phaseolus vulgaris*) cell suspension cultures. In: Secondary Metabolism in Plant Cell Cultures, (P. Morris, A.H. Scragg, A. Stafford, M.W. Fowler eds.), Cambridge University Press, Cambridge, pp. 89-102.

93. BLOCK, A., DANGL, J.L., HAHLBROCK, K., SCHULZE-LEFERT, P. 1990. Functional borders, genetic fine structure, and distance requirements of *cis* elements mediating light responsiveness of the parsley chalcone synthase promoter. Proc. Natl. Acad. Sci. USA 87:5387-5391.

94. RADEMACHER, W., FRITSCH, H., GRAEBE, J.E., SAUTER, H., JUNG, J. 1987. Tetcyclasis and triazole-type plant growth retardants: their influence on the biosynthesis of gibberellins and other metabolic processes. Pesticide Sci. 21:241-252.

95. BINGHAM, E.T., HURLEY, C.V., KAATZ, D.M., SAUNDERS, J.W. 1975. Breeding alfalfa which regenerates from callus tissue in culture. Crop Sci. 15:719-721.

96. RAY, I.M., BINGHAM, E.T. 1989. Breeding diploid alfalfa for regeneration from tissue culture. Crop Sci. 29:1545-1548.

97. NOLAN, K.E., ROSE, R.J., GORST, J.R. 1989. Regeneration of
     Medicago truncatula from tissue culture: Increased somatic embryo-
     genesis using explants from regenerated plants. Plant Cell Rep.
     8:278-281.
98. WINICOW, I., MAKI, D.H., WATERBORG, J.H., RIEHM, M.R.,
     HARRINGTON, R.E. 1988. Characterization of alfalfa (*Medicago
     sativa*) genome by DNA reassociation. Plant Mol. Biol. 10:369-371.
99. BINGHAM, E.T., CLEMENT, W.M. 1989. Alfalfa transposable
     elements and variegation. Developmental Genetics 10:552-560.
100. DIXON, R.A., BLYDEN, E.R., ELLIS, J.S. 1990. Biochemistry and
     molecular genetics of plant-pathogen systems. In: "Biochemical
     Aspects of Crop Improvement," (K.R. Khanna, ed.), CRC Press,
     Boca Raton, Fl., pp. 179-222.
101. VANETTEN, H.D., MATTHEWS, D.E., MATTHEWS, P.S. 1989.
     Phytoalexin detoxification: Importance for pathogenicity and practical
     implications. Annu. Rev. Phytopathol. 27:143-164.
102. KEPHART, K.D., BUXTON, D.R., HILL, R.R., Jr. 1990. Digestibility
     and cell-wall components of alfalfa following selection for divergent
     herbage lignin concentration. Crop Sci. 30:207-212.

Chapter Five

BIOSYNTHESIS AND METABOLISM OF ISOFLAVONES AND PTEROCARPAN PHYTOALEXINS IN CHICKPEA, SOYBEAN AND PHYTOPATHOGENIC FUNGI*

Wolfgang Barz[#] and Roland Welle[^]

[#] Department of Plant Biochemistry and Biotechnology, Westfälische, Wilhelms-Universität, D-4400 Münster, FRG

[^] Department of Plant Biochemistry, Albert Ludwigs-Universität, D-7800 Freiburg, FRG

INTRODUCTION

Among the almost unlimited number of plant secondary constituents, isoflavonoids are especially suitable to demonstrate physiological functions and ecological roles for such compounds in the producing plant. Isoflavonoids (i.e., isoflavones, isoflavanones, isoflavans, pterocarpans, coumestans, rotenoids) are

---

*This manuscript is dedicated to the memory of Professor Hans Grisebach (1926-1990).

Phenolic Metabolism in Plants, Edited by H.A. Stafford
and R.K. Ibrahim, Plenum Press, New York, 1992

139

characteristic constituents of the Leguminosae and they are known to be involved in the interaction of the individual plant species with other organisms. Furthermore, they provide plants with a means of adjustment to changing circumstances such as microbial infection. These functions are a consequence of their bioactive properties which may be antibiotics, repellents, attractants or signal compounds.[1-4]

The constitutively produced isoflavones ("preinfectional inhibitors") and the infection-induced, *de novo* synthesized pterocarpans ("phytoalexins") are of special interest due to their pronounced antimicrobial properties.[5] Pterocarpans presently represent the largest group among the nearly 300 known phytoalexins.[5,6] Recent aspects of the enzymology and molecular biology of isoflavone and pterocarpan biosynthesis and the metabolism of their conjugates in chickpea and soybean will be discussed. Furthermore, isoflavone and pterocarpan phytoalexin degradation by phytopathogenic fungi deserves attention because such metabolic routes may be essential traits of parasite pathogenicity.

PTEROCARPAN PHYTOALEXINS AS PART OF PLANT ANTIMICROBIAL DEFENCE

Plants have developed sophisticated active defence mechanisms against pathogens. These reactions involve inhibition of microbial growth in the plant tissue, isolation of the microbes in lesions by the erection of special permeability barriers, and death of the invading pathogen by the accumulation of antibiotic compounds. Most of the defence reactions (i.e., formation of chitinases, β–1,3-glucanases or inhibitors of fungal hydrolases, reinforcement of plant cell walls by integration of glycoproteins, callose, lignin or other phenolic polymers as well as secretion of antimicrobial peroxidases) appear to start simultaneously upon infection.[7,8] The principle of differential gene activation is an essential element in defence expression. Phytoalexin induction and accumulation is, therefore, integrated into a complex network of several antimicrobial reactions so that the relative importance of the plant phytoalexin response needs to be elucidated. For example, in the chickpea *(Cicer arietinum)* - *Ascochyta rabiei* interaction, pterocarpan accumulation was shown to be accompanied by the formation of various groups of pathogenesis related proteins[9] and increased polyphenol deposition.[10]

Chickpea plants and cell cultures synthesize the (6aR:11aR)-pterocarpans, medicarpin and maackiain (see Fig. 2). Incompatible interactions between resistant plant cultivars and virulent *A. rabiei* isolates are characterized by a hypersensitive response followed by rapid accumulation of much higher phytoalexin concentrations in comparison to compatible interactions.[11]

In soybean *(Glycine max.)* tissues infected with the stem and root rot pathogen *Phytophthora megasperma* f. sp. *glycinea* (Pmg), the pterocarpan phytoalexins glyceollin I - III (see Fig. 3) are induced. Intensive investigations by Grisebach and his associates on the compatible and the incompatible interaction, using different Pmg races and soybean cultivar Harosoy 63, have unequivocally shown the decisive role of the glyceollins in plant resistance and in the inhibition of fungal growth in the plant tissue.[12]

Incompatible interactions lead rapidly to such high phytoalexin concentrations that the *in vitro* EC_{50} value of the inhibitor is surpassed. Furthermore, in comparison to the compatible interaction, the rate of glyceollin formation in the incompatible interaction continues much longer. This rapid increase in, and continued duration of, the biosynthesis of the glyceollins were positively correlated with a concomitant induction of several enzymes of phytoalexin formation[12] (see Figs, 1,3). Finally, inhibition of phenylalanine ammonia lyase, the initial enzyme of phenylpropanoid biosynthesis, with molecular inhibitors prevented glyceollin accumulation so that an incompatible interaction turned into a compatible one[13]. These and other data clearly indicate that phytoalexins represent an essential element in the plant antimicrobial defence system.

ENZYMATIC SYNTHESIS OF ISOFLAVONES AND THEIR CONJUGATES

Pterocarpan phytoalexin biosynthesis proceeds through the stage of an isoflavone intermediate. Isoflavones themselves are prominent constitutive compounds in chickpea and soybean. *C. arietinum* plants and cell cultures mainly accumulate formononetin and biochanin A, whereas soybean tissues are rich in daidzein and genistein (Fig. 1). In both systems these isoflavones predominantly occur in the form of 7-*O*-glucoside 6"-*O*-malonate conjugates.[14,15] As shown for other polar hydrophilic plant constituents,[16] the isoflavone conjugates are located in vacuoles, as recently demonstrated with chickpea cells (U. Mackenbrock, W. Barz, unpublished).

The biosynthesis of isoflavones has been studied extensively in both chickpea and soybean, so that the essential enzymes of the general phenylpropanoid pathway and chalcone synthase, chalcone isomerase as well as isoflavone synthase are well known[6,12] (Fig. 1). In the chickpea system, the additional enzymes for 4'-*O*-methylation of isoflavones and for the addition of the

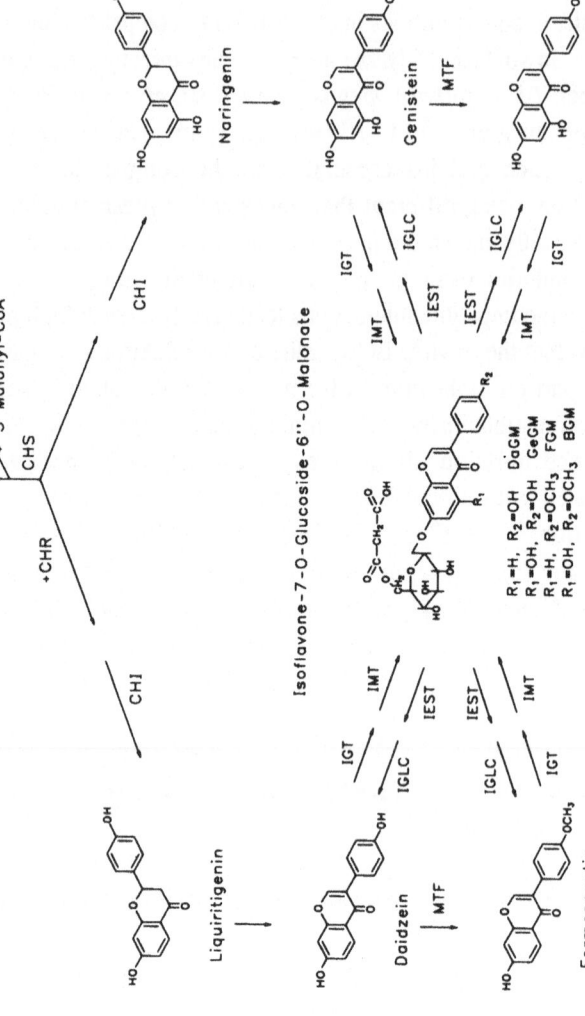

Figure 1. Biosynthetic pathway to isoflavones and their malonylglucosides in soybean (diadzein: DaGM; genistein: GeGm) and chickpea (formononetin: FGM; biochanin A: BGM). The enzymes are: PAL, phenylalanine ammonia lyase; CA4H cinnamic acid 4-hydroxylase; 4CL, p-coumaric acid CoA-ligase; CHS, chalcone synthase; CHR, chalcone reductase; CHI, chalcone isomerase; MTF, isoflavone methyltransferase. IGT, isoflavone 7-O-glucosyltransferase; IMT, isoflavone glucoside malonyltransferase; IEST, isoflavone malonylglucoside esterase; IGLC, isoflavone glucoside glucosidase.

conjugating moieties have also been characterized.[16] During pterocarpan phytoalexin induction in both plants, all enzymes from PAL to isoflavone synthase including the isoflavone O-methyltransferase in chickpea are subject to *de novo* formation resulting in a pronounced increase of enzyme activities above the constitutively expressed levels.[12,13,17] However, tissue-specific differences in the localization of constitutive enzyme activity and infection-increased levels have to be considered. Thus, the constitutive chalcone synthase in chickpea leaves is found in epidermal layers whereas the newly induced enzyme activity representing two isoforms mainly appears in mesophyll cells (L. Beerhues, R. Wiermann, W. Barz, unpublished).

Several recent developments in the enzymology of isoflavone biosynthesis require more detailed descriptions. Chalcone synthase (CHS), the key enzyme of flavonoid biosynthesis and carefully studied in numerous plant systems,[18] catalyzes the formation of a chalcone with a phloroglucinol substitution pattern in ring A. This 6'-hydroxychalcone (not shown) is isomerized to naringenin, the intermediate for genistein and biochanin A formation (Fig. 1). In soybean, a CHS gene family comprising six members has been described. The various CHS isoforms are differently regulated according to physiological situation and enviromental stimuli.[19] Similarly, five CHS-isoforms appear to be involved in chickpea flavonoid metabolism.[9]

The long sought explanation for the enzyme mechanism operating in the formation of 6'-deoxychalcones (DOC, i.e. 4,2',4'-trihydroxychalcone) as the intermediate via liquiritigenin for daidzein, formononetin and the pterocarpan phytoalexins has recently been found by Welle and Grisebach.[20] An independent enzyme, chalcone reductase (CHR), coacts with chalcone synthase and NADPH as cofactor in the formation of DOC. Prior to the closing of ring A in the CHS-reaction at the stage of an enzyme-bound polyketide the oxygen function is removed. Such a mechanism requires a functional complex between CHS and CHR during the enzyme reaction. Thus, the CHR protein shows a high activity for CHS (K_m = 0.75μM) and the ratio of 6'-hydroxy- to 6'-deoxyproducts depends on the CHS:CHR molar ratio. However, under *in vitro* conditions where an optimal ratio of CHS:CHR of 2:1 was found, only 60% of the total CHS products are DOC. The *in vivo* regulation of this substrate shuttle for the channelling of the two competing pathways (Fig. l) and the cellular localization remain to be investigated. In general, CHR may be considered as the key enzyme in the biosynthesis of 5-deoxyisoflavones and l-deoxypterocarpans which represent the majority of all known isoflavonoid phytoalexins.[5]

At times of massive phytoalexin accumulation in soybean cell cultures, the CHR (one polypeptide chain, Mr 34 kD) was found to be strongly induced in parallel with CHS, both at the level of enzyme and mRNA activities.[21] Using a polyclonal antibody raised against the soybean CHR, a very similar situation has been observed for the chickpea system (W. Bless, W. Barz, unpublished). This enzyme (M_r 34 kD, consisting of four isoforms with IEP between 5.55 and 6.10) is rapidly induced upon elicitation of chickpea cell cultures, with the same kinetics observed for both CHS and the accumulation of medicarpin and maackiain.

The recent report on the DNA sequences of reductase cDNA clones,[22] the amino acid composition of the enzyme and the synthesis of the CHR protein in an *E. coli* expression system, forms the basis for investigations on CHR gene regulation and the CHS-CHR enzyme mechanism. An interesting feature of the CHR amino acid sequence derived from the various cDNA clones and their proteins is the observation that the first 14 amino acids of the N-terminus are essential for full enzyme activity.[22] Furthermore, hybridization experiments of the soybean CHR cDNA with genomic DNA from other plants revealed homologies in bean and peanut. Surprisingly, no such homologies were found for pea, although this plant is known to produce a 1-deoxypterocarpan.

A central step in isoflavone biosynthesis is the molecular rearrangement catalyzed by isoflavone synthase in which a (2S)-flavanone is converted to an isoflavone such as daidzein or genistein (Fig. 1). The enzyme reaction for this oxidative aryl migration, first reported by Hagmann and Grisebach,[23] consists of two enzyme steps. First, a membrane-bound cytochrome P-450 monoxygenase produces a 2-hydroxyisoflavanone with the oxygen atom of the hydroxyl function being derived from O_2. This intermediate is then converted by a soluble dehydratase to the isoflavone.[24] Although recent studies on this rearrangement reaction using microsomal preparations of *Pueraria lobata* cell cultures favoured a different reaction mechanism (radical versus anionic incorporation of the 2-hydroxyfunction) the essential elements of this biosynthetic step are clear. The soluble dehydratase, however, remains to be characterized.

4'-O-Methylation of isoflavones yielding formononetin and biochanin A (Fig. 1) appeared to be catalyzed by a SAM-dependent methyltransferase characterized from chickpea roots and cell cultures.[26] Conflicting data on the expression of this enzyme activity in plant and cell culture material required reexamination. Recent studies (C. Weidemann, W. Barz, unpublished) have led to the discovery of both an isoflavone 7-O-methyltransferase (7-OMT) and an 4'-O-methyltransferase (4'-OMT). In roots and leaves the ratio of these two enzyme activities is approximately 1:8, whereas in cell cultures the reverse situation has been measured. The bulk of the cell culture O-methyltransferase is the 7-OMT (pH optimum 8-9; K_m daidzein = 6.5μM; K_m SAM = 26μM). However, the expected 7-0-methylated isoflavones (isoformononetin, prunetin) do not accumulate in these cultures which form predominantly biochanin A and formononetin. Furthermore, upon elicitation of the chickpea cell cultures which leads to medicarpin and maackiain accumulation, the activity of the 7-0MT is highly increased. Similar confusing observations have been made for 7-0MT and 4'-0MT in alfalfa plants and cell cultures (R. Edwards, R.A. Dixon, personal communication). These data require future investigations on the regulation and the expression of isoflavone 4'-0MT in these Leguminosae because such a reaction is necessary for phytoalexin formation.

Daidzein and genistein in soybean and formononetin and biochanin A in chickpea occur predominantly as 7-0-glucoside 6"-0-malonate conjugates (Fig. 1). Investigations in the Münster laboratory using chickpea plants and cell cultures have thoroughly elucidated the enzymology of the formation and the breakdown of such conjugates.[6,16]

The metabolic cycle indicated in Figure 1 (see also Fig. 5) consists of isoflavone 7-0-glucosyltransferase (IGT), malonyl-transferase (IMT), a highly substrate-specific malonylesterase (IEST) and a set of isoflavone glucoside specific β-glucosidases (IGLC). The IEST enzyme appears to be associated with the tonoplast membrane (U. Mackenbrock, W. Barz, unpublished). IMT and IEST are present in all organs and in cell cultures of chickpea. Therefore, a strict metabolic and spatial separation of these two enzyme activities must exist to control the pool sizes of malonylglucosides in vacuoles. Since vacuolar influx and efflux of such conjugates occur simultaneously[16] the regulation and substrate funnelling during this bidirectional tonoplast transport should be elucidated with high priority; this work is presently under investigation.

THE PTEROCARPAN-SPECIFIC BRANCH OF PHYTOALEXIN BIOSYNTHESIS

Pterocarpan phytoalexins are biosynthesized from isoflavone inter-mediates, which are formononetin in chickpea (Fig. 2) and daidzein in soybean (Fig. 3). Cell suspension cultures treated with fungal elicitors to induce phytoalexin accumulation were especially helpful in the elucidation of the pathways and the isolation of the enzymes.[6,12,13,17]

The pterocarpan-specific branch leading to medicarpin and maackiain in chickpea starts with two cytochrome P-450 monooxygenases for isoflavone 2'-

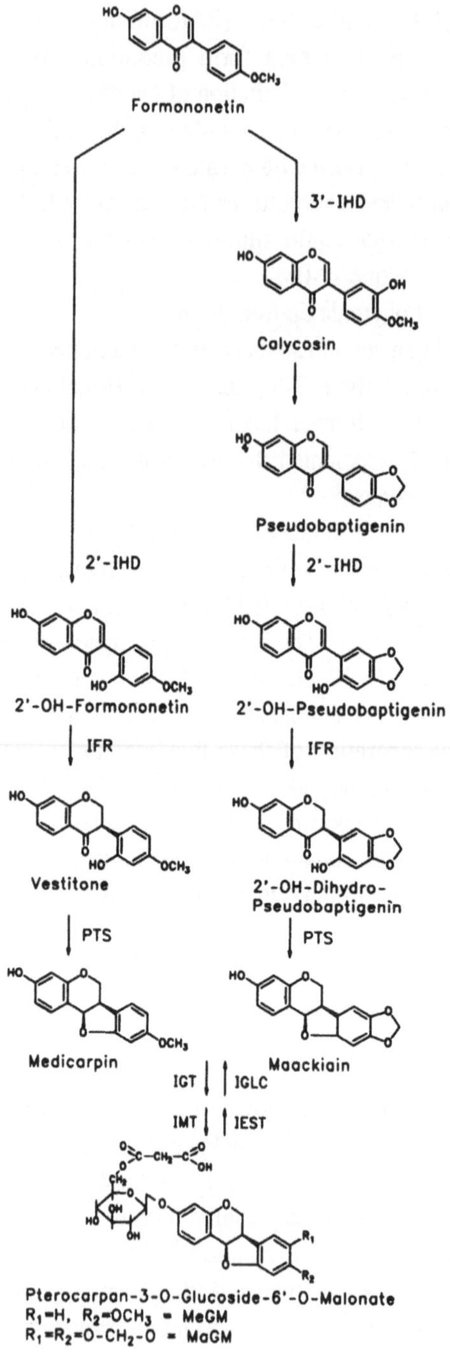

Formononetin

3'-IHD

Calycosin

Pseudobaptigenin

2'-IHD   2'-IHD

2'-OH-Formononetin   2'-OH-Pseudobaptigenin

IFR   IFR

Vestitone   2'-OH-Dihydro-
Pseudobaptigenin

PTS   PTS

Medicarpin   Maackiain

IGT ⇅ IGLC

IMT ⇅ IEST

Pterocarpan-3-O-Glucoside-6'-O-Malonate
R_1=H, R_2=OCH$_3$ = MeGM
R_1=R_2=O-CH$_2$-O = MaGM

and 3'- hydroxylation (2'-IHD, 3'-IHD). These inducible, microsomal enzymes together with the other P-450 monooxygenases depicted in Figures 2 and 3 were well characterized by CO-inhibition of enzyme activities, reversal of this inhibition with blue light and studies with recognized cytochrome P-450-specific inhibitors. The chickpea enzymes are highly specific for 4'-0-methylisoflavones. The regulatory importance of the 2'-IHD enzyme is indicated by the observation that the very low induction of this enzyme step appears to be essential for the moderate accumulation of the pterocarpans in *A. rabiei* suspectible chickpea cultivars.[17,27]

The subsequent biosynthetic step is catalyzed by a NADPH:isoflavone oxidoreductase (IFR) which reduces a 2'-hydroxyisoflavone to the corresponding isoflavanone (Figs. 2,3). This enzyme creates the first chiral compound which has a $3R$-configuration in the case of chickpea and soybean. The enzyme (one polypeptide chain, M_r 36 kDa, IEP 6.3, K_m NADPH = 20µM) is highly specific for the two substrates shown in Figure 2. The cDNA cloning of the IFR gene led to a 1183 bp clone with an open reading frame of 954 bases, encoding a polypeptide of 318 amino acids (calculated M_r 35.4 kDa).[28] Comparison of the IFR amino acid sequence with other known protein sequences revealed no significant degree of homology, not even with other oxidoreductases. Hybridization experiments with chickpea leaves and cell cultures using the cDNA clone and RNA preparations revealed that the enzyme is under strict control of infection/elicitor induction and gene activation (K. Tiemann, W. Barz, unpublished.)

The terminal enzyme reaction for the formation of the dihydrofuran ring

---

Figure 2. Pterocarpan-specific pathway from the isoflavone formononetin to medicarpin and maackiain and their malonylglucosides (GM) in chickpea. The enzymes are: 2'-IHD, isoflavone 2'-hydroxylase; 3'-IHD, isoflavone 3'-hydroxylase; IFR, 2'-hydroxyisoflavone oxidoreductase; PTS, pterocarpan synthase. For the enzymes IGT, IMT, IGLC and IEST, see Figure 1. (Note that the numbering system is confusing in the case of the 2'-hydroxylation leading to maackiain because a 3'-hydroxylation step precedes it. However, the 3',4' positions occupied by the methylenedioxy moiety can be considered equivalent to those at 4',5'. The same 2'-hydroxylase (2'-IHD) is involved in both the medicarpin and maackiain pathways, and the 2'-hydroxyl group introduced is *para* to the one formed by the 3'-hydroxylase).

of pterocarpans (Fig. 2) is catalyzed by pterocarpan synthase (PTS). The enzyme purified from chickpea (one polypeptide chain, M_r 27 kDa, K_m NADPH = 40μM) only accepts the $3R$-configurated isoflavanone intermediates shown in Figure 2, whereas the corresponding 2'-hydroxy intermediate from the soybean pterocarpan pathway (Fig. 3) is not converted.[29] Elicitation experiments on the chickpea pterocarpan pathway revealed that all enzymes are coordinately induced and reach maximum activities prior to maximum pterocarpan aglycone accumulation.[17]

After the complete elucidation of the chickpea pathway[6,16,17] (Fig. 2), reports from other laboratories demonstrated an identical sequence of reaction steps and the involvement of closely related enzymes for other pterocarpan-producing plants. The pathway operating in soybean (Fig. 3), pisatin biosynthesis in pea and medicarpin production in alfalfa (see Dixon, Chapter 4) revealed a high degree of homology.[30,31,32]

The terminal part of glyceollin biosynthesis (Fig. 3) was elucidated by Grisebach and his associates as a series of membrane-bound enzymes.[33,34,35] First, (6aR:11aR)-3,9-dihydroxypterocarpan is hydroxylated at position 6a, the product is prenylated at carbon atoms 2 or 4 and finally, glyceollidin cyclase catalyzes the formation of glyceollins I - III. This cyclase explains the biosynthesis of 2,2-dimethylchromen and 2-isopropyl-dihydrofuran rings often found in secondary products. The enzyme yields the glyceollin isomers I - III in a ratio of 15:1:5 starting with a mixture of the isomeric 2- and 4-glyceollidins of 8:92. The enzymatic data for the preferential formation of glyceollin I is in agreement with the *in vivo* distribution of the phytoalexin isomers.

Except for prenyltransferase all of these membrane-associated enzymes are cytochrome P-450 monooxygenases. It is especially noteworthy that the cytochrome P-450 6a-hydroxylase has been solubilized and successfully purified to homogeneity.[33] These investigations also provided evidence for the elicitor induction of multiple cytochrome P-450 enzyme species at the level of individual proteins.[33] The very pronounced differential elicitor induction of such enzyme species has also been demonstrated with microsomes and their solub-ilized proteins from elicited chickpea cells (W. Gunia, W. Barz, unpublished).

A surprising new aspect of pterocarpan phytoalexin formation in chickpea plants and cell cultures was the isolation of medicarpin and maackiain 3-0-glucoside 6'-0-malonate conjugates[36] (Fig. 2). These compounds (MeGM, MaGM) occur constitutively in non-elicited cell cultures and have also been observed in noninfected roots. In infected aerial parts, the accumulation of the aglycones is accompanied and sometimes quantitatively exceeded by the conjugates MeGM and MaGM.

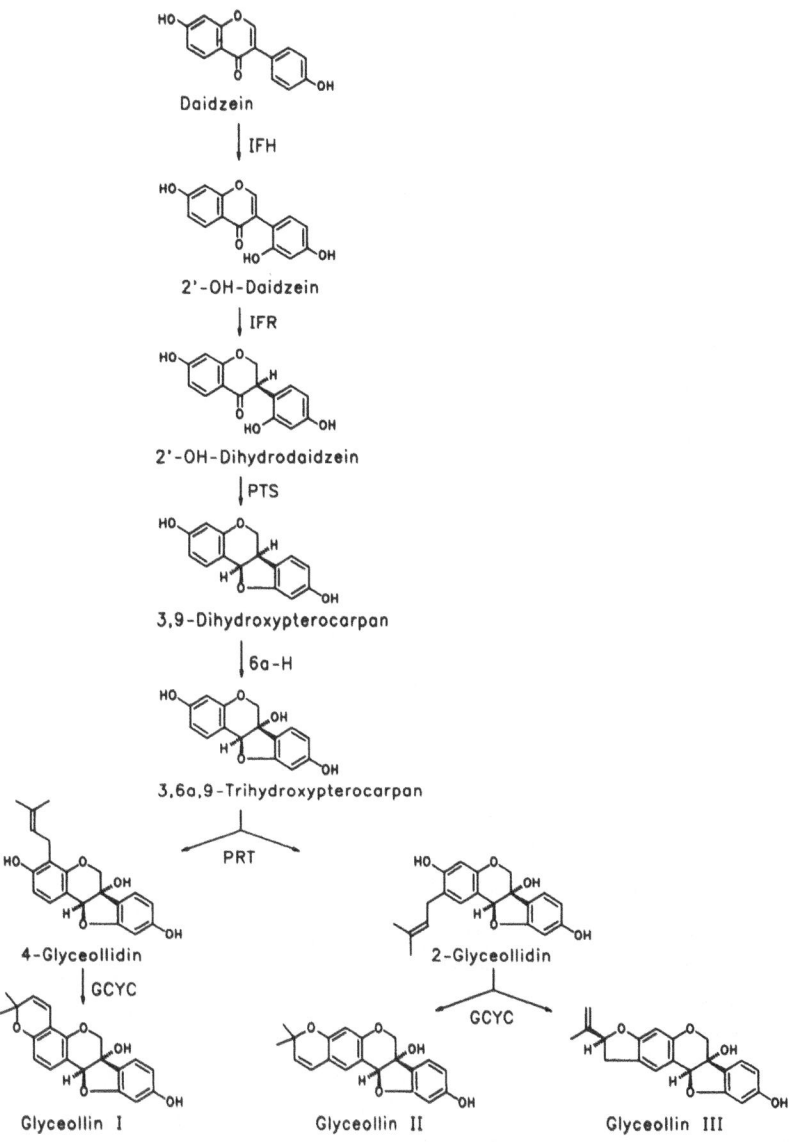

Figure 3: Pterocarpan-specific pathway from the isoflavone daidzein to glyceollins I - III in soybean. The enzymes are: IFH, isoflavone hydroxylase; IFR, 2'-hydroxyisoflavone oxidoreductase; PTS, pterocarpan synthase; 6a-H, pterocarpan 6a-hydroxylase; PRT, prenyltransferase; GCYC, glyceollidin cyclase.

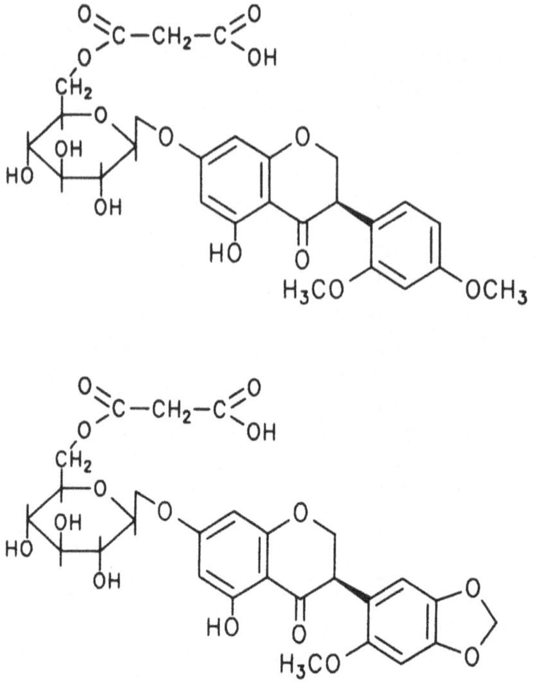

Figure 4. Structures of malonylglucosides of 2'-methoxyiso-flavanones homoferreirin (upper structure) and cicerin newly isolated from chickpea plants and cell cultures.

The constitutive formation of these pterocarpan conjugates requires that under normal conditions of cell culture and plant growth sufficient levels of the biosynthetic enzymes leading from formononetin to the pterocarpans (Fig. 2; 2'-IHD/3'-IHD to PTS) are expressed. Present investigations have provided evidence for such basic levels of these enzymes (W. Gunia, U. Mackenbrock, W. Barz, unpublished). With regard to the biosynthesis and the possible breakdown of the malonylglucosides of medicarpin and maackiain the same enzymes characterized in connection with FGM and BGM (Fig. 1) are thought to be involved.[36] Aspects of the metabolic function of MeGM and MaGM will be discussed in the subsequent section.

Medicarpin and maackiain are 1-deoxypterocarpans synthesized from 5-deoxyisoflavone formononetin (Fig. 2). Although 1-hydroxypterocarpans are known to occur in plants,[5] there has been no evidence for the 5-hydroxyisoflavone biochanin A being involved in a pathway analogous to the pterocarpan-specific branch used for l-deoxypterocarpans (Fig. 2). However, the isolation and structural identification of the 5-hydroxy-2'-methoxyisoflavanones, homoferreirin and cicerin (Fig. 4), from chickpea cell cultures and plants now points to a closely related pathway. These new compounds occur in substantial amounts, exclusively as 7-0-glucoside 6"-0-malonates, and their biosynthetic sequence, starting with biochanin, has now been shown to be strictly analogous to the pathway shown in Figure 2 (C. Weidemann, R. Tenhaken, W. Barz, unpublished). The highly active 2'-0-methyltransferase for the formation of homoferreirin and cicerin possibly prevents a terminal PTS-reaction. The possible differences in the substrate specificities of the enzymes used for medi-carpin/maackiain and homoferreirin/cicerin, respectively remain to be clarified.

CONSTITUTIVE ISOFLAVONE AND PTEROCARPAN CONJUGATES AS PRECURSORS FOR INDUCIBLE PTEROCARPAN PHYTOALEXINS

The prevailing assumption in the literature holds that phytoalexins are quantitatively synthesized *de novo* from early precursors of primary metabolism.[7,12,18] In the case of chickpea, the constitutive biosynthetic route to formononetin and its 7-0-glucoside 6"-0-malonate was thus viewed as being metabolically separated from the inducible medicarpin and maackiain isoflavonoid pathway. Recent investigations[16,37] have now shown, however, that a metabolic linkage between these pathways exists in chickpea. The isolation of constitutively formed medicarpin and maackiain malonylglucosides from chickpea further points to a high degree of regulatory complexity in the production of pterocarpan aglycones. Figure 5 illustrates that pterocarpan aglycones, the actual phytoalexins released from the producing cells, may be synthesized partly *de novo* and partly from the aglycone moieties of the conjugates of formononetin (FGM) and of pterocarpans (MeGM, MaGM).

A direct proof that formononetin released from the conjugate pool may act as a precursor for pterocarpan biosynthesis was obtained by using the competitive PAL inhibitor L-AOPP[37] (Fig. 5). In chickpea cell suspension cultures, the pool of FGM labelled with ^{14}C-formononetin, was shown to be quantitatively converted to pterocarpans upon elicitation and PAL inhibition. In addition, other data support the assumption that also under normal conditions FGM contributes to phytoalexin formation.[37]

MeGM and MaGM may be regarded as a reservoir from which the

aglycones can be rapidly released upon demand, i.e. a pronounced elicitation signal. Low or moderate amounts of elicitor applied to chickpea cell cultures led to a substantial increase in pterocarpans but almost quantitatively in form of the

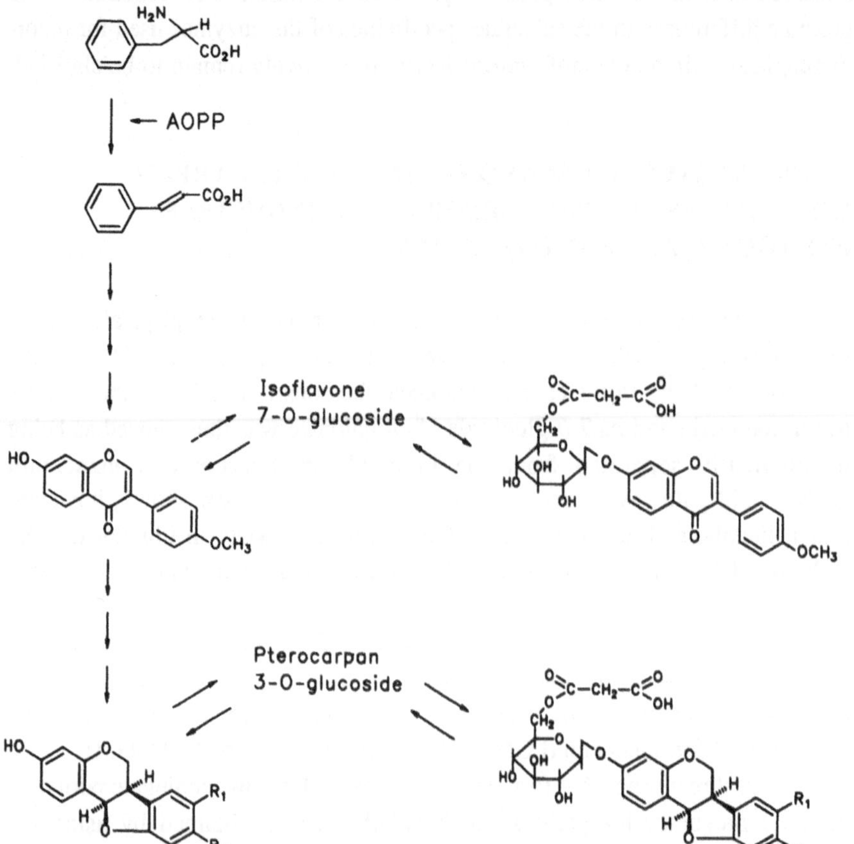

Figure 5. Metabolic grid indicating medicarpin ($R_1 = H$; $R_2 = OCH_3$) and maackiain ($R_1 = R_2 = -OCH_2O-$) formation from isoflavone or pterocarpan malonylglucosides in chickpea. AOPP: L-α- aminooxy -β-phenylpropionic acid, a competitive inhibitor of phenylalanine ammonia lyase. For explanation see text.

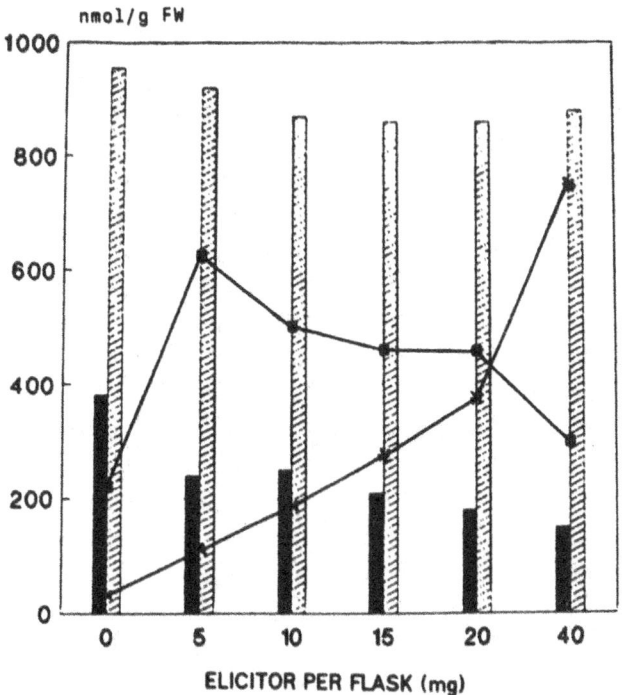

Figure 6. Dose-response of elicitor effect on the metabolism of formononetin 7-*0*-glucoside 6"*0*-malonate (FGM) (solid bars) and biochanin A 7-*0*-glucoside 6"-*0*-malonate (BGM) (hatched bars) and the accumulation of pterocarpan phytoalexin aglycones (∗—∗) and pterocarpan malonylglucosides (•—•) in chickpea cell suspension cultures. See Figure 5 for pathways involved.

conjugates. Furthermore, cell cultures enriched in MeGM and MaGM by low doses of elicitor rapidly produced the pterocarpan aglycones at the expense of the conjugates when a more pronounced elicitation signal was applied (U. Mackenbrock, W. Barz, unpublished).

The metabolic grid shown in Figure 5 is derived from various experiments which are partly summarized in Figure 6. With increasing amounts of elicitor applied to chickpea cell suspension cultures, FGM steadily decreased whereas the pool of the 5-hydroxyisoflavone conjugate, BGM, remained practically unaltered. The high levels of MeGM and MaGM found with low amounts of elicitor gradually decreased, while the aglycones that are the predominant compounds at high levels of elicitor increased. Quantitative evaluation of the data indicates that under massive elicitor impact the pools of FGM, MeGM and MaGM substantially contribute to phytoalexin aglycone accumulation. *De novo* synthesis of pterocarpans not funnelled through the pools of the conjugates is also evident but it amounts to only 40% of total pterocarpan accumulation. These data show a significant selectivity with regard to the differential mobilization of FGM and BGM and reflects a highly dynamic pattern of regulation of pools spatially separated in the cytoplasm (i.e., biosynthetic enzymes and aglycones) and in vacuoles (i.e., conjugates).

The direction of substrate funnelling into and out of the conjugate pools indicated in Figure 5 appears to be regulated by alternating changes in the activities of the enzymes involved in conjugate formation. With increasing levels of elicitor, a parallel and steady rise in the activities of pterocarpan bio-synthetic enzymes (i.e. PAL, CA4H, CHS, 2'-IHD) has been measured. However, isoflavone glucosyltransferase (IGT) and malonyltransferase (IMT) which were found to be highly induced at low elicitor levels, became strongly suppressed when elicitor amounts were raised. An inverse situation was observed for malonylesterase (IEST) and isoflavone glucoside β-glucosidase (IGLC) (U. Mackenbrock, W. Barz, unpublished). These findings indicate the regulatory importance of the enzymes involved in conjugate metabolism.

Mobilization of constitutive isoflavone malonylglucosides concomitant with pterocarpan phytoalexin accumulation upon an infection stimulus has also been observed for the soybean-*Phytophthora megasperma* interaction.[15] In Pmg-infected seedlings of soybean, the accumulation of the glyceollin phytoalexins is preceded by a substantial turnover of daidzein 7-0-glucoside 6"-0-malonate. The transient accumulation of the 5-deoxyisoflavone during glyceollin formation may be interpreted as an indication of its consumption for phytoalexin biosynthesis. In these soybean seedlings, genistein is also liberated from its conjugate pool, although a metabolic function for the 5-hydroxy compound has not yet been found.

In general, the consumption of vacuolar, constitutively formed conjugates of pterocarpans for phytoalexin formation can be regarded as an anaplerotic sequence and thus a protective measure for the very rapid accumu-lation of defence compounds.

ISOFLAVONE AND PTEROCARPAN DEGRADATION BY PHYTOPATHOGENIC FUNGI

Constitutive antimicrobial isoflavones may be regarded as universal protective barriers against microbes, whereas pterocarpan phytoalexins are active plant defences directed at specific invading pathogens. Sucessful invasion of the plant may, in turn, involve the ability of the fungal pathogen to detoxify the defence compounds of the host.[6] A variety of data indicate that pathogens are able to alter plant chemical defences by producing metabolites less toxic to microbes, conferring to the pathogen a higher degree of virulence. This potential represents an important genetic trait of pathogenicity.[1,38] This is best documented by genetic transformation of fungi with a gene for a phytoalexin detoxifying enzyme.[39]

In comprehensive studies on fungal catabolism of chickpea isoflavones and pterocarpan phytoalexins, the degradative pathways of these compounds have been identified in various *Fusarium* species and in *Ascochyta rabiei*. [40-42] Similar investigations with *Phytophthora megasperma* and the glyceollins are not known.

The degradative sequence of biochanin A determined with *Fusarium javanicum* (Fig. 7) first involves the formation of dihydrobiochanin A and then

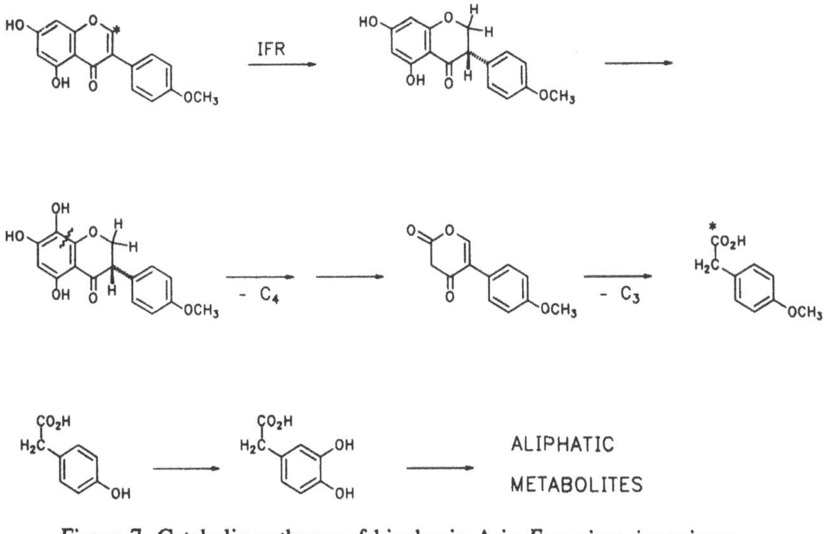

Figure 7. Catabolic pathway of biochanin A in *Fusarium javanicum*. The asterisk (*) indicates ^{14}C-label.

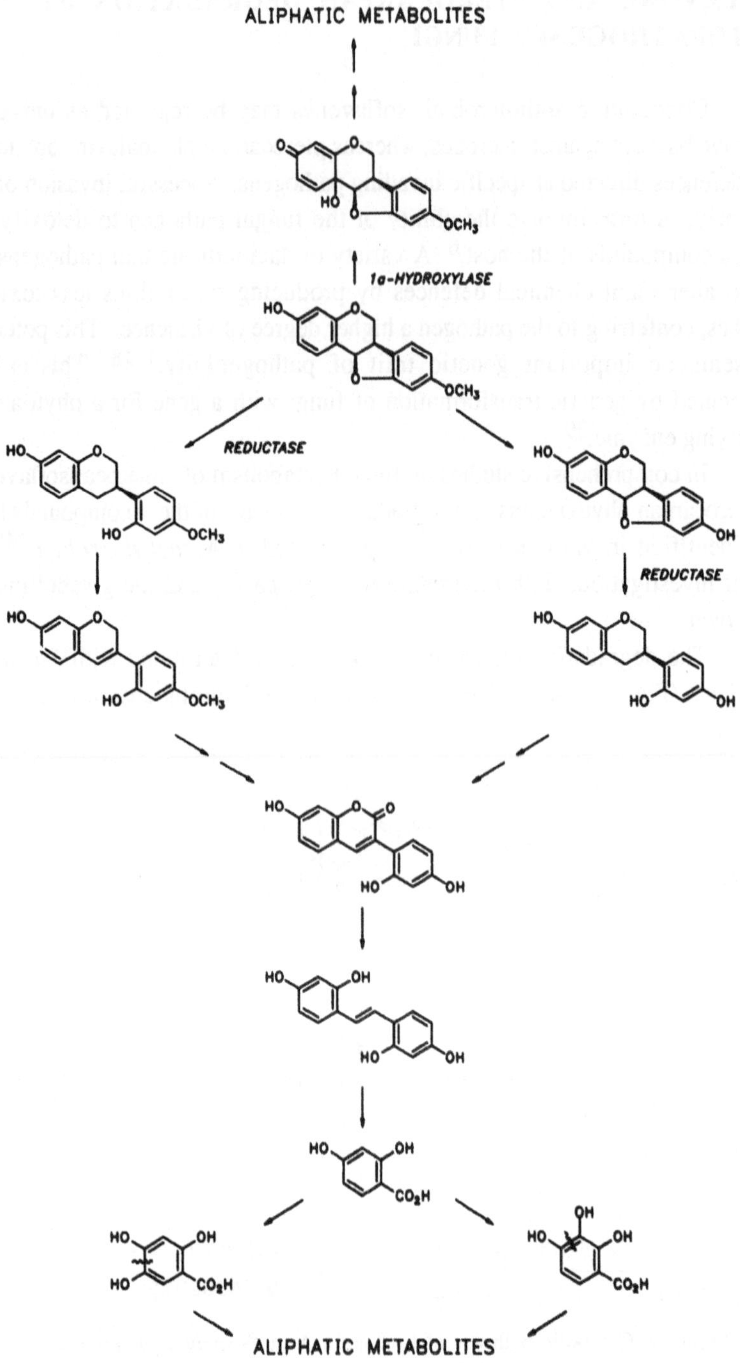

ALIPHATIC METABOLITES

1α-HYDROXYLASE

REDUCTASE

REDUCTASE

ALIPHATIC METABOLITES

leads via 8-hydroxydihydrobiochanin A through a dioxygenase type of cleavage reaction to a diketodihydropyran. The C_4-unit removed in this step is most likely oxaloacetate. The terminal aromatic catabolites consist of a series of phenylacetic acids derived from the side chain phenyl ring of the isoflavone and carbon atoms 2 and 3 of the heterocycle.[43] The NADPH:isoflavone oxido-reductase (Fig. 7; IFR) involved in this catabolic pathway belongs to the A-series of pyridine nucleotide linked enzymes and produces the 3S-stereoisomer.[44] Thus, this enzyme leads to the opposite configuration than the chickpea IFR (Fig. 1), but is identical in this respect with the pea enzyme.[32]

All *A. rabiei* isolates investigated so far readily degrade 5-hydroxyisoflavones through a pathway probably similar to the sequence depicted in Figure 7.[41] Quite remarkably, in this fungus as well as in all *Fusarium* species, 5-deoxyisoflavones (i.e. daidzein, formononetin) are highly recalcitrant to fungal degradation.[43]

The chickpea pathogen *A. rabiei* dissimilates the phytoalexins medi-carpin and maackiain by two parallel routes which are started by either oxidative (pterocarpan:FAD hydroxylase; M_r 58 kDa) or reductive (pterocarpan: NADPH reductase; M_r 29 kDa) conversion[42] (Fig. 8). These two initial enzymes have been thoroughly characterized, and are constitutively expressed by the *A. rabiei* isolates.[46] An important catabolite of the pterocarpans, 2,4-dihydroxybenzoic acid, (Fig. 8) is derived from a stilbene intermediate. Among the two subsequent trihydroxybenzoic acids, the 2,4,5-trihydroxy substituted compound appears to be the more prominent intermediate for dioxygenase ring fission.

The high degree of adaptation of a fungal pathogen to the phytoalexins of its host is illustrated by the observation[45] that *A. rabiei* will convert only the chickpea (6a*R*:11a*R*)-pterocarpans (Fig. 8) and not the (6a*S*:11a*S*)-isomers which also occur as natural products in pea or peanut plants.[38] Possible consequences of such specificity for the genetic transformation of more resistant plants have been evaluated by Van Etten and his associates.[38] Present investigations in the Münster laboratory aim at cloning the genes for the two pterocarpan modifying enzymes of *A. rabiei* (Fig. 8), to enable subsequent gene disruption experiments with the chickpea pathogen to reveal more clearly the relative importance of this metabolic trait for pathogenesis in the fungal-host interaction.

Figure 8. Degradative pathways of medicarpin in *Ascochyta rabiei*. Pterocarpan reductase and pterocarpan 1a-hydroxylase only convert the (6a*R*: 11a*R*)-pterocarpans occurring in chickpea but not the 6a*S*: 11a*S*--isomers.

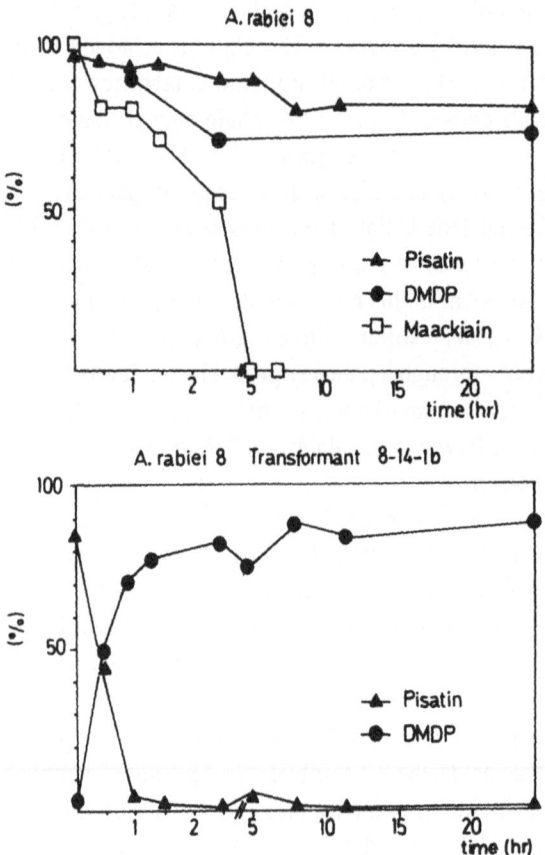

Figure 9. Comparative metabolism of maackiain, pisatin and 6a-
hydroxymaackiain (DMDP) in *Ascochyta rabiei*, wild type isolate
8 and transformant 8-14-1b, after transfer of pda gene from *Nectria*
haematococca. Concentrations of compounds applied: $100\% = 5$ x
10^{-5} M.

Pisatin 3-*O*-demethylation represents the initial detoxification reaction
of this pea pterocarpan phytoalexin in *Nectria haematococca*.[38] The gene for
this cytochrome P-450 monooxygenase/*O*-demethylase (pda-gene) has recently
been cloned[46] and expressed in non-phytopathogenic and phytopathogenic
fungi.[39] The results of such transformation studies have shown that the ability
to degrade pisatin confers to the transformed pathogen a certain degree of hitherto

lacking pathogenicity towards pea plants.[39]

Similar transformation experiments in the senior author's laboratory were recently performed with the 8,8 kb pUCH I/pda plasmid and various *A. rabiei* isolates which lack the capability to *O*-demethylate or otherwise metabolize pisatin. Southern analyses of various stable *A. rabiei* transformants revealed that the integration of the heterologous plasmid in the genome had occurred both in single and in multiple copies at different loci. The majority of the transformants readily *O*-demethylated pisatin with the stoichiometric accumulation of the 6a-hydroxymaackiain (DMDP) product (Fig. 9). Furthermore, the *A. rabiei* transformants also showed a measurable virulence in pea plants. Although such results provide direct evidence that in some host-parasite interactions phytoalexin degradation is essential for pathogenicity, additional presently unknown genetic elements are required for the complete determination of fungal virulence.[47]

CONCLUSION

Biochemical and recent molecular biological investigations of constitutive and inducibly formed isoflavonoids have led to a better understanding of the function of these plant secondary products. The various interrelated pathways show a considerable complexity, and the regulatory network involved still warrants more detailed investigations. These should aim especially at understanding the cellular and spatial compartmentation in the producing plant. The dynamics of the isoflavone and pterocarpan pathways provide to the plant a high degree of adaptation to environmental stimuli. On the other hand, phytopathogenic microbes have evolved strategies to combat the defence mechanisms of their hosts. Future investigations should preferentially be devoted to the analysis of microbial pathways that interact with the plant defence system.

ACKNOWLEDGEMENTS

The research work conducted in the Münster and the Freiburg laboratories have generously been supported by Deutsche Forschungsgemeinschaft and Fonds der Chemischen Industrie. R.W. acknowledges with thanks a postdoctoral fellowship by Deutsche Forschungsgemeinschaft. ICARDA, Aleppo, Syria, has reliably provided seed material of chickpea cultivars.

REFERENCES

1. MANSFIELD, J.W. 1982. The role of phytoalexins in disease resistance. In: Phytoalexins, (J. A. Bailey, J. W. Mansfield, eds.), Blackie & Son, Glasgow, pp. 253-288.

2. HARBORNE, J.B. 1988. Introduction to Ecological Biochemistry. 3rd ed. Academic Press, London.

3. WINK, M. 1988. Plant breeding: importance of plant secondary metabolites for protection against pathogens and herbivores. Theor. Appl. Genet. 75:225-233.

4. KOSSLAK, R.M., BOOKLAND, R., BARKAI, J., PAAREN, H.E., APPELBAUM; E.R. 1987. Induction of *Bradyrhizobium japonicum* common nod genes by isoflavones isolated from *Glycine max*. Proc.Natl. Acad. Sci. USA 84:7428-7432.

5. BAILEY, J.A., MANSFIELD, J.W., eds. 1982. Phytoalexins. Blackie & Son, Glasgow.

6. BARZ, W., BLESS, W., BÖRGER-PAPENDORF, G., GUNIA, W., MACKENBROCK, U., MEIER, D., OTTO, C.H., SÜPER, E. 1990. Phytoalexin as part of induced defence reactions in plants: their elicitation, function and metabolism. In: Bioactive Compounds From Plants. (Ciba Foundation Symposium 154), Wiley, Chichester, pp. 140-156.

7. DIXON, R.A. 1986. The phytoalexin response: elicitation, signalling and control of host gene expression. Biol. Rev. 61:239-291.

8. LAMB, C.J., LAWTON, M.A., DRON, M., DIXON, R.A. 1989. Signal and transduction mechanisms for activation of plant defences against microbial attack. Cell 56:215-224.

9. DANIEL, S., BARZ, W. 1990. Elicitor-induced metabolic changes in cell cultures of chickpea *(Cicer arietinum* L.) cultivars resistant and susceptible to *Ascochyta rabiei*. II. Differential induction of chalcone synthase mRNA activity and analysis of *in vitro* translated protein patterns. Planta 182:279-286.

10. HÖHL, B., PFAUTSCH, M., BARZ, W. 1990. Histology of disease development in resistant and susceptible cultivars of chickpea (*Cicer arietinum* L.) inoculated with spores of *Ascochyta rabiei*. J. Phytopathol. 129:31-45.

11. KESSMANN, H., BARZ, W. 1987. Accumulation of isoflavones and pterocarpan phytoalexins in cell suspension cultures of different cultivars of chickpea *(Cicer arietinum* L.). Plant Cell Rep. 6:55-59

12. EBEL, J., GRISEBACH, H., BONHOFF, A., GRAB, D., HOFFMANN, CH., KOCHS, G., MIETH, H., SCHMIDT, W., STÄB, M. 1986. Phytoalexin synthesis in soybean following infection of roots with *Phytophthora megasperma* or treatment of cell cultures with fungal elicitors. In: Recognition in Microbe-Plant Symbiotic and Pathogenic Interactions, (B. Lugtenberg, ed.) NATO ASI Series, Vol. H4, Springer-Verlag, Berlin, Heidelberg, pp. 345-361.

13. EBEL, J., GRISEBACH, H. 1988. Defense strategies of soybean against the fungus *Phytophthora megasperma* f. sp. *glycinea:* a molecular analysis. Trends Biochem. Sci. 13:23-27.

14. KÖSTER, J., STRACK, D., BARZ, W. 1983. High performance liquid chromatographic separation and structural elucidation of isoflavone 7-0-glucoside 6"-malonates from *Cicer arietinum*. Planta Med. 48:131-135.

15. GRAHAM, T.L., KIM, J.E., GRAHAM, M.Y. 1990. Role of constitutive isoflavone conjugates in the accumulation of glyceollin in soybean infected with *Phytophthora megasperma*. Molec. Plant-Microbe Interact. 3:157-166.

16. BARZ, W., BEIMEN, A., DRÄGER, B., JAQUES, U., OTTO, CH., SÜPER, E., UPMEIER, B. 1990. Turnover and storage of secondary products in cell cultures. In: Secondary Products From Plant Tissue Culture. (B.V. Charlwood, M.J.C. Rhodes, eds.) Proc. Phytochem. Soc. Europe 30:79-102.

17. DANIEL, S., TIEMANN, K., WITTKAMPF, U., BLESS, W., HINDERER, W., BARZ, W. 1990. Elicitor-induced metabolic changes in cell cultures of chickpea *(Cicer arietinum* L.) cultivars resistant and susceptible to *Ascochyta rabiei*. I. Investigations of enzyme activities involved in isoflavone and pterocarpan phytoalexin biosynthesis. Planta 182:270-278.

18. HAHLBROCK, K., SCHEEL, D. 1989. Physiology and molecular biology of phenylpropanoid metabolism. Annu. Rev. Plant Physiol. 40:347-369.

19. WINGENDER, R., RÖHRIG, H., HÖRICKE, CH., WING, D., SCHELL, J. 1989. Differential regulation of soybean chalcone synthase genes in plant defence symbiosis and upon environmental stimuli. Mol. Gen. Genet. 218:315-322.

20. WELLE, R., GRISEBACH, H. 1988. Isolation of a novel NADPH-dependent reductase which coacts with chalcone synthase in the biosynthesis of 6'-deoxychalcone, FEBS Lett. 236:221-225.

21. WELLE, R., GRISEBACH, H. 1989. Phytoalexin synthesis in soybean cells: elicitor induction of reductase involved in the biosynthesis of 6'deoxychalcone. Arch. Biochem. Biophys. 272:97-102.

22. WELLE, R., SCHRÖDER, G., SCHILTZ, E., GRISEBACH, H., SCHRÖDER, J. 1991. Induced plant responses to pathogen attack. Analysis and heterologous expression of the key enzyme in the biosynthesis of phytoalexin in soybean (Glycine max Merr. cv Harosoy 63). Eur. J. Biochem. 196:423-430.

23. HAGMANN, M., GRISEBACH, H. 1984. Enzymatic rearrangement of flavanone to isoflavone. FEBS Lett. 175:199-202.

24. KOCHS, G., GRISEBACH, H. 1986. Enzymic synthesis of isoflavones. Eur. J. Biochem. 155:311-318.

25. HASHIM, M.F., HAKAMATSUKA, T., EBIZUKA, Y., SANKAWA; U. 1990. Reaction mechanism of oxidative rearrangement of flavanone in isoflavone biosynthesis. FEBS Lett. 271:219-222

26. WENGENMAYER, H., EBEL, J., GRISEBACH, H. 1974. Purification and properties of a S-adenosylmethionine: isoflavone 4'-O-methyl-transferase from cell suspension cultures of Cicer arietinum L. Eur. J. Biochem. 50:135-143.

27. GUNIA, W., HINDERER, W., WITTKAMPF, U., BARZ, W. 1991. Elicitor induction of cytochrome P_{450} monooxygenases in cell suspension cultures of chickpea (Cicer arietinum L.) and their involvement in pterocarpan phytoalexin biosynthesis. Z. Naturforsch. 46c:58-66.

28. TIEMANN, K., INZE, D., VAN MONTAGU, M., BARZ, W. 1991. Pterocarpan phytoalexin biosynthesis in elicitor-challenged chickpea (Cicer affetinum L.) cell cultures; purification, characterization and cDNA cloning of NADPH: isoflavone oxidoreductase. Eur. J. Biochem., in press.

29. BLESS, W., BARZ, W. 1988. Isolation of pterocarpan synthase, the terminal enzyme of pterocarpan phytoalexin biosynthesis in cell suspension cultures of Cicer arietinum L. FEBS Lett. 235:47-50.

30. FISCHER, D., EBENAU-JEHLE, CH., GRISEBACH, H. 1990. Phytoalexin synthesis in soybean: purification and characterization of NADPH: 2'-hydroxydaidzein oxidoreductase from elicitor-challenged soybean cell cultures. Arch. Biochem. Biophys. 276:390-395.

31. FISCHER, D., EBENAU-JEHLE, CH., GRISEBACH, H. 1990. Purification and characterization of pterocarpan synthase from elicitor-challenged soybean cell cultures. Phytochemistry 29:2879-2882.

32. SUN, Y., WU, Q., VANETTEN, H.D., HRAZDINA, G. 1991. Stereoisomerism in plant disease resistance: Induction and isolation of the 7,2'-dihydroxy-4'-5'-methylenedioxyisoflavone oxidoreductase, an enzyme introducing chirality during synthesis of isoflavonoid phytoalexins in pea (Pisum sativum L.). Arch. Biochem. Biophys. 284:167-173.

33. KOCHS, G., GRISEBACH, H. 1989. Phytoalexin synthesis in soybean: Purification and reconstitution of cytochrome P450 3,9-dihydroxy-pterocarpan 6a-hydroxylase and separation from cytochrome P450 cinnamate 4-hydroxylase. Arch. Biochem. Biophys. 273:543-553.

34. WELLE, R., GRISEBACH, H. 1988. Induction of phytoalexin synthesis in soybean: enzymatic cyclization of prenylated pterocarpans to glyceollin isomers. Arch. Biochem. Biophys. 263:191-198.

35. WELLE, R., GRISEBACH, H. 1991. Properties and solubilization of the prenyltransferase of isoflavonoid phytoalexin biosynthesis in soybean. Phytochemistry 30:479-484.

36. WEIDEMANN, C., TENHAKEN, R., HÖHL, U., BARZ, W. 1991. Medicarpin and maackiain 3-0-glucoside 6'-0-malonate conjugates are constitutive compounds in chickpea (Cicer arietinum L.) cell cultures. Plant Cell Rep., in press.

37. MACKENBROCK, U., BARZ, W. 1991. Elicitor-induced formation of pterocarpan phytoalexins in chickpea (Cicer arietinum L.) cell suspension cultures from constitutive isoflavone conjugates upon inhibition of phenylalanine ammonia lyase. Z. Naturforsch. 46c:43-50.

38. VANETTEN, H.D., MATTHEWS, D.E., MATTHEWS, P.S. 1989. Phytoalexin detoxification: importance for pathogenicity and practical implications. Annu. Rev. Phytopathol. 27:143-164.

39. SCHÄFER, W., STRANEY, D., CIUFFETTI, L., VANETTEN, H.D., YODER, O.C. 1989. One enzyme makes a fungal pathogen, but not a saprophyte, virulent on a new host plant. Science 246:247-249.

40. WILLEKE, U., BARZ, W. 1982. Catabolism of 5-hydroxyisoflavones by fungi of the genus Fusarium. Arch. Microbiol. 132:266-269.

41. KRAFT, B., BARZ, W. 1985. Degradation of the isoflavone biochanin A and its glucoside conjugates by Ascochyta rabiei. Appl. Environ. Microbiol. 50:45-48.

42. HÖHL, B., ARNEMANN, M., SCHWENEN, L., STÖCKL, D., BRINGMANN, G., JANSEN, J., BARZ, W. 1989. Degradation of the pterocarpan (-)-maackiain by Ascochyta rabiei. Z. Naturforsch. 44c:771-776.

43.   MACKENBROCK, U., 1986. Untersuchungen zum Isoflavon-
      stoffwechsel in *Cicer arietinum*, Doctoral thesis, University of
      Munster.
44.   SCHLIEPER, D., TIEMANN, K., BARZ, W. 1990. Stereospecificity of
      hydrogen transfer by fungal and plant NADPH: isoflavone
      oxidoreductases. Phytochemistry 29:1519-1524.
45.   TENHAKEN, R., SALMEN, H.CH., BARZ, W. 1991. Purification and
      characterization of pterocarpan hydroxylase, a flavoprotein mono-
      oxygenase from the fungus *Ascochyta rabiei* involved in pterocarpan
      phytoalexin metabolism. Arch. Microbiol. 155:353-359.
46.   WELTRING, K.M., TURGEON, B.G., YODER, O.C., VANETTEN,
      H.D. 1988. Isolation of a phytoalexin-detoxifying gene from the
      plant pathogenic fungus *Nectria haematococca* by detecting its
      expression in *Aspergillus nidulans*. Gene 68: 335-344.
47.   LEONG, S.A., HOLDEN, D.W. 1989. Molecular genetic approaches to
      the study of fungal pathogenesis. Ann. Rev. Phytopath. 27:463-481.

Chapter Six

FLAVONOID SYNTHESIS IN *PETUNIA HYBRIDA*; GENETICS AND MOLECULAR BIOLOGY OF FLOWER COLOUR

Anton G.M. Gerats[#] and Cathie Martin[*]

[#]Laboratorium voor Genetika, Rijksuniversiteit Gent, Ledeganckstraat 35, B-9000, Ghent, Belgium

[*]Department of Genetics, John Innes Institute, Colney Lane, Norwich NR4 7UH, United Kingdom

Phenolic Metabolism in Plants, Edited by H.A. Stafford
and R.K. Ibrahim, Plenum Press, New York, 1992

INTRODUCTION

Except for yellow colours due to carotenoids, the major flower pigments are flavonoids, more precisely anthocyanins and flavonol glycosides. Under natural conditions, coloured flowers attract pollinators and, as such, flavonoids can be considered to perform a vital function in the life cycle of the plant. Besides contributing to floral pigmentation, flavonoids have been shown to play a role in a number of phenomena: defence against phytopathogens[1,2] and predators[3] and nodule induction in the *Rhizobium*-legume symbiosis.[4,5,6] Because flavonoids are phenolic compounds they can act as metal chelators and antioxidants, and because they are aromatic compounds they might provide protection against damage by UV light. Flavonoids are widely used in medicine as therapeutic drugs, although they are also known to be causative agents of some diseases.[7]

Changes in pigment accumulation are easy to observe, and therefore have been widely used in genetic studies since Mendel first worked on flower colouration of pea.[8] The identification of single genes that alter the nature of anthocyanins produced in flowers[9,10] can be viewed as an early basis for the formulation of the one gene—one enzyme concept.[11] In 1922 Vavilov[12] developed the concept of homologous variation on studies of flower colour, stating that similar genes occur in different plants.

With the development of paper chromatography for flavonoid separation[13] and later tracer experiments and enzymology, especially by the laboratories of Grisebach and Hahlbrock, the biochemical steps involved in flavonoid production were extensively analyzed. Flavonoid biosynthesis can now be regarded as the best understood plant secondary metabolic pathway. The available genetic and biochemical knowledge has led to the molecular isolation and analysis of many of the genes involved in flavonoid biosynthesis, constituting a major contribution to the understanding of gene structure, function and expression in plants.

In this chapter, we will focus on aspects of flavonoid biosynthesis in flowers of *Petunia hybrida* We will attempt to summarize the genetic, biochemical, physiological and molecular data to illustrate the complexity of flavonoid biosynthesis. Some of these and other aspects will be covered in more detail by other contributions in this volume (see Chapters 1-4 for details of the pathways involved).

BIOCHEMISTRY AND GENETICS OF FLAVONOID SYNTHESIS

General phenylpropanoid metabolism

The first step in the general phenylpropanoid metabolism is the *trans*-elimination of ammonia from phenylalanine to form *trans*-cinnamate by the enzyme phenylalanine ammonia-lyase.[14,15] Cinnamate 4-hydroxylase transforms *trans*-cinnamate to *trans*-4-coumarate.[16,17] The latter is then ligated to CoA to form a 4-coumaroyl CoA ester by the enzyme 4-coumarate:CoA ligase.[18,19]

Hydroxycinnamic CoA esters form a precursor pool common to a number of diverging pathways involved in the biosynthesis of compounds such as lignin, stilbenes, cyanogenic compounds, phytoalexins and flavonoids. Isozymes of 4-coumarate:CoA ligase have been detected in a number of species: *Glycine max* cell cultures,[20] *Pisum sativum* shoots[21,22] and *Petunia hybrida* leaves.[23,24] These different isoforms are thought to provide the precursors for the biosynthesis of specific phenylpropanoid branches.[22,24,25] Excellent reviews on flavonoid biosynthesis are available.[26,27,28] Therefore, we will describe in this paper the genetic, physiological and molecular aspects of flavonoid biosynthesis, mainly in *Petunia hybrida* (Fig. 1).

Chalcone synthase (CHS)

The formation of the C_{15} skeleton, typical for all flavonoids, is catalyzed by the enzyme chalcone synthase and leads to the formation of chalcones.[26] Chalcone synthase condenses three malonyl-CoA residues with the CoA ester of a hydroxycinnamic acid. In some species (*Matthiola incana, Antirrhinum majus, Phaseolus vulgaris*) 4-coumaroylCoA is exclusively used as a substrate,[29,30,31] whereas the enzyme from other species also accepts caffeoyl-CoA or feruloyl-CoA as a substrate (for a detailed review see Heller and Forkmann[28]). There are varieties of *Verbena* that accumulate cyanidin, even in the absence of a 3'-hydroxylase. It has been suggested that in these varieties chalcone synthase will use caffeoyl-CoA in the condensation reaction to give a product with *o*-dihydroxyl groups in the B-ring. However, the condensation using caffeoyl-CoA must be relatively inefficient since the basic level of cyanidin accumulated can be considerably enhanced in such varieties by the subsequent hydroxylation of the 3'-position of the *p*-coumaroyl-CoA product, naringenin by a 3'-hydroxylase.[32] In *Silene dioica*, the *P* gene has been shown to be required for the hydroxylation of 4-coumaroyl-CoA to caffeoyl-CoA,

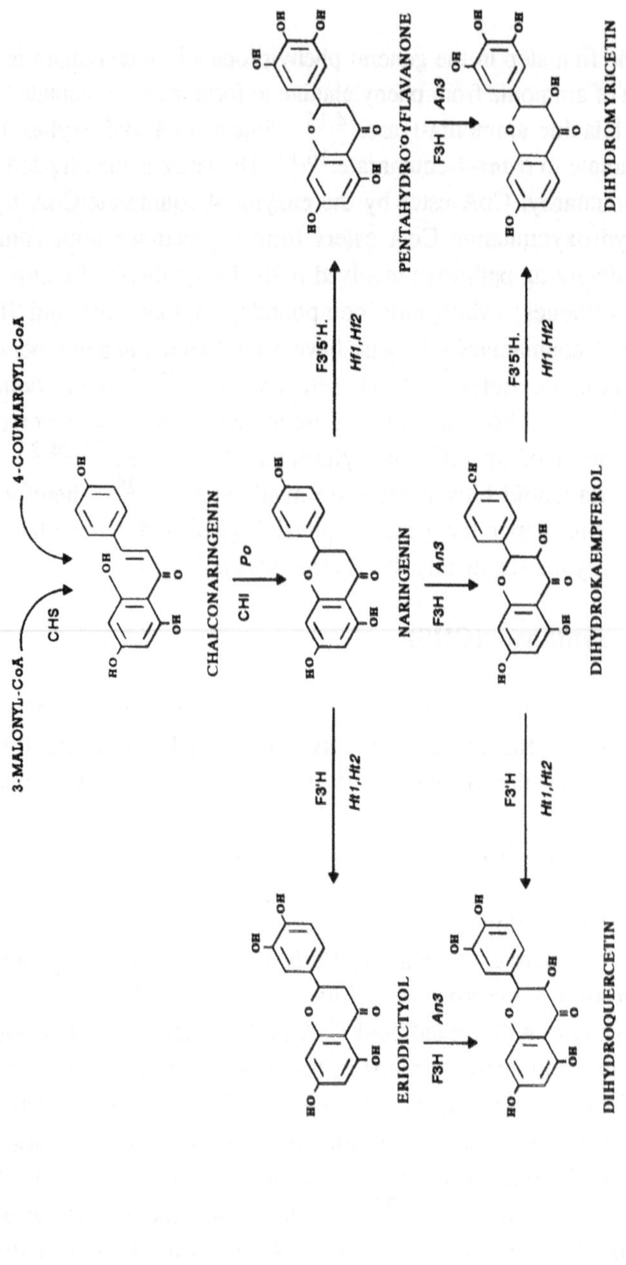
3-MALONYL-CoA
4-COUMAROYL-CoA
CHS
CHALCONARINGENIN
CHI
Po
NARINGENIN
F3H
An3
DIHYDROKAEMPFEROL
F3'H
Ht1,Ht2
ERIODICTYOL
DIHYDROQUERCETIN
F3'5'H.
Hf1,Hf2
DIHYDROMYRICETIN

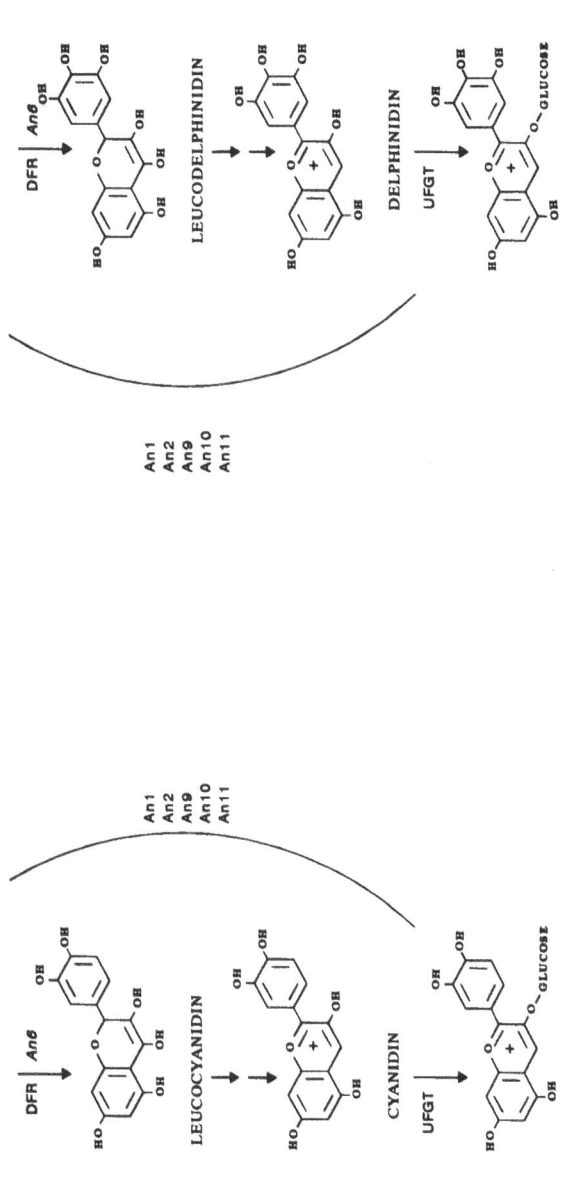

Fig. 1. Representation of the central part of the anthocyanin biosynthetic pathway in *Petunia hybrida*. Enzyme symbols are placed on the left of or above arrows, gene symbols on the right of or below arrows. CHS: chalcone synthase; CHI: chalcone flavanone isomerase; F3H: flavanone-3-hydroxylase; F3'H: flavonoid 3'-hydroxylase; F3'5'H: flavonoid 3',5'-hydroxylase; DFR: dihydroflavonol 4-reductase; UFGT: UDPglucose: flavonoid 3-O- glucosyltransferase. Further modifications of the glycosides are possible. For an explanation of the gene symbols see text. *Po, An3, Ht1, Ht2, Hf1, Hf2* and *An6* are structural genes, coding for a specific flavonoid enzyme; *An1, An 2, An10 and An11* are regulatory genes in the sense that they influence the expression of a number of structural genes. The action of the *An9* gene remains unclear (see text).

which is used by chalcone synthase in the condensation reaction.[33]

Genes controlling chalcone synthase activity have been described for a number of species: *Nivea (Antirrhinum majus)*, *F (Matthiola incana)*[29,30] and the *C2* gene for *Zea mays*.[34] Chalcone synthase genes have been cloned from several species: *Antirrhinum majus*,[35] *Petunia* hybrida[36,37] *Zea mays*,[39] *Petroselinum crispum*,[38] *Phaseolus vulgaris*,[41] *Arabidopsis thaliana*,[42] *Ranonculus acer*,[40] *Magnolia liliflora*[40] and *Hordeum vulgare*.[40]

In *Petunia*, no mutations of the structural gene(s) for chalcone synthase are known. However, some *Petunia* cultivars exhibit a pattern of flower colour in which coloured sectors alternate with white sectors (Red Star). This phenotype is very sensitive to environmental changes[43,44] and at least four genes are believed to be involved in the pattern formation (Gerats, unpublished). Interestingly, the white sectors lack chalcone synthase enzyme activity due to the absence of active CHS mRNA. While there is some effect on chalcone isomerase, the expression of the enzyme activity of structural genes further along the pathway is less affected.[45]

A number of chalcone synthase genes have been cloned from *Petunia hybrida*. Some varieties express two genes more or less equally in flower petals,[36] whereas others show predominant expression of a single gene.[46] At least eight complete CHS genes are present in the *Petunia* variety V30, as well as gene fragments consisting of one of the two CHS exons. The CHS gene family can be divided into subfamilies, based on sequence homology and chromosomal position: a tightly linked and highly homologous subfamily is localized in chromosome II, whereas a less homologous and more loosely linked subfamily is found in chromosome V. The number of gene copies varies between different varieties.[37] Both clusters are present in the parental species for the *Petunia hybrida* cultivars.

Chalcone flavanone isomerase (CHI)

The enzyme chalcone flavanone isomerase (CHI) catalyzes the formation of flavanones from chalcones. Flavanones are the intermediates for a number of flavonoids, for example flavones, flavonols, isoflavones and anthocyanins. The isomerization catalyzed by CHI occurs spontaneously under *in vitro* conditions, but this can be repressed by the addition of proteins like bovine serum albumin.[47] This might explain the accumulation of chalcones in pollen of various plants in the absence of chalcone flavanone isomerase activity.[48] In all cases described, CHI exhibits a high substrate specificity for trihydroxychalcones.[48,49,50]

Genes coding for CHI have been cloned from *Phaseolus vulgaris*,[51] *Petunia hybrida*[52,53] and *Antirrhinum majus*.[66] In anthers of *Petunia*, CHI expression is controlled by the gene *Po*, since chalcones accumulate in the pollen of *po/po* plants. CHI activity is absent in anthers of *po/po* plants, although they show a wildtype CHI activity in flower petals.[54,55] Two *CHI* genes (*CHI-A* and *CHI-B*) have been found in *Petunia* and these are located on different chromosomes. The *CHI-A* gene is intronless, whereas the *CHI-B* gene contains three introns.[52] RFLP analysis of the *CHI-A gene* shows complete linkage to the *Po* gene and the *po/po* phenotype can be molecularly complemented by transformation of a mutant line with a genomic *CHI-A* clone. Therefore, the *Po* gene encodes *CHI-A*.[57]

Both *CHI-A and CHI-B* are expressed. A 1Kbp mRNA transcribed from the *CHI-B* gene is found exclusively in immature anthers; *CHI-B* segregates independently from *Po* and does not produce a functional isomerase that converts chalcones to flavanones.[56,57] The expression of the *CHI-A* gene is more complicated: a 1 Kbp *CHI-A* transcript is detected in, among other tissues, flower tubes, flower petals and immature anthers. This transcript is lacking in immature anthers of *po/po* lines, suggesting that the *po* mutation involves the inactivation of anther-specific expression.[57] Furthermore, a 1.3 Kbp RNA is produced in mature pollen. This RNA is extended at the 5'-site of the above mentioned 1Kbp *CHI-A* mRNA. The function of this larger RNA remains to be elucidated.[53]

Flavanone 3-hydroxylase (F3H)

The enzyme flavanone 3-hydroxylase (F3H) catalyzes the conversion of flavanones to dihydroflavonols. F3H is a soluble enzyme that belongs to the 2-oxoglutarate-dependent dioxygenases.[58] In *Petunia*, the nature of the substrate (naringenin, eriodictyol or 5,7,3',4',5'-pentahydroxyflavanone) is determined by the allelic status for the genes *Ht1* and *Hf1*.[59] Apparently, the *Petunia* F3H enzyme can use all three substrates for conversion to the respective dihydroflavonol.

The gene encoding F3H is located at the *An3* locus,[60,61] based on differences in kinetic parameters of the enzyme as encoded by the wildtype and the *an3-1* mutant allele. The *an3-1* allele exhibits a 90% reduction in F3H activity, resulting in a wildtype colouration of petals and anthers, but an almost uncoloured tube.[60] As for all flavonoid genes analyzed so far in *Petunia*, mRNA levels and enzyme activity appear to be lower in tubes then in petals (although there might simply be less pigmented tissue in the tube, due to the presence of relatively more mesoderm cells). Thus, either the *an3-1* mutation lowers the overall level of gene expression, rendering the resultant enzyme activity rate limiting in tubes but not in petals, or the mutation may involve sequences that influence the spatial-specific expression pattern of the *An3* gene within the *Petunia* flower.

A further peculiarity of the *an3-1* allele is that while anthocyanin production is almost normal in petals, flavonol production is severely reduced. The F3H enzyme converts flavanones into dihydroflavonols, which serve as a common precursor for both anthocyanin and flavonol synthesis. The conclusion from kinetic criteria is that the *an3-1* allele involves a structural alteration in the gene encoding F3H. How such a structural alteration can lead to a differential influence on anthocyanin- and flavonol-synthesis remains to be elucidated.

Acyanic lines of a number of species have been shown to lack F3H activity. In addition, biochemical complementation with dihydroquercetin led to anthocyanin production, thus identifying the block to be most likely in the F3H step. In this way, mutant *F3H* loci have been defined in *Antirrhinum (Inc)*,[62] *Gerbera (Fht)*, *Streptocarpus (Fht)*, *Zinnia (Fht)* and *Verbena (A)*.[63,64]

The *Incolourata* gene from *Antirrhinum* has been cloned.[66] The wildtype allele is incompletely dominant, causing a paler flower colour when heterozygous with a recessive allele. This suggests that in fully-pigmented *Antirrhinum* lines, F3H activity might be the rate limiting step for anthocyanin biosynthesis.

Hydroxylation of the B-ring: 3'- and 3'-5'-hydroxylases

Dihydrokaempferol can be converted to dihydroquercetin by a 3'-hydroxylase. Two genes, *Ht1* and *Ht2*, have been described in *Petunia* that control 3'-hydroxylation of naringenin and dihydrokaempferol[67,68] but not of cinnamic acids.[59] Interestingly, these genes exhibit a difference in expression pattern: *Ht1* is expressed in both petal and tube, whereas *Ht2* is expressed only in the tube. Furthermore, two genes have been described that control 3',5'-hydroxylation, *Hf1* and *Hf2*.[67] These also show a differential expression pattern; *Hf1* is expressed in petals and tube and leads to the exclusive production of delphinidins, whereas *Hf2* is expressed only in petals and converts part of the precursor pool to dihydromyricetin, leading to a mixture of cyanidins and delphinidins.[69] The 3',5'-hydroxylase activity can convert both dihydrokaempferol and dihydroquercetin to dihydromyricetin.

Feeding experiments and genetic data suggest that the 3-, 3'- and 3',5'-hydroxylases can act in a grid fashion. Naringenin can either be converted to eriodictyol and subsequently to dihydroquercetin, or to dihydrokaempferol and then to dihydroquercetin and even pentahydroxyflavanone. The latter can form dihydromyricetin, whereas dihydromyricetin can also be formed from dihydroquercetin.[70] Apparently, the addition of a hydroxyl group at the 3-position does not interfere with the addition of either one or two hydroxyl groups in the B-ring of the flavonoid molecule. While F3H is thought to be a soluble enzyme, the 3'- and the 3',5'-hydroxylases appear to be localized in the microsomal fraction.[62,70]

The nature of the anthocyanins produced depends on the dihydroflavonols available. There is no evidence for further hydroxylation of the B-ring following conversion of dihydroflavonols to leucoanthocyanidins. The different anthocyanins give rise to different colours, according to the degree of hydroxylation of the B-ring. As a rule of thumb, flower colour shifts from orange to red to purple/blue when the anthocyanin present is pelargonidin, cyanidin or delphinidin, respectively. However, the actual colour present also depends on a number of other factors. The absence of blue shades in species such as *Tulipa*, *Rosa* and *Chrysanthemum* might be explained by the lack of a 3',5'-hydroxylase activity. 3'-Hydroxylase activity is controlled by the *B* gene in *Matthiola incana*[50] and the *Eosinea* gene in *Antirrhinum*.[62]

Whereas in most species studied so far the hydroxylation reactions described above appear to be carried out at the flavanone-dihydroflavonol stage, the situation in *Silene* is quite different. Gene *P* has been shown to control the formation of caffeoyl-CoA from *p*-coumaroyl-CoA. Caffeoyl-CoA is then used as a precursor for both cyanidin biosynthesis and the acylation of cyanidin 3-rhamnoside.[33]

Dihydroflavonol 4-reductase (DFR)

The next step in the biosynthesis of anthocyanins is the reduction of dihydroflavonols to 3,4-*cis*-diols (leucoanthocyanidins) by DFR. The enzyme activity was first described by Stafford and Lester[72,73,74] with enzyme preparations extracted from cell suspension cultures of *Pseudotsuga mensiezii* and by Kristiansen[75,76] in barley.

The *DFR* gene from *Petunia hybrida* has been cloned by using part of the *DFR* gene (*Pallida*) from *Antirrhinum majus*[77,78] as a probe. Three copies of this gene are found in the *Petunia* genome, each on different chromosomes.[78] The *DFR-A* gene, located in chromosome IV was identified as the *An6* locus, based on RFLP analysis and molecular complementation of a white flowered mutant with a genomic construct harbouring the *DFR-A* gene under the control of its own promoter (Huits *et al.*, in preparation). This system allows the visual analysis of floral activity of any promoter sequence, hooked up to the *DFR-A* gene and transformed into the *an6* mutant. The expression from the promoters analyzed can be followed on a cell by cell basis in living material.

The substrate specificity of the *Petunia* DFR enzyme has been analyzed by Forkmann and Ruhnau.[79] It preferentially converts dihydromyricetin to leucodelphinidin; dihydroquercetin is a poor substrate, and dihydrokaempferol is not converted to leucopelargonidin. This distinct substrate specificity explains the preferential accumulation of delphinidin and the lack of pelargonidin synthesis in *Petunia*.[80] Further evidence that the lack of pelargonidin synthesis is due to the substrate specificity of the DFR enzyme comes from experiments in which the *DFR* gene (*A1*) from *Zea mays* was transformed into a *Petunia* line that accumulates dihydrokaempferol. The *A1* gene that readily converts dihydro-kaempferol in corn[81] continued to do so in *Petunia*, leading to the synthesis of significant amounts of pelargonidin in transformants and the appearance of a hitherto unknown brick red colouration.[82,83]

Genes controlling the expression of DFR have been reported for *Matthiola* (*e*),[84,85] *Callistephus* (*f*) and *Dianthus* (*a*).[64] The structural gene has been cloned from *Zea mays* (*A*),[86] *Antirrhinum maius* (*Pallida*)[77] and *Petunia hybrida* (*An6*).[78] The *Pallida* gene from *Antirrhinum* has been studied in great detail at the molecular level.[87] Analysis of an allelic series induced by a transposable element, including lines with coloured flowers of varying anthocyanin content, suggests that *Pallida* expression can be reduced to about 30% before DFR activity starts to limit anthocyanin biosynthesis.

For the role of leucoanthocyanidins in the biosynthesis of catechin and proanthocyanidin the reader is referred to the review by Heller and Forkmann.[28]

Conversion of leucoanthocyanidins to anthocyanidins

Feeding experiments of a *Matthiola* line blocked in chalcone synthase with radioactively labeled leucopelargonidin led to the detection of labeled pelargonidin derivatives, thus confirming the role of leucoanthocyanins in antho-cyanin biosynthesis.[84] It is assumed that the 3,4-*cis*-diol (leucoanthocyanidin) is the actual precursor since it is formed by the dihydroflavonol 4-reductases of all plants studied so far.[28] Anthocyanin aglycones (anthocyanidins) are rarely found in nature; the molecule is stabilized by 3-*O*-glycosylation. Before leucoanthocyanidins can be glycosylated, however, three more modifications of the flavonoid molecule seem to be required: a hydroxylation of the 2 position and two dehydration steps.[28] Some mutants have been described for the steps in the conversion of leucoanthocyanidins to anthocyanidins. However, accumulation of leucoanthocyanidins or postulated later intermediates has never been observed.

Two acyanic *Antirrhinum* mutants do not produce colour on feeding with leucoanthocyanidin or earlier precursors. These are *Candica* (previously known as *Incolourata(III)* and *Incolouratal*.[88] Mutants of the *A2* gene from maize are blocked at the same stage.[89] Both *A2* and *Candica* have recently been cloned.[90,66] Sequence comparison indicates that the *A2* and *Candica* gene code for the same enzyme. Surprisingly, the enzyme shows high sequence homology to the 3-hydroxylase encoded by *Incolourata* and other oxoketoglutarate-dependent dioxygenases. The *Incolourata I* gene might encode a dehydratase.

In vitro formation of anthocyanin 3-*O*-glucosides was first demonstrated in enzyme extracts from maize pollen.[91] It has been shown that the enzyme catalyzes the glucosylation of the hydroxyl group at the 3-position of both flavonols and anthocyanidins.[92] Glycosylation confers greater stability[93] and is suggested to prevent leakage out of the vacuole.[94] Alternatively, glycosylation might be a prerequisite for transport of the anthocyanidin through the vacuolar membrane.

Flavonoid 3-*O*-glucosylation

Flavonoid 3-*O*-glucosylation in *Petunia hybrida* was analyzed by Kho *et al*.[95] and later by Gerats *et al*.[96] Based on the detection of allelic differences in thermal inactivation properties, the structural gene encoding *UFGT* appears to be located in chromosome VII, close to the gene *An4*.[97] The enzyme has been studied in a number of species. In corn, the structural gene is *Bz*.[98,99] Like the *Petunia* enzyme, the corn enzyme appears to be under intensive genetic control (see below). *UFGT* activity in *Matthiola* is controlled by the genes *E*, *G* and *Z*. Single recessive mutants of these genes exhibit 10 to 40% of the wildtype activity;[100] none of these genes appears to be the structural *UFGT* gene. [84,85,100] Moreover, gene *E* has also been shown to affect the activity of dihydroflavonol 4-reductase[85] and thus is presumably a regulatory gene.

Further modifications of anthocyanidin 3-glucosides

Further modifications of the anthocyanidin 3-glucosides occur in *Petunia*. The gene *Rt* controls the addition of a rhamnose residue to the glucose at the 3-position of both cyanidin and delphinidin to form 3-rutinoside.

The gene *Gf* affects the subsequent addition of glucose at the 5-position and an acyl group, usually *p*-coumaric acid, to the rutinose (apparently a mutant for either one of both steps is not yet available). Griesbach *et al.*[101] described the isolation of anthocyanidin 3-rutinoside-5-glucosides acylated with caffeic acid. They suggested that these compounds had escaped detection because of their lower stability compared to the previously described *p*-coumaroyl acylated anthocyanins. The activity of the UDP-glucose: anthocyanin 5-*O*-glucosyltransferase has been studied by Jonsson *et al.*[102] This enzyme exhibits a very strong substrate specificity for anthocyanidin 3-*p*(coumaroyl)rutinosides and does not seem to be controlled by the *Gf* gene. This might imply that the *Gf* gene controls the acylation of anthocyanidin 3-rutinosides and affects the production of anthocyanidin 5-glucosides through the supply of substrates to the 5-*O*-glucosyltransferase.

In contrast, the 5-*O*-glucosyltransferase from *Silene dioica* glucosylates the 3-glucoside and 3-rutinoside derivatives of both pelargonidin and cyanidin but not the acylated glucosides.[104-106] Moreover, a number of significant differences in kinetic parameters were found between the 5-*O*-glucosyltransferases of *Petunia* and *Silene*, suggesting that different and specialized 5-*O*-glucosyltransferases have developed during evolution in different plant families.[102]

A soluble acyltransferase has been characterized in *Matthiola* that catalyzes the acylation of 3-*O*-glucosides and of 3-*O*-xylosylglucosides with 4-coumaric and caffeic acid respectively. In contrast, 5-*O*-glucosides were not acylated.[100]

The last step in anthocyanin modification is the methylation of the 3'-OH group of cyanidin (peonidin), the 3'-OH of delphinidin (petunidin), or the 3',5'-OH groups of delphinidin (malvidin). Genes that specifically control the methylation of anthocyanins have been identified in *Primula malacoides*,[107] *Solanum tuberosum*[108] and *Medicago sativa*.[109] In *Petunia*, anthocyanin methylation has been extensively analyzed, both at the genetic and the biochemical level.[110-114] The SAM:anthocyanidin 3-(*p*-coumaroyl)rutinoside-5-*O*-glucoside 3',5'-*O*-methyltransferase (OMT) has a very high affinity for the acylated rutinoside-5-*O*-glucoside derivatives of anthocyanidins, and does not readily accept caffeic acid, coumaric acid, aglycones, 3-glucosides or 3-rutinosides. An interesting aspect of the *Petunia* OMT is the involvement of two pairs of duplicated genes for anthocyanin methylation. The genes *Mt1* and *Mf1* are localized in chromosome III, whereas the genes *Mt2* and *Mf2* are found in chromosome V. The four genes specify four different OMT's. All of which catalyze both 3'- and 5'-*O*-methylation, although double methylation is more pronounced with the isozymes encoded by the *Mf* genes.

REGULATORY ASPECTS OF FLAVONOID GENE EXPRESSION

General remarks

The genetic determination of flower colour is not limited to the structural genes involved in flavonoid synthesis. A number of genes that affect the pattern and intensity of anthocyanin biosynthesis are known to regulate the expression of the structural genes. The separation of genes that affect colour into structural and regulatory types depends partly on the identification of a subset of genes encoding the flavonoid biosynthetic enzymes. In the past, gene identification has relied heavily on a biochemical correlation between enzyme activity and allelic status of the gene. However, this is often insufficient for identification of structural genes. For example, the genes *An1* and *An2* of *Petunia hybrida* exhibit a dosage effect for flavonoid:3-*O*-glucosyltransferase activity in crosses between a dominant line and a line recessive for either of the genes. However, the structural gene encoding this enzyme appears to be located on a different chromosome close to the gene *An4*.[97] Final proof that a specific gene encodes a specific enzyme requires either molecular complementation of a mutant with a clone harbouring the gene in question or, even more rigorously, demonstration of enzyme activity of the gene product *in vitro*. For example, a full-length *A1* cDNA clone (*DFR* of *Zea mays*) transcribed and translated *in vitro* yielded a protein that converted dihydroquercetin to leucocyanidin.[81]

High homology at the DNA level does not preclude that the two encoded proteins might have clearly discrete functions: resveratrol synthase from *Arachis hypogaea* exhibits high regional homology with chalcone synthase.[160] Furthermore, sequence comparison between the *Candica* gene, acting after formation of leucoanthocyanidins and the *Incolourata* gene (*F3H*) of *Antirrhinum majus* revealed strong sequence homology between these genes and of these with other dioxygenases.[115]

In general, it can be stated that mutations in anthocyanin structural genes only change the expression of the encoded step and not that of other structural genes. To our knowledge, there is so far no exception to this rule. This would imply that the *E* gene (*Matthiola incana*) which controls the activity of *DFR* and *UFGT* is a regulatory gene.[85,100]

Regulatory genes

A number of genes which do not encode the structural gene products also influence the biosynthesis of anthocyanins. Among these we perceive a discrete subset of genes, the regulatory genes. These genes influence anthocyanin biosynthesis because they are involved in controlling the expression of structural genes. Such genes have been recognized genetically and their mutant phenotypes suggest that they control structural gene expression with respect to developmental parameters. For example, they influence tissues or organs in which the anthocyanidin structural genes are expressed. However, this generalization does not exclude the possibility that environmental influences such as light may induce anthocyanin biosynthesis, at least in part, through the operation of these regulatory genes.

The greatest understanding of the action of regulatory genes lies in maize. Two primary regulatory genes, R(S) and C1 have been shown to be necessary for the expression of the structural genes CHS, DFR, A2 and UFGT in the aleurone layer of the kernel.[81,99,116-122] Both regulatory genes are required for the transcription of these structural genes. C1 is homologous to the myb-family of transcriptional activators in animals,[123] and R(S) is homologous to the myc family of animal transcription factors.[118] This implies that their influence on the transcription of structural genes is direct and both have been shown to regulate expression of the UFGT promoter in maize tissue. [119,120,122] R(S) and C1 control the transcription of structural genes in the aleurone tissue of the maize kernels, and are not expressed in other tissues. The pigmentation of tissues within other parts of the maize plant depends upon other genes: R(P), Lc, B, Sn and Pl.[118,124-133,158,159] Molecular analysis has shown R(P), Lc, B and Sn to be highly homologous to R(S), and Pl to be homologous to C1. Therefore, duplicate genes perform the same regulatory roles as R(S) and C1 in other parts of the plant,[131] and the specificity of pigmentation of different organs reflects the specificity of expression of the regulatory genes.

Light will stimulate the production of anthocyanin in maize and light can substitute for Pl if it is inactive.[134] Both R and C1 expression are induced by light, suggesting that it affects anthocyanin biosynthesis partly through these regulatory genes and partly through an independent mechanism.[135,136]

An appreciation of the role of regulatory genes in controlling the expression of the anthocyanin biosynthetic pathway, leads to a further requirement of a higher order of regulators that control the expression of these regulatory genes. One example of such a gene, Viviparous-1 (Vp1) will be discussed later.

In maize kernels, R(S) and C1 control the expression of the first (CHS) and later (DFR and UFGT) steps of the biosynthetic pathway; enzymological measurements imply that genes such as CHI and F3H are regulated in the same way.[128,137] Thus, changing the activity of regulatory genes can modify the expression of the entire anthocyanin pathway as a single unit. The situation in other parts of the maize plant is not so clear, since *R(P)* does not appear to regulate CHS expression to the same degree as *R(S)*.[136]

Regulatory genes that affect the production of anthocyanin in flowers have been identified in other species also. Six genes are known to be involved in the conversion of dihydroflavonols to anthocyanins in *Petunia hybrida*: *An1*, *An2*, *An6*, *An9*, *An10* and *An11*. Of these, *An6* has been shown to harbour the structural *DFR* gene.[78] The genes *An1*, *An2*, *An10* and *An11* influence the steady state level of *DFR-A* mRNA and the enzymatic activity of flavonoid 3-*O*-glucosyltransferase.[78] Since single mutants for all these genes accumulate dihydroflavonols,[80] they apparently do not significantly influence the CHS, CHI or F3H activity. Both *An1* and *An2* have also been shown to control later steps in the pathway, based on *in vitro* enzyme assays; in a multiple allelic series for both genes, a correlation was found between anthocyanin accumulation, *UFGT* activity and OMT activity.[138] The gene *An1* exhibits the "strongest" regulatory effects since in lines recessive for *An1*, *Hf1* and *Gf*, gene expression is also inhibited, whereas this is not the case for *an2/an2* lines.[80,102]

Mutation of the *Delila* (*Del*) gene of *Antirrhinum majus* blocks anthocyanin production in flower tubes, whereas the colouration of the lobes remains unaffected.[139] Analysis of the expression of the structural genes in lobes and tubes has shown that loss of an active *Del* gene product prevents the expression of *F3H*, *DFR*, *Candi* and *UFGT* in tubes, whereas *CHS* and *CHI* activites remain unaffected.[66,140] This implies that *Del* is a transcriptional activator required for expression of some of the structural genes in a specific part of the flower. In this case, the pathway is regulated in at least two discrete blocks; one comprising *CHS* and *CHI*, and the other *F3H*, *DFR*, *Candi* and *UFGT*. This may allow metabolic flexibility since the production of flavonols and anthocyanins can be regulated separately. A second gene, *Eluta*, reduces the overall anthocyanin content of the flower 8-10 fold, and limits its production in the central region of the face, the back of the lobes and the base of the tube, and has similar effects on the expression of *F3H*, *DFR*, *Candi* and *UFGT*.[66] Whereas the *Eluta* allele reduces expression of these genes, it increases expression of *CHI* in lobes and has no effect on *CHS*. Thus, the regulatory action of *Eluta* separates the biosynthetic genes into at least two groups. The dominant *Eluta* allele, which reduces pigmentation in the flower, is found in a number of wild species; [141] it may encode a transcriptional repressor rather than an activator like *Del*.

Potential regulatory genes have been described in other species. The *E* gene in *Matthiola* controls the expression of *DFR* and *UFGT*.[64] In pea, the genes *A* and *A2* regulate the expression of *CHS* in flowers;[142] it is not known whether this reflects a specific *CHS* regulation or whether other genes are also affected.

Many regulatory genes have been described that influence the expression of at least a number of structural genes in the anthocyanin pathway. There are some interesting differences in regulation in different species: in corn, regulatory genes influence expression of all flavonoid genes tested so far. In *Antirrhinum majus*, a key regulatory point seems to involve the conversion of flavanone to dihydroflavonol, whereas in *Petunia hybrida* the regulatory mutants accumulate dihydroflavonols, suggesting that the conversion of dihydroflavonols to leucoanthocyanidins is the key regulatory point.[66,80,81,99,116-122,140]

The action of regulatory genes is presumed to involve binding of their gene products to the DNA sequences upstream of the transcribed regions of the structural genes. When bound, these proteins would interact with the general transcription complex to stimulate transcriptional activity. The regulation of structural gene expression will depend on the concentration of the regulator, its affinity for the binding site motif, and its interactions with other proteins that may affect the affinity for the binding site, or the ability of the regulatory gene product to activate or repress transcription. *R* and *C1* are believed to interact in this way to bring about transcriptional activation in maize.[120]

In most cases where promoter sequences controlling the expression of a specific anthocyanin biosynthetic gene have been examined, specificity has been found to lie in sequences a few hundred base pairs upstream of the transcription start. Flower-specific expression can be obtained using just 67 bp upstream of the *CHS* promoter in *Petunia*.[143] In *CHS* genes from *Antirrhinum* and parsley, on the other hand, a CACGTG motif (lying between -129 bp and -135 bp upstream of the transcription start) has been shown to bind nuclear factors and may be involved in light induction of gene expression in at least some tissues.[144-147] A region lying between -90 bp and -170 bp of the start of transcription of the *DFR* promoter in *Antirrhinum* has been implicated in binding to the *Del* gene product.[140]

It is likely that genes similar to *R* and *C1* are primary regulators of anthocyanin biosynthetic genes in many species. *Myb* homologous genes have been isolated from several plant species, although they have not yet been shown to be involved in the control of anthocyanin biosynthetic genes.[148,149] The promoter sequences of the biosynthetic genes may, therefore, include the target motifs for these regulators. If these *R*- and *C1*-type proteins interact intimately, their binding motifs may also be closely juxtapositioned within the structural gene promoters. A motif in the *UFGT* promoter, conforming to the *myc* consensus (NCANNTGN) has been shown to be important in R-induced expression in maize.[120,121] This motif lies within 6 bp of a potential *myb*-binding site that may be the target of the *C1* gene. Similar closely positioned potential *myb* and *myc* target motifs have also been reported to be present in the promoters of anthocyanin biosynthetic genes in *Antirrhinum majus*.[66]

Fluxes through the pathway in *Petunia*

Dihydroflavonols form a precursor pool for the formation of anthocyanins and flavonols, both of which are accumulated as glycosides. In colourless mutants, dihydroflavonols are accumulated as the 7- or 4'-glucosides. Feeding experiments, using limbs of flower buds of anthocyanin-synthesizing mutants as acceptors and radioactively labeled malonic acid as a substrate, indicate that glucosylation of dihydroflavonols occurs concomitantly with anthocyanin synthesis.[150] The glucosylated dihydroflavonols can no longer serve as precursors for anthocyanin synthesis.[151] In the coloured line R27, dihydroflavonol 7-glucoside accumulates as well as cyanidin 3-glucoside,[152] so that part of the precursor pool is not used for anthocyanin or flavonol synthesis.

A genotype, dominant for all known flavonoid biosynthetic genes would theoretically be able to produce the flavonols kaempferol, quercetin and myricetin and the anthocyanins pelargonidin, cyanidin, peonidin, petunidin and malvidin. Actually, malvidin and quercetin accumulate predominantly.[80] In order to obtain appreciable amounts of cyanidins, the synthesis of delphinidins has to be suppressed. Moreover, when delphinidin production is independent of flavonol synthesis, cyanidin synthesis is greatly reduced in a genotype that leads to a high flavonol synthesis.[80] This phenomenon can be explained partly by the high affinity of DFR for dihydromyricetin and a much lower affinity for dihydroquercetin.[79]

A similar shift in anthocyanin levels can be obtained if either one of the six described pH genes is made homozygous recessive. In strains dominant for all pH genes, the pH of a floral extract is about 5.5; whereas in strains recessive for one of these genes, the pH in such an extract increases to about 6.0.[153] In a delphinidin genetic background, there is no change in the amount of delphinidins, although there is a change in colour to a bluer hue. However, cyanidin production is significantly reduced in a genetic background with a recessive pH gene. The combination of high flavonol production and high pH leads to an almost completely acyanic phenotype in genotypes that are still capable of cyanidin synthesis. While the pH genes have a marked effect on (probably) vacuolar pH in coloured *Petunia* lines, no difference in pH values for dominant and recessive plants is observed in an *anl/anl* background; this suggests that either *Anl* controls the expression of the pH genes, or that there is a direct relationship between pH value and the presence of anthocyanins.[153]

Genes have been described that influence the quantity or localization of pigment,[154-156] a fading of flower colour[157] or intensity of colour.[152] Very little is known about the mechanisms by which these genes exert their influence.

Subcellular localization of flavonoids and their enzymes

This subject is treated elsewhere in this volume, but warrants some specific remarks on the situation in *Petunia*. Floral anthocyanins are known to acumulate in the vacuole of epidermal cells[161] or in the outer layers of the pollen wall.[162,163] The primary metabolism of the cell is protected thereby against the reactive phenolic groups of the anthocyanins. As has been shown by Jonsson et al.,[112] the last modifiyng enzyme in the pathway, a methyltransferase, is a cytosolic enzyme which might be loosely bound to the endoplasmic membrane (ER). This indicates that none of the steps involved in anthocyanin biosynthesis take place in the vacuole; Jonsson supports the concept that flavonoids are synthesized in the cytosol outside the large central vacuole.

As stated before, the glucosylated anthocyanidins are found in the vacuole, whereas dihydroflavonols are produced in the cytoplasm. Six genes have been defined in *Petunia* that control the conversion of dihydroflavonols to anthocyanins. Four of these, *Anl, An2, An10* and *An11*, appear to control the expression of at least DFR and the flavonoid:3-*O*-glucosyltransferase and, therefore, may be regarded as regulatory genes.

The fact that anthocyanidin 3-glucosides are vacuolar, whereas the enzymes that carry out further modifications of the 3-glucosides are presumably cytosolic (so-called soluble enzymes), poses another intriguing question. An explanation might be that the late enzymes in anthocyanin synthesis are organized in a multienzyme complex, which might be close to or losely bound with the vacuolar membrane. Reactions leading to further modifications of anthocyanins would then compete with direct transport into the vacuole.

Hormonal Effects On Flavonoid Synthesis

Flavonoid biosynthesis is influenced by external factors such as temperature, light and nutritional status as well as by internal factors such as hormone balance and developmental programs. Clearly, the endogenous factors are the most significant regulatory influences of the biosynthesis in developing flowers. Now that a number of structural and regulatory genes of the flavonoid pathway have been cloned, it is possible to analyze these types of controls. One example is the analysis of the viviparous-1 (*vp1*) mutant of *Zea mays*.[135] The wild type gene is expressed specifically in the embryo and endosperm of developing seeds, whereas expression is lacking in seeds of the mutant. Mutant seeds germinate precociously on the ear and exhibit no anthocyanin synthesis in aleurone and embryo tissues. Some *vp1* alleles, however, permit near normal embryo dormancy, but still block anthocyanin synthesis completely, suggesting a separate control of both properties.[164] The *vp1* mutant embryos exhibit a reduced sensitivity to the phytohormone abscisic acid.[165] It has been shown that the *Vp1* product is required for *C1* expression, since no *C1* mRNA could be detected in embryo or endosperm of *vp1* kernels.[135] *C1* expression, in turn, is one of the regulatory signals needed for activation of the structural flavonoid genes.

Both gibberellic acid and abscisic acid have been reported to inhibit anthocyanin synthesis in radish seedlings.[166] In carrot cell suspension cultures, addition of gibberellic acid to the culture medium leads to suppression of anthocyanin synthesis, possibly through the induction of a nucleotidase activity that catalyzes the formation of 3'-dephospho-CoA derivatives, which in turn inhibit CHS expression. Subsequent biochemical complementation with naringenin and dihydroquercetin restored the synthesis of cyanidin glycosides, indicating that CHS is specifically inhibited by the addition of gibberellic acid.[167]

Petunia stamens contain high levels of gibberellins.[168] Removal of the anthers at a stage of flower development before the beginning of anthocyanin synthesis, inhibits both growth and anthocyanin synthesis *in vivo*. Both effects can be reversed by the application of GA3; the effect on corolla growth is independent of that on anthocyanin synthesis.[169] This suggests that during normal flower development GA3 is synthesized in anthers, and is subsequently transported to other parts of the flower to induce anthocyanin synthesis. Besides the activity of phenylalanine ammonia lyase,[169] it has been shown that CHS and CHI proteins and steady state mRNA levels are affected by the removal of stamens and application of GA3.[170] In experiments with transgenic plants, containing a construct in which the β-glucuronidase gene was fused with the *CHI* promotor, a sixfold increase in β-glucuronidase activity was measured following GA3 application to destaminated flower buds.[170] The hormone apparently induces transcription of flavonoid genes, either directly or indirectly, or it influences the stability of the mRNA's.

GENETIC ENGINEERING OF FLOWER COLOUR

Based on the quite detailed biochemical and genetic knowledge of how floral pigments are synthesized, it has become possible to genetically (or more precisely, molecularly) engineer flower colour. The first modified plants in this area were produced by Meyer *et al.*[82] who showed that a new colour could be obtained in *Petunia* by transforming a variety that produced low amounts of cyanidin by itself with a construct of the corn *A1* gene. This led to the synthesis of appreciable amounts of pelargonidin, an anthocyanin that is not found normally in *Petunia*.

Other examples are the sense and anti-sense experiments carried out by van der Krol *et al.*[171-174] and Napoli *et al.*[175] White flowered plants, as well as plants showing new patterns of colour, were obtained with both sense and antisense constructs of CHS or DFR genes. The results with sense constructs pose intriguing questions concerning the mechanisms by which gene expression may be suppressed and may lead to new insights in gene regulation.[176]

The analysis of structural genes in transgenic plants could be augmented by the engineering of regulatory genes to either inhibit or activate the expression of blocks of structural genes in new tissues or under different conditions. Whereas the flavonoid system provides an excellent model system for testing the more general aspects of genetic engineering in plants, these experiments also provide considerable commercial interest because flowers are a high value commodity, particularly susceptible to fashion.

CONCLUSION

With the wealth of biochemical and classical genetic data available and the rapidly developing input from molecular genetics, flavonoid synthesis has again become a field of research that offers tremendous opportunities. Molecular techniques now enable us to analyze gene expression on the inter- and intratissue level and to analyze the location of gene products at the subcellular level. Such research will undoubtedly further clarify the subcellular localization of the enzymes and the routing of intermediates and products.

The availability of expression vectors should enable the isolation of flavonoid biosynthetic enzymes in milligram quantities, providing the opportunity to analyze the structure of the proteins. An exciting opportunity to analyze structure-function relationships is offered by the different DFR genes cloned so far, as they show clear differences in substrate specificity. This is also true for the diverse CHS and CHI genes. The analysis of these genes on the protein level can contribute to the elucidation of the rules governing the formation of secondary and tertiary protein structures and provide insight into the processes of secondary metabolism in general. This, after all, is the best understood plant secondary metabolic pathway, but just the tip of the iceberg in terms of the biosynthetic capacity of plants.

On a different level, transformation experiments will not only provide information on the requirements for gene expression, but also tell us more about the influence of the topology of the genome on gene expression. Confocal Laser Scanning Microscope analysis might tell us more about the actual position of transcribed genes in the nucleus and of the gene products in the cell.[177,178]

REFERENCES

1. DERNO, R.F., McCLURE, M.S. 1983. Variable Plants And Herbivores in Natural and Managed Systems. Academic Press, New York.
2. DIXON, R.A. 1986. The phytoalexin response: elicitation signalling and control of host gene expession. Biol. Rev. 61:239-291.
3. HEDIN, P.A., WAAGE, S.K. 1986. In: Plant Flavonoids in Biology and Medicine, (V. Cody, E. Middleton, J.B. Harborne, eds.) Alan R. Liss Inc., New York, pp. 87-106.
4. FIRMIN, J.L., WILSON, K.E., ROSSEN, L., JOHNSTON, A.W.B. 1986. Flavonoid activation of nodulation genes in *Rhizobium* reversed by other compounds present in plants. Nature 324:90-92.
5. PETERS, N.K., FROST, J.N., LONG, S.R. 1986. A plant flavone, inteolin, induces expression of *Rhizobium meliloti* nodulation genes. Science 233:977-980.
6. REDMOND, J.W., BATLEY, M., DJORDJEVIC, M.A., INNES, M.W., KUEMPEL, P.W., ROLFE, B.G. 1986. Flavones induce expression of nodulation genes in *Rhizobium*. Nature 323, 632-635.
7. CODY, V., MIDDLETON, E., HARBORNE, J.B. 1986. Plant Flavonoids in Biology and Medicine: Biochemical, Pharmacological and Structure-Activity Relationships. Alan R. Liss Inc., New York.
8. MENDEL, G. 1865. Versüche über Pflanzen-Hybriden. Verh. Naturf. Verein. Brünn, IV, 3-47.
9. WHELDALE, M.W. 1909. The colours and pigments of flowers with special reference to genetics. Proc. Roy. Soc. B, 81:44-60.
10. WHELDALE, M.W. 1925. The anthocyanin pigments of plants.

University Press, Cambridge.

11. BEADLE, G.W., TATUM, E.C. 1941. Genetic control of biochemical reactions in *Neurospora*. Proc. Natl. Acad. Sci. USA, 27:499-506.

12. VAVILOV, N.I. 1930. The Linnean series as a system. Proc. 5th Int. Bot. Cong. Cambridge. 213-216.

13. BATE-SMITH, E.C. 1948. Paper chromatography of anthocyanins and related substances in petal extracts. Nature 161:835.

14. HANSON, K.R., HAVIR, E.A. 1972. The enzymatic elimination of ammonia. In: The Enzymes. (P.D. Boyer, ed.) Academic Press, New York, Vol. 7, pp. 577-625.

15. HANSON, K.R., HAVIR, E.A. 1981. Phenylalanine Ammonia Lyase. In: The Biochemistry of Plants. (P.K. Stumpf, E.E. Conn, eds.) Academic Press, New York, Vol 7, pp. 577-625.

16. NAIR, P.M., VINING, L.C. 1965. Cinnamic acid hydroxylase in spinach. Phytochemistry 4:161-168.

17. RUSSELL, D.W., CONN, E.E. 1967. The cinnamic acid 4-hydroxylase of pea seedlings. Arch. Biochem. Biophys. 122:256-258.

18. HAHLBROCK, K., GRISEBACH, H. 1970. Formation of coenzyme A esters of cinnamic acid with an enzyme preparation from cell suspension cultures of parsley. FEBS Lett. 11:62-64.

19. WALTON, E., BUTT, V.S. 1970. The activation of cinnamate by an enzyme from leaves of spinach beet (*Beta vulgaris* L. ssp. *vulgaris*). J. Exp. Bot. 21:887-891.

20. KNOBLOCH, K.H., HAHLBROCK, K 1975. Isoenzymes of *p*-Coumarate: CoA ligase from cell suspension cultures of *Glycine max*. Eur. J. Biochem. 52:311-320.

21. WALLIS, P.J., RHODES, M.J.C. 1977. Multiple forms of hydroxy-cinnamate:CoA ligase in etiolated pea seedlings. Phytochemistry 16:1891-1894.

22. BUTT, V.S., WILKINSON, E.M. 1979. In: Regulation of Secondary Product and Plant Hormone Metabolism. (M. Luckner, K. Schreiber, eds.) Pergamon Press, Oxford, Vol. 55, pp. 147-154.

23. RANJEVA, R., FAGGION, R., BOUDET, A. 1975. Physiol. Veg. 13:725-734.

24. RANJEVA, R., BOUDET, A., FAGGION, R. 1976. Phenolic metabolism in petunia tissues. IV. Properties of *p-coumarate*: coenzyme A ligase isoenzymes. Biochimie 58:1255-1262.

25. RANJEVA, R., BOUDET, A.M., ALIBERT, G. 1979. In: Regulation of Secondary Product and Plant Hormone Metabolism. (M. Luckner, K.

Schreiber, eds.) Pergamon Press, Oxford, Vol. 55, pp. 91-100.

26. EBEL, J., HAHLBROCK, K. 1982. Biosynthesis. In: The Flavonoids, Advances in Research. (J.B. Harborne, T.J. Mabry, eds.) Chapman and Hall, London, pp. 641-679.

27. HELLER, W. 1986. Flavonoid biosynthesis, an overview. In: Plant Flavanoids in Biology and Medicine. (V. Cody, E. Middleton, J.B. Harborne, eds.) Alan R. Liss, New York, pp. 25-42.

28. HELLER, W., FORKMANN, G. 1988. Biosynthesis. In: The Flavonoids. (J.B. Harborne, ed.) Chapman and Hall, London, pp.399-425.

29. SPRIBILLE, R., FORKMANN, G. 1981. Genetic control of chalcone synthase activity in flowers of *Matthiola incana* R.Br. Z. Naturforsch. 36c:619-624.

30. SPRIBILLE, R., FORKMANN, G. 1982. Genetic control of chalcone synthase activity in flowers of *Antirrhinum majus*. Phytochemistry 21:2231-2234.

31. WHITEHEAD, J.M., DIXON, R.A. 1983. Chalcone synthase from cell suspension cultures of *Phaseolus vulgaris* L. Biochim. Biophys. Acta 747:298-303.

32. STOTZ, G., SPRIBILLE, R., FORKMANN, G. 1984. Flavonoid synthesis in flowers of *Verbena hybrida*. J. Plant Physiol. 116:173-183.

33. KAMSTEEG, J., VAN BREDERODE, J., VERSCHUREN, P.M., VAN NIGTEVECHT, G. 1981. Z. Pflanzenphysiol. 102:435.

34. DOONER, H.K. 1983. Co-ordinate genetic regulation of flavonoid biosynthetic enzymes in maize. Mol. Gen. Genet. 189:136-141.

35. SOMMER, H., SAEDLER, H. 1986. Structure of the chalcone synthase gene of *Antirrhinum majus*. Mol. Gen. Genet. 202:429-434.

36. REIF, H.J., NIESBACH, U., DEUMLING, B., SAEDLER, H. 1985. Cloning and analysis of two genes for chalcone synthase from *Petunia hybrida*. Mol. Gen. Genet. 199:208-215.

37. KOES, R.E., SPELT, C.E., MOL, J.N.M., GERATS, A.G.M. 1987. The chalcone synthase multigene family of *Petunia hybrida*: sequence homology, chromosomal localization and evolutionary aspects. Plant Mol. Biol. 10:159-169.

38. HERMANN, A., SCHULZ, W., HAHLBROCK, K. 1988. Two alleles of the single-copy chalcone synthase gene in parsley differ by a transposon-like element. Mol. Gen. Genet. 212:93-98.

39. WIENAND, U., WEYDEMANN, U., NIESBACH-KLOESGEN, U., PETERSON, P.A., SAEDLER, H. 1986. Molecular cloning of the C2 locus of *Zea mays*, the gene coding for chalcone synthase. Mol.

Gen. Genet. 203:202-207.

40.   NIESBACH-KLOESGEN, U., BARZEN, E., BERNHARDT, J.,
      ROHDE, W., SCHWARZ-SOMMER, Z.S., REIF, H.J.,
      WIENAND, U., SAEDLER, H. 1987. Chalcone synthase genes in
      plants: a tool to study evolutionary relationships. J. Mol. Evol.
      26:213-225.

41.   RYDER, T.B., HEDRICK, S.A., BELL, J.N., LIANG, X., CLOUSE,
      S.D., LAMB, C.J. 1987. Organization and differential activation of a
      gene family encoding the plant defense enzyme chalcone synthase in
      Phaseolus vulgaris. Mol. Gen. Genet. 210:219-233.

42.   FEINBAUM, R.L., AUSUBEL, F.M. 1988. Transcriptional regulation of
      the *Arabidopsis thaliana* chalcone synthase gene. Mol. Cell. Biol.
      8:1985-1992.

43.   HARDER, R., MARHEINEKE, J. 1935. Weitere untersuchungen über
      die Musterbildung an Petunienblüten. Nachr. Biol. 2:97-105.

44.   HARDER, R. 1938. Ueber Farb- und Muster änderungen bei Blüten.
      Naturwissenschaften 26:713-728.

45.   MOL J.N.M., SCHRAM, A.W., DE VLAMING, P., GERATS,
      A.G.M., KREUZALER, F., HAHLBROCK, K., REIF, H.J.,
      VELTKAMP, E. 1983. Regulation of flavonoid gene expression in
      Petunia hybrida: description and partial characterization of a
      conditional mutant in chalcone synthase gene expression. Mol. Gen.
      Genet. 192:424-429.

46.   KOES, R.E., SPELT, C.E., REIF, H.J., VAN DEN ELZEN, P.,
      VELTKAMP, E., MOL, J.N.M. 1986. Floral tissue of *Petunia
      hybrida* (V30) expresses only one member of the chalcone synthase
      multigene family. Nucl. Acids Res. 14:5229-5239.

47.   MOL, J.N.M., ROBBINS, M.P., DIXON, R.A., VELTKAMP, E. 1985.
      Spontaneous and enzymic rearrangement of naringenin chalcone to
      flavanone. Phytochemistry 24:2267-2269.

48.   CHMIEL, E., SUETFELD, R., WIERMANN, R. 1983. Conversion of
      phlorolgucinol-type chalcones by purified chalcone isomerase from
      Tulip anthers and from Cosmos petals. Biochem. Physiol. Pflanzen
      178:139-146.

49.   KUHN, B., FORKMANN, G., SEYFFERT, W. 1978. Genetic control of
      chalcone flavanone isomerase activity in *Callistephus chinensis*.
      Planta 138:199-203.

50.   FORKMANN, G., DANGELMAYER, B. 1980. Genetic control of
      chalcone isomerase activity in flowers of *Dianthus caryophyllus*.

Biochem. Genet. 18:519-527.

51. MEHDY, M.C., LAMB, C.J. 1987. Chalcone isomerase cDNA cloning and mRNA induction by fungal elicitor, wounding and infection. EMBO J. 6:1527-1533.

52. VAN TUNEN, A.J., KOES, R.E., SPELT, C.E., VAN DER KROL, A.R., STUITJE, A.R., MOL, J.N.M. 1988. Cloning of the two chalcone flavanone isomerase genes from *Petunia hybrida*: Coordinate, light regulated and differential expression of flavonoid genes. EMBO J. 7:1257-1263.

53. VAN TUNEN, A.J., HARTMAN, S.A., MUR, L.A., MOL, J.N.M. 1989. Regulation of chalcone flavanone isomerase (CHI)gene expression in *Petunia hybrida*: The use of alternative promoters in corolla, anthers and pollen. Plant Mol. Biol. 12:539-551.

54. FORKMANN, G., KUHN, B. 1979. Genetic control of chalcone isomerase activity in anthers of *Petunia hybrida*. Planta 144:189-192.

55. VAN WEELY, S., BLEUMER, A., SPRUYT, R., SCHRAM, A.W. 1983. Chalcone isomerase in flowers of mutants of *Petunia hybrida*. Planta 159:226-230.

56. VAN TUNEN, A.J., MUR, L.A., BROUNS, G.S., RIENSTRA, J.D., KOES, R.E., MOL, J.N.M. 1990. Pollen- and anther-specific *chi* promotors from petunia: Tandem promoter regulation of the *chiA* gene. Plant Cell 2:393-401.

57. VAN TUNEN, A.J., MUR, L.A., RECOURT, K., GERATS, A.G.M., MOL, J.N.M. 1991. Regulation and manipulation of flavonoid gene expression in anthers of *Petunia*: the molecular basis of the *Po* mutation. Plant Cell 3:39-48.

58. FRITSCH, H., GRISEBACH, H., 1975. Biosynthesis of cyanidin in cell cultures of *Haplopappus gracilis*. Phytochemistry 14:2437-2442.

59. DOODEMAN, M., TABAK, A.J.H., SCHRAM, A.W., BENNINK, G.J.H. 1982. Hydroxylation of cinnamic acids and flavonoids during biosynthesis of anthocyanins in *Petunia hybrida* Hort. Planta 154:546-549.

60. FROEMEL, S., DE VLAMING, P., STOTZ, G., WIERING, H., FORKMANN, G., SCHRAM, A.W. 1985. Genetic and biochemical studies on the conversion of flavanones to dihydroflavonols in flowers of *Petunia hybrida*. Theor. Appl. Genet. 70:561-568.

61. BRITSCH, L.H., GRISEBACH, H. 1986. Purification and characterization of (2S) flavanone 3-hydroxylase from *Petunia hybrida*. Eur. J. Biochem. 135:569-577.

62.   FORKMANN, G., STOTZ, G. 1981. Genetic control of flavanone 3-
      hydroxylase and flavonoid 3'-hydroxylase activity in *Antirrhinum
      majus* (Snapdragon). Z. Naturforsch. 36C:411-416.
63.   FORKMANN, G., STOTZ, G. 1984. Selection and characterisation of
      flavanone 3-hydroxylase mutants in *Dahlia, Streptocarpus, Verbena*
      and *Zinnia*. Planta 161:261-265.
64.   FORKMANN, G. 1989. Gene-enzyme relations and genetic manipulation
      of anthocyanin biosynthesis in flowering plants. In: The Genetics of
      Flavonoids. (E.D. Styles, G.A. Gavazzi, M.L. Racchi, eds.) Edizioni
      Unicopli., Milan, pp. 49-60.
65.   LARSON, R.L. 1989. Genetics, precursors and enzymes in flavonoid bio-
      synthesis in maize. In: The Genetics of Flavonoids. (E.D. Styles,
      G.A. Gavazzi, M.L. Racchi, eds.) Edizioni Unicopli., Milan. pp. 71-
      78.
66.   MARTIN, C., PRESCOTT, A., MACKAY, S., BARTLETT, J.,
      VRIJLANDT, E. 1991. Control of anthocyanin biosynthesis in
      flowers of *Antirrhinum majus*. THe Plant Journal 1:37-49
67.   WIERING, H. 1974. Genetics of flower colour in *Petunia hybrida* hort.
      Gen. Phaenen 17:117-134.
68.   TABAK, A.J.H., MEYER, H., BENNINK, G.J.H. 1978. Modification of
      the B-ring during flavonoid biosynthesis in *Petunia hybrida*:
      introduction of the 3' hydroxyl group is regulated by the gene *Ht1*.
      Planta 139:67-71.
69.   WIERING, H., DE VLAMING, P. 1984. Genetics of flower and pollen
      colours. In: Petunia. (K.C. Sink, ed.) Monographs on Theoretical and
      Applied Genetics, Springer Verlag, Berlin, pp. 49-67.
70.   STOTZ, G., DE VLAMING, P., WIERING, H., SCHRAM, A.W.,
      FORKMANN, G. 1985. Genetic and biochemical studies on
      flavonoid 3'-hydroxylation in flowers of *Petunia hybrida*. Theor.
      Appl. Genet. 70:300-305.
71.   FORKMANN, G., HELLER, W., GRISEBACH, H. 1980. Anthocyanin
      biosynthesis in flowers of *Matthiola incana*:flavanone 3- and
      flavonoid 3'-hydroxylases. Z. Naturforsch. 35c:691-695.
72.   STAFFORD, H.A., LESTER, H.H. 1982. Enzymatic and non-enzymatic
      reduction of (+) dihydroquercetin to its 3,4-diol. Plant Physiol.
      70:695-698.
73.   STAFFORD, H.A., LESTER, H.H. 1984. Flavan-3-ol biosynthesis. The
      conversion of (+)-dihydroquercetin and flavan-3,4-cis-diol (leucoantho-
      cyanidin) to (+)-catechin by reductases extracted from cell suspension

cultures of Douglas fir. Plant Physiol. 76:184

74. STAFFORD, H.A., LESTER, H.H. 1985. Flavan-3-ol biosynthesis. The conversion (+)-dihydroquercetin and its flavan-3,4-cis-diol (leucodelphinidin) and to (+)-gallocatechin by reductases extracted from tissue cultures of *Ginkgo biloba* and *Pseudotsuga menziesii*. Plant Physiol. 78:791-794.

75. KRISTIANSEN, K.N. 1984. Biosynthesis of proanthocyanidins in barley: Genetic control of the conversion of dihydroquercetin to catechin and procyanidins. Carlsberg Res. Commun. 49:503-524.

76. KRISTIANSEN, K.N. 1986. Conversion of (+)-dihydroquercetin to (+)-2,3-*trans*-3,4-*cis*-leucocyanidin and (+)catechin with an enzyme extract from maturing grains of barley. Carlsberg Res. Commun. 51:51-60.

77. MARTIN, C., CARPENTER, R., SOMMER, H., SAEDLER, H., COEN, E.S. 1985. Molecular analysis of instability in flower pigmentation in *Antirrhinum majus*, following the isolation of the *pallida* locus by transposon tagging. EMBO J 4:1625-1630.

78. BELD, M.G.H.M., MARTIN, C., HUITS, H., STUITJE, A.R., GERATS A.G.M. 1989. Flavonoid synthesis in *Petunia hybrida*: partial characterisation of dihydroflavonol 4-reductase genes. Plant Mol. Biol. 13:491-502.

79. FORKMANN, G., RUHNAU, B. 1987. Distinct substrate specificity of dihydroflavonol 4-reductase from flowers of *Petunia hybrida*. Z. Naturforsch. 42C:1146-1148.

80. GERATS, A.G.M., DE VLAMING, P., DOODEMAN, M., AL, B., SCHRAM, A.W. 1982. Genetic control of the conversion of dihydroflavonols into flavonols and anthocyanins in flowers of *Petunia hybrida*. Planta 155:364-368.

81. REDDY, G.M., BRITSCH, L., SALAMINI, F., SAEDLER, H., ROHDE, W. 1987. The *A1* (*anthocyanin-1*) locus in *Zea mays* encodes dihydroquercetin reductase. Plant Sci. 52:7-13.

82. MEYER, P., HEIDMANN, I., FORKMANN, G., SAEDLER, H. 1987. A new *Petunia* flower colour generated by transformation of a mutant with a maize gene. Nature 330:677-678.

83. LINN, F., HEIDMANN, I., SAEDLER, H., MEYER, P. 1990. Epigenetic changes in the expression of the maize A1 gene in *Petunia hybrida*: Role of numbers of integrated gene copies and state of methylation. Mol. Gen. Genet. 222:329-336.

84. HELLER, W., BRITSCH, L., FORKMANN, G., GRISEBACH, H.

1985. Leucoanthocyanidins as intermediates in anthocyanidin biosynthesis in flowers of *Matthiola incana* R. Br. Planta 163:191-196.

85.   HELLER, W., FORKMANN, G., BRITSCH, L., GRISEBACH, H. 1985. Enzymatic reduction of the (+)-dihydroflavnols to flavan-3,4-*cis*-diols with flower extracts from *Matthiole incana* and its role in anthocyanin biosynthesis. Planta 165:284-287.

86.   O'REILLY, C., SHEPHERD, N.S., PEREIRA, A., SCHWARZ-SOMMER, Z.S., BERTRAM, I., ROBERTSON, D.S., PETER-SON, P.A., SAEDLER, H. 1985. Molecular cloning of the *a1* locus of *Zea mays* using the transposable elements En and Mu. EMBO J. 44:877-882.

87.   COEN, E.S., CARPENTER, R., MARTIN, C. 1986. Transposable elements generate novel patterns of gene expression in *Antirrhinum majus*. Cell 47:285-296.

88.   BARTLETT, N.J.R. 1989. The genetic control of anthocyanin biosynthesis in *Antirrhinum majus*. Ph.D. thesis, U. of East Anglia, Norwich.

89.   REDDY, A.R., COE, E.H. 1962. Inter-tissue complementation: A simple technique for direct analysis of gene action sequence. Science 138:149-150.

90.   MENSSEN, A., HOEHMANN, S., MARTIN, W., SCHANLE, P.S., PETERSON, P.A., SAEDLER, H., GIERL, A. 1990. The En/Spm transposable element of *Zea mays* contains splice sites at the termini generating a novel intron from a dSpm element in the *A2* gene. EMBO J. 9:3051-3058.

91.   LARSON, R.L., COE, E.H. 1968. Enzymatic action of the *Bz* anthocyanin factor in maize. Proc. XII Int. Cong. Genet. 1:131.

92.   JONSSON, L.M.V., AARSMAN, M.E.G., BASTIAANNET, J., DONKER-KOOPMAN, W., GERATS, A.G.M., SCHRAM, A.W. 1984. Common identity of UDP-Glucose:anthocyanidin-3-*O*-glucosyltransferase and UDP-Glucose:flavonol-3-*O*-glucosyltransferase in flowers of *Petunia hybrida*. Z. Naturforsch. 39c:559-567.

93.   SWAIN, T. 1976. Flavonoids. In: Chemistry and Biochemistry of Plant Pigments. (T.W. Goodwin, ed.) Academic Press, New York, Vol. 1, pp. 425-463.

94.   MARTY, F., BRANTON, D., LEIGH, R.A. 1980. Plant vacuoles. In: The Biochemistry of Plants. (P.K. Stumpf, E.E. Conn, eds.) Academic Press, New York, Vol. 1, pp. 625-658.

95.   KHO, K.F.F., KAMSTEEG, J., VAN BREDERODE, J. 1978. Identification, properties and genetic control of UDP Glucose:

cyanidin 3-*O*-glucosyltransferase in *Petunia hybrida*. Z. Pflanzen-physiol. 88:449-464.

96. GERATS, A.G.M., WALLROTH, M., DONKER-KOOPMAN, W., GROOT, S., SCHRAM, A.W. 1983. UDPGlucose: 3-*O*-flavonoid glucosyltransferase in flowers of *Petunia hybrida*. Theor. Appl. Genet. 65:349-352.

97. GERATS, A.G.M., VRIJLANDT, E., WALLROTH, M., SCHRAM, A.W. 1985. The influence of the genes An1, An2 and An4 on the activity of the enzyme UDP glucose: flavonoid 3-o-glucosyltransferase in flowers of *Petunia hybrida*. Biochem. Genet. 23:591-598.

98. LARSON, R.L., COE, E.H. 1977. Gene-dependent flavonoid glucosyltransferase in maize. Biochem. Genet. 15:153-156.

99. DOONER, H.K., NELSON O.E. 1977. Genetic control of UDP-Glucose: flavonol 3-*O*-glucosyltransferase in the endosperm of maize. Biochem. Genet. 15:501-519.

100. TEUSCH, M., FORKMANN, G., SEYFFERT, W. 1986. UDP-Glucose:anthocyanidin/flavonol 3-*O*-glucosyltransferase in enzyme preparation from flower extracts of genetically defined lines of *Matthiola incana* R.Br. Z. Naturforsch. 41c:699-706.

101. GRIESBACH, R.J., ASEN, S. 1990. Characterization of the flavonol glycosides in *Petunia*. Plant Sci. 70:49-56.

102. JONSSON, L.M.V., AARSMAN, M.E.G., VAN DIEPEN, J., DE VLAMING, P., SCHRAM, A.W. 1984. Properties and genetic control of anthocyanin 5-*O*-glucosyltransferase in flowers of *Petunia hybrida*. Planta 160:341-347.

103. TEUSCH, M., FORKMANN, G., SEYFFERT, W. 1986. Genetic control of UDP-glucose: anthocyanin 5-*O*-glucosyltransferase from flowers of *Matthiola incana* R. Br. Planta 168:586-591.

104. KAMSTEEG, J., VAN BREDERODE, J., VAN NIGTEVECHT, G. 1978. Identification, properties and genetic control of UDP-Glucose: Cyanidin 3-rhamnosyl (1-6)-glucoside, 5-*O*-glucosyl-transferase isolated from petals of the Red Campion (*Silene dioica*). Biochem. Genet. 16:1059-1071.

105. KAMSTEEG, J., VAN BREDERODE, J., VAN NIGTEVECHT, G. 1980. Identification, properties and genetic control of UDP-*l*-rhamnose: anthocyanidin 3-*O*-glucoside, 6"-*O*-rhamnosyltransferase from petals of the Red Campion (*S. dioica*). Z. Naturforsch. 35C:249-257.

106. KAMSTEEG, J., VAN BREDERODE, J., VAN NIGTEVECHT, G. 1980. The pH-dependent substrate specificity of UDP-Glucose:

anthocyanidin 3-rhamnosylglucoside, 5-*O*-glucosyltransferase in petals of *Silene dioica* . Z. Pflanzenphysiol 96:87-93.

107. SEYFFERT, W. 1959. Untersuchungen Über interallele Wechsel-wirkungen III. Der Dosiseffect eines die Methylierung von Anthocyanen kontrollierenden Gene. Naturwissenschaften 46:271.

108. HARBORNE, J.B. 1967. Inheritance and biosynthesis of flavonoids in plants. In: Comparative Biochemistry of Flavonoids, Academic Press, London, pp. 250-280.

109. GUPTA, S.B. 1970. Biochemical aspects of the inheritance of floral anthocyanins in diploid alfalfa. Genetics 65:267-278.

110. WIERING, H., DE VLAMING, P. 1977. Glycosylation and methylation patterns of anthocyanins in *Petunia hybrida*:. The genes Mf1 and Mf2. Z. Pfl. Zücht. 78:113-123.

111. JONSSON, L.M.V., AARSMAN, M.E.G., SCHRAM, A.W., BEN-NINK G.J.H. 1982. Methylation of anthocyanins by cell-free extracts of flower buds of *Petunia hybrida*. Phytochemistry 21:2457-2459.

112. JONSSON, L.M.V., DONKER-KOOPMAN, W.E., UITSLAGER, P., SCHRAM, A.W. 1983. Subcellular localization of anthocyanin methyltransferase in flowers of *Petunia hybrida*. Plant Physiol. 72:287-290.

113. JONSSON, L.M.V., DE VLAMING, P., WIERING, H., AARSMAN, M.E.G., SCHRAM, A.W. 1983. Genetic control of anthocyanin *O*-methyltransferase activity in flowers of *Petunia hybrida*.. Theor. Appl. Genet. 66:349-355.

114. JONSSON, L.M.V., AARSMAN, M.E.G., POULTON, J.E., SCHRAM, A.W. 1984. Properties and genetic control of four methyltransferases involved in methylation of anthocyanins in flowers of *Petunia hybrida*. Planta 160:174-179.

115. MARTIN, C., GERATS, A.G.M. 1991 The control of flower colouration. In: The Molecular Biology of Flowering. (B. Jordan, ed.) CAB Internat., in press.

116. DOONER, H.K. 1983. Co-ordinate genetic regulation of flavonoid biosynthetic enzymes in maize. Mol. Gen. Genet. 189:136-141.

117. CONN, K.C., BURR, F.A., BURR, B. 1986. Molecular analysis of the maize regulatory locus *C1*. Pro. Nat. Acad. Sci. USA 83:9631-9635.

118. LUDWIG, S.R., HABERA, L.F., DELLAPORTA, S.L., WESSLER, S.R. 1989. *Lc* a member of the maize *R* gene family responsible for tissue specific anthocyanin production encodes a protein similar to transcriptional activators and contains the *myc*-homology region.

Proc. Natl. Acad. Sci. USA 86:849-851.

119. GOFF, S.A., CONE, K.C., FROMM, M.E. 1991. Identification of functional domains in the maize transcriptional activator *C1*: comparison of wildtype and dominant inhibitor proteins. Genes and Develop. 5:298-309.

120. ROTT, B.A., GOFF, S.A., KLEIN, T.M., FROMM, M.E.1991. *C1* and *R* dependent expression of the maize *Bz1* gene requires sequences with homology to mammalian *myb* and *myc* binding sites. Plant Cell 3:317-325.

121. GOFF, S.A., KLEIN, T.M., ROTH, B.A., FROMM, M.E., CONE, K.C., RADICELLA, J.P., CHANDLER, V.L. 1990. Transactivation of anthocyanin biosynthetic genes following transfer of of *B* regulating genes into maize tissues. EMBO J. 9:2517-2522.

122. KLEIN, T.M., ROTH, B.A., FROMM, M.E. 1989. Regulation of anthocyanin biosynthetic genes introduced into intact maize tissues by microprojectiles. Proc. Natl. Acad. Sci. USA 86:6681-6685.

123. PAZ-AREZ, J., GHOSAL, D., WIENAND, U., PETERSON, P.A., SAEDLER, H. 1987. The regulatory *C1* locus of *Zea mays* encodes a protein with homology to *myb* protooncogene products and with structural similarities to transcriptional activators. EMBO J. 6:3553-3558.

124. CHEN, S-M. H. 1973. Anthocyanins and their Control by the *C* locus in Maize. Ph.D. Thesis, Univ. of Missouri, Columbia.

125. CHEN, S-M, COE, E.H. 1977. Control of anthocyanin synthesis by the *C* locus in maize. Biochem. Genet. 15:333-346.

126. COE, E.H. 1985. Phenotypes in corn: control of pathways by alleles, time and place. In: Plant Genetics. (M. Freeling, ed.) Alan R. Liss Inc., New York, pp 509-521.

127. CONE, K.C., BURR, B. 1989. Molecular and genetic analyses of light requirement for anthocyanin synthesis in maize. In: The Genetics of Flavonoids. (E.D. Styles, G.A. Gavazzi, M.L. Racchi, eds.) Edizione Unicopli, Milano, pp. 143-146.

128. CONSONNI, G., RACCHI, M.L., SHAMMAH, S., GAVAZZI, G.A. 1987. The role of *Sn* in the light-regulated activity of enzymes of flavonoid biosynthesis. Maize Genet. Coop. Newsl. 61:35.

129. DOONER, H.K., KERMICLE, J.L. 1971. Structure of the R^r tandem duplication in maize. Genetics 67:427-436.

130. EMERSON, R.A. 1921. The genetic relations of plant colours in maize. Cornell Univ. Agric. Exp. Sta. 39:3-156.

131. LUDWIG, S.R., WESSLER, S.R. 1990. Maize *R* gene family: Tissue specific helix-loop-helix proteins. Cell 62:849-851.

132. CHANDLER, V.L., RADICELLA, J.P., ROBBINS, T.P., CHEN, J., TRUKS, D. 1989. Two regulating genes of the maize anthocyanin pathway are homologous: isolation of *B* using *R* genomic sequences. Plant Cell 1:1175-1183.

133. STADLER, L.D. 1946. Spontaneous mutation at the *R* locus in maize 1. The aleurone colour and plant colour effects. Genetics 31:377-394.

134. COE, E.H., NEUFFER, M.G., HOISINGTON, D.A. 1988. Corn and corn improvement. Agronomy monograph 18:81-258. Amer. Soc. Agron. Inc., Madison, WI USA.

135. McCARTY, J.P., HATTORI, T., VASIL, V., VASIL, I.K. 1991. A regulatory hierarchy in seed development:interaction of *viviparous-1* and abscisic acid in the regulation of the *C1* gene in maize. J. Cell. Biochem., supp. 15A:25.

136. TAYLOR, L.P., BRIGGS, W.R. 1990. Genetic regulation and photocontrol of anthocyanin accumulation in maize seedlings. Plant Cell 2:115-127.

137. LARSON, R.L. 1989. Genetics, precursors and enzymes in flavonoid biosynthesis in maize. In: The Genetics of Flavonoids. (E.D. Styles, G.A. Gavazzi, M.L. Racchi, eds.) Edizione Unicopli, Milano, pp. 71-78.

138. GERATS, A.G.M., FARCY, E., WALLROTH, M., GROOT, S.P.C., SCHRAM, A.W. 1984. Control of anthocyanin synthesis in *Petunia hybrida* by multiple allelic series of the genes *An1* and *An2*. Genetics 106:501-508.

139. DE VRIES, H. 1903. Die Mutationstheorie, Band II. Verlag von Veit und Co., Leipzig, pp. 194-206.

140. ALMEIDA, J., CARPENTER, R., ROBBINS, T.P., MARTIN, C., COEN, E.S. 1989. Genetic interactions underlying flower colour patterns in *Antirrhinum majus*. Genes Dev. 3:1758-1767.

141. STUBBE, H. 1966. Genetik und Zytologie von *Antirrhinum* L. sec. *Antirrhinum*. Veb. Gustaf Fischer Verlag, Jena.

142. HARKER, C.L., ELLIS, T.H.N., COEN, E.S. 1990. Identification and genetic regulation of the chalcone synthase multigene family in pea. Plant Cell 2:185-194.

143. VAN DER MEER, I.M., SPELT, C.E., MOL, J.N.M., STUITJE, A.R. 1990. Promoter analysis of the chalcone synthase (CHSA) gene of *Petunia hybrida*:A 67 bp promoter region directs flower-specific

expression. Plant Mol. Biol. 15:95-109.

144. STAIGER, D., KAULEN, H., SCHELL, J. 1989. A CACGTG motif of the *Antirrhinum majus* chalcone synthase promoter is recognised by an evolutionary conserved nuclear protein. Proc. Natl. Acad. Sci. USA 86:6930-6934.

145. LIPPHARDT, S., BRETTSCHNEIDER, R., KREUZALER, F., SCHELL, J., DANGL, J. 1988. UV-inducible transient expression in parsley protoplasts identifies regulatory *cis* elements of a chimeric *Antirrhinum majus* chalcone synthase gene. EMBO J. 7:4027-4033.

146. SCHULZE-LEFERT, P., BECKER-ANDRE, M., SCHULZ, W., HAHLBROCK, K., DANGL, J. 1989. Functional architecture of light-responsible chalcone synthase promoter from parsley. Plant Cell 1:707-714.

147. SCHULZE-LEFERT, P., DANGL, J., BECKER-ANDRE, M., HAHL-BROCK, K., SCHULZ, W. 1989. Inducible *in vivo* DNA footprints define sequences necessary for U.V. light activation of the parsley chalcone synthase gene. EMBO J. 8:651-656.

148. MAROCCO, A., WISSENBACH, M., BECKER, D., PAZ-ARES, J., SAEDLER, H., SALAMINI, F., ROHDE, W. 1989. Multiple genes are transcribed in *Hordeum vulgare* and *Zea mays* that carry the DNA-binding domain of the Myb oncoprotein. Mol. Gen. Genet. 216:183-187.

149. JACKSON, D., CULIANEZ-MACIA, F., PRESCOTT, A., ROBERTS, K., MARTIN, C. 1991. Expression patterns of *myb* genes from *Antirrhinum* flowers. Plant Cell 3:115-125.

150. TABAK, A.J.H., SCHRAM, A.W., BENNINK, G.J.H. 1981. Modification of the B-ring during flavonoid synthesis in *Petunia hybrida*: effect of the hydroxylation gene *Hf1* on dihydroflavonol intermediates. Planta 153:462-465.

151. SCHRAM, A.W., TIMMERMAN, A.W., DE VLAMING, P., JONSSON, L.M.V., BENNINK, G.J.H. 1981. Glucosylation of flavonoids in petals of *Petunia hybrida*. Planta 153:459-461.

152. GERATS, A.G.M., CORNELISSEN, R.T.J., HOGERVORST, J.M.W., SCHRAM, A.W., BIANCHI, F. 1982. A gene controlling rate of anthocyanin synthesis and mutation frequency of the gene *Anl* in *Petunia hybrida*. Theor. Appl. Gen. 62:199-203.

153. DE VLAMING, P., SCHRAM, A.W., WIERING, H. 1983. Genes affecting flower colour and pH of flower limb homogenates in *Petunia hybrida*. Theor. Appl. Genet. 66:271-278.

154. SCOTT-MONCRIEFF, R.A. 1936. A biochemical survey of some mendelian factors for flower colour. J. Genet. 32:117-170.

155. HARBORNE, J.B. 1967. The anthocyanin pigments. In: Comparative Biochemistry of Flavonoids, Academic Press, London, pp. 1-30.

156. CORNU, A., FARCY, E., MAIZONNIER, D., HARING, M., VEERMAN, W., GERATS, A.G.M. 1990. *Petunia hybrida* (2n=14). In: Genetic Maps: Locus Maps of Complex Genomes. Book 6, Plants. (S.J. O'Brien, ed.) Cold Spring Harbor Lab. Press, fifth ed., pp. 6.113-6.124.

157. DE VLAMING, P., VAN EEKERES, J.E.M., WIERING, H. 1982. A gene for flower colour fading in *Petunia hybrida*. Theor. Appl. Genet. 61:41-46.

158. GERATS, A.G.M., BUSSARD, J., COE, E.H., LARSON, R. 1984. Influence of *B* and *Pl* on UDPG flavonoid 3-*O*-glucosyltransferase in *Zea mays* L. Biochem. Genet. 22:1161-1169.

159. GAVAZZI, G., MIKEREZI, I., PAPINUTTI, P., TONELLI, C. 1985. Light induced effects on tissue specific gene expression in *Zea mays* L. Maydica 30:309-319.

160. SCHROEDER, G., BROWN, J.W.S., SCHROEDER, J. 1988. Molecular analysis of resveratrol synthase. cDNA, genomic clones and relationship with chalcone synthase. Eur. J. Biochem. 172:161-169.

161. WAGNER, G.J. 1979. Content and vacuole/extravacuole distribution of neutral sugars, free amino acids and anthocyanin in protoplasts. Plant Physiol. 64:88-93.

162. WIERMANN, R., BUTH-WEBER, M. 1980. The distribution and localization of an UDP-Glucose:flavonol 3-*O*-glucosyl transferase activity in pollen. Protoplasma 104:307-313.

163. WIERMANN, R., VIETH, K. 1983. Outer pollen wall, an important accumulation site for flavonoids. Protoplasma. 118:230-233.

164. ROBERTSON, D.S. 1965. A dormant allele of *vp1*. Maize Gen. Coop. Newsl. 39:104.

165. ROBICHAUD, C.S., SUSSEX, I.M. 1986. The response of *viviparous-1* and wildtype embryos of *Zea mays* to culture in the presence of abscisic acid. J. Plant Physiol. 126:235-242.

166. GURUPRASAD, K.N., LALORAYA, M.M. 1980. Effect of pigment precursors on the inhibition of anthocyanin biosynthesis by GA and ABA. Plant Sci. Lett. 19:73-79.

167. HINDERER, W., PETERSEN, M., SEITZ, H.U. 1984. Inhibition of flavonoid biosynthesis by gibberellic acid in cell suspension cultures

of *Daucus carota* L. Planta 160:544-549.

168. BARENDSE, G.W.M., PEREIRA, R.A.S., BARKERS, P.A., DRIESSEN, F.M., VAN EYDEN, E.A., LINSKENS H.F. 1970. Growth hormones in pollen, styles and ovaries of *Petunia hybrida* and *Lilium* species. Acta Bot. Neerl. 19:175-186.

169. WEISS, D., HALEVY, A.H. 1989. Stamens and gibberellin in the regulation of corolla pigmentation and growth in *Petunia hybrida*. Planta 179:89-96.

170. WEISS, D., VAN TUNEN, A.J., HALEVY, A.H., MOL, J.N.M., GERATS, A.G.M. 1990. Stamens and gibberellic acid in the regulation of flavonoid gene expression in the corolla of *Petunia hybrida*. Plant Physiol. 94:511-515.

171. VAN DER KROL, A.R., LENTING, P.J., VEENSTRA, J.G., VAN DER MEER, I.M., KOES, R.E., GERATS, A.G.M., MOL, J.N.M., STUITJE, A.R. 1988. An antisense chalcone synthase gene in transgenic plants inhibits flower pigmentation. Nature 333:866-869.

172. VAN DER KROL, A.R., MUR, L.A., DE LANGE, P., GERATS, A.G.M., MOL, J.N.M., STUITJE, A.R. 1990. Antisense chalcone synthase genes in *Petunia*: visualisation of variable transgene expression. Mol. Gen. Genet. 220:204-212.

173. VAN DER KROL, A.R., MUR, L.A., BELD, M., MOL, J.N.M., STUITJE, A.R. 1990. Flavonoid genes in *Petunia*: addition of a limited number of gene copies may lead to a suppression of gene expression. Plant Cell 2:291-299.

174. VAN DER KROL, A.R., MUR, L.A., MOL, J.N.M., STUITJE, A.R. 1990. Inhibition of flower pigmentation by antisense CHS genes: promoter and minimal sequence requirements for the antisense effects. Plant Mol. Biol. 14:457-466.

175. NAPOLI, C., LEMIEUX, C., JORGENSEN, R. 1990. Introduction of a chimeric chalcone synthase gene into *Petunia*: results in reversible co-suppression of homologous genes *in trans*. Plant Cell 2:279-289.

176. JORGENSEN, R. 1990. Altered gene expression in plants due to *trans* interactions between homologous genes. TIBtech. 8:340-344.

177. BRAKENHOFF, G.J., VAN DER VOORT, H.T.M., OUD, J.L. 1990 Three dimensional image representation in confocal microscopy. In: Confocal Microscopy. (T.Wilson, ed.) Acad. Press, London, pp. 185-197.

178. NANNINGA, N., OUD, J.L. 1990. Analysis of chromosomes by CSLM. Eur. Microscopy and Analysis, pp. 23-25.

Chapter Seven

FLAVONOIDS: PLANT SIGNALS TO SOIL MICROBES

Donald A. Phillips
Department of Agronomy & Range Science
University of California, Davis, CA 95616

INTRODUCTION

Recent discoveries have established that some flavonoids and several simple phenolics function as transcriptional signals from plants to bacteria. These findings open an important new chapter for understanding plant—microbe interactions and offer a potentially powerful intellectual foundation for analyzing factors controlling rhizosphere ecology.

This new area of study is based on demonstrations that certain flavonoids from legumes induce transcription of nodulation (*nod*) genes in N$_2$-

Phenolic Metabolism in Plants, Edited by H.A. Stafford
and R.K. Ibrahim, Plenum Press, New York, 1992

fixing *Rhizobium*[1-3] and *Bradyrhizobium*[4] bacteria (rhizobia) and that acetosyringone (3',5'-dimethoxy-4'-hydroxyacetophenone) induces transcription of virulence (*vir*) genes in *Agrobacterium tumefaciens*,[5,6] a common plant pathogen. Discussion in this review will be limited to flavonoids, but a concise comparison of the two systems is available elsewhere.[7] Isoflavonoid phytoalexins are well established as plant compounds that are produced in response to microbes,[8,9] and like many phenolics,[7,10] they may induce catabolic microbial genes to promote their degradation in some bacteria. However, because there is little evidence that those compounds function as transcriptional signals, rather than as toxins or substrates for microbes, they will not be discussed here.

Traces of flavonoids in root exudate were first reported in 1944 for wheat.[11] That observation has not been ignored by rhizosphere biologists, but investigations have centered more on the total flux of organic compounds from roots[12] than on the identity and function of phenolics, which are present in exudates at low concentrations.[13,14] Various workers have reported flavonoids in root extracts for many years,[15] but accurate identification and quantification of flavonoids in root exudates is a recent accomplishment.[16] The subsequent discovery that some flavonoids are transcriptional signals to bacteria emphasizes the importance of those initial observations and is the primary subject of this review. Understanding how plants release flavonoid signals that affect microbes is a first step toward managing these molecular triggers in agricultural and natural ecosystems.

FLAVONOIDS AS INDUCERS OF RHIZOBIAL NODULATION GENES

Specificity of Rhizobia

Rhizobia traditionally have been grouped on the basis of the legume plants with which they form effective root nodules[17,18] (Table 1). Evolutionary relationships among legumes and rhizobia are extremely complex,[19] and interest in these symbioses as an agriculturally important model of plant—microbe interactions led to the discovery of new rhizobia.[20-27] Flavonoids are crucial components in the signaling process that determines these relationships because the flavonoid released by the plant interacts with an unknown, but somewhat specific, rhizobial receptor before the *nod* genes responsible for infection are transcribed. In some instances, incompatible interactions occur between closely related plants and a single rhizobial species. *R. meliloti*, for example, normally is viewed as the organism that nodulates alfalfa (*Medicago sativa*). Nevertheless, the host specificities of *R. meliloti* strains isolated from soils vary considerably on other leguminous species (Table 2).[28] These types of barriers to the effective symbiotic pairing of rhizobia and legumes can be important in agricultural systems and may result from ineffective signaling between those organisms. Understanding which flavonoids are released by a particular plant and how rhizobia perceive those signals may facilitate the establishment of desired *Rhizobium*—legume associations.

Table 1. Symbiotic N_2-fixing associations between rhizobial bacteria and plants. Whereas taxonomy of the *Rhizobiaceae* was clarified recently,[17,18] several new designations currently in use or proposed for use are listed.[20-27]

Bacterium	Representative Host	Notable Trait
Rhizobium leguminosarum		
biovar *viceae*	*Pisum sativum*	Fast-growing
biovar *trifolii*	*Trifolium repens*	Fast-growing
biovar *phaseoli*	*Phaseolus vulgaris*	Fast-growing
Rhizobium meliloti	*Medicago sativa*	Fast-growing
Rhizobium loti	*Lotus corniculatus*	Fast-growing
Rhizobium fredii	*Glycine max*	Fast-growing
(*Sinorhizobium fredii*)		
(*Sinorhizobium xinjiangensis*)		
Rhizobium sp. (*Galega*)	*Galega orientalis*	Fast-growing
Bradyrhizobium japonicum	*Glycine max*	Slow-growing
Bradyrhizobium sp. (*Vigna*)	*Vigna unguiculata*	Slow-growing
Bradyrhizobium sp. (*Lupinus*)	*Lupinus albus*	Slow-growing
Bradyrhizobium sp. (*Parasponia*)	*Parasponia rigida*	Non-legume host
Azorhizobium caulinodans	*Sesbania rostrata* nodules	Stem & root
Photorhizobium thompsonianum	*Aeschynomene indica*	'photosynthetic' stem nodules

Table 2. Morphological characteristics of root nodules formed on various plant species by *R. meliloti* strains isolated from Australian soils. Nodulation responses distinguished three groups of rhizobia on various *Medicago* (*M.*) and *Melilotus* (*Mel.*) species, and bacteria were grouped accordingly.[28]

Test Host	Bacterial Group		
	I	II	III
M. falcata, M. lupulina, M. praecox, M. rigidula, M. sativa, M. truncatula	Normal	Normal	Normal
M. littoralis, M. orbicularis, Melilotus alba, Mel. officinalis	Odd	Normal	Normal
M. arabica, M. minima	Odd/None	Odd/None	Normal
M. laciniata	Normal	Normal	Odd/None

Nodulation Genes

Formation of root nodules by rhizobia requires expression of bacterial nodulation (*nod*) genes. Most rhizobia contain a modest number of *nod* genes, such as the 13 described recently in *R. leguminosarum* bv. *viceae*.[29] Rhizobial *nod* genes have been studied extensively by modern molecular techniques, and their complexity has been addressed in recent reviews.[30-32] Three simplified categories include host-range *nod* genes, regulatory *nodD* genes, and the highly conserved "common" *nod* genes *nodA, B,* and *C,* which in many rhizobia are contiguous in a single operon. Mutations in *nodA, B,* or *C* prevent root nodule formation, but such Nod⁻ phenotypes can be corrected with the corresponding gene from other rhizobial species.[33-36]

The biochemical functions of gene products from *nodABC* in *R. meliloti* have not been defined, but they operate together to produce NodRm-1, a sulfated β-1,4-tetrasaccharide of D-glucosamine in which three amino groups are acetylated and one is acylated with a C_{16} fatty acid.[37] Nanomolar concentrations of NodRm-1 deform root hairs on alfalfa, and mimic the morphological response of those cells to the actual presence of *R. meliloti*. A strain of *R. meliloti* mutated in *nodH* produces NodRm-2, a molecule similar to NodRm-1 except that it lacks the sulfate.[38] NodRm-2 has no effect on alfalfa root hairs, but it does deform root hairs on vetch (*Vicia sativa* subsp. *nigra*), a plant which is not infected by *R. meliloti*. NodRm-1 and NodRm-2 may facilitate nodulation by binding to plant lectins,[37] which control specific *Rhizobium* plant interactions,[39] but that possibility remains to be proven. Taken as a whole, these findings indicate that transcription of *nodABC* is crucial for normal rhizobial infection, and thus controlling expression of these genes is a biologically significant process.

The protein product of the *nodD* gene is an important factor that regulates transcription of the *nodABC* operon by interacting with flavonoids. Rhizobia normally have one *nodD* gene, which is contiguous to *nodABC*, but many rhizobial species have additional *nodD* genes.[40-44] Most rhizobial genomes show physical homology to *nodD* at one or two loci,[45] but *R. meliloti*[40,41,43] and *R.l.* bv. *phaseoli*[44] are especially interesting because each has three different *nodD* genes, *nodD1*, *nodD2*, and *nodD3*. In *R. meliloti*, successive mutation of each *nodD* gene slows the rate of root nodule formation on alfalfa, and bacteria mutated in all three *nodD* genes fail to produce nodules.[41]

The mechanism by which *nodD* genes control transcription of *nodABC* is unclear. One working hypothesis is that the protein product of the *nodD* gene (NodD) binds to conserved sequences (*nod* boxes) that precede *nod* genes and probably function as transcriptional control regions. Flavonoids then "activate" the NodD protein into a configuration that allows *nod* gene transcription. Although gel retardation studies suggest that flavonoids bind to the NodD protein,[46] there is as yet no evidence that flavonoids affect binding of the NodD protein to DNA.[31] Most *nodD* genes are constitutive, but in several cases flavonoids enhance their transcription.[47,48]

The First Flavonoid *nod* Inducers

Beginning in 1985, several research groups showed that, in addition to an active *nodD* gene, transcription of *nodABC* required an unknown plant factor.[49-53] Those results were based on the use of a reporter gene, *lacZ*, whose transcription can be detected easily by assaying for β-galactosidase activity. Rhizobia containing genetic constructions such as that shown in Figure 1 produced β-galactosidase in the presence of crude plant extracts and thus demonstrated that *nod* genes were being transcribed.

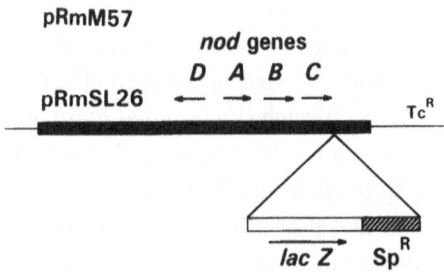

Fig. 1. One reporter gene system that helped identify plant flavonoids which induce *nod* genes in *Rhizobium* bacteria.[50] Plasmid pRmM57 was produced by inserting a gene responsible for β-galactosidase synthesis (*lacZ*) from *Escherichia coli* bacteria into the *nodC* gene of *R. meliloti* on the plasmid pRmSL26. Active flavonoids function with the product of a regulatory *nodD* gene to induce transcription of *nodABC-lacZ*, which is measured as β-galactosidase activity. Additional *nodD* genes were supplied on multiple copies of a separate plasmid to maximize the sensitivity of the assay system. Genetic markers: Tc[R]-tetracycline resistance; Sp[R]-spectinomycin resistance.

Flavonoids that were active *nod*-gene inducers in *Rhizobium* and *Bradyrhizobium* were quickly described (Table 3). Luteolin (5,7,3',4'-tetrahydroxyflavone) from alfalfa and 7,4'-dihydroxyflavone from white clover were identified as *nod*-gene inducers in *R. meliloti*[1] and *R.l.* bv. *trifolii*,[2] respectively, by UV/visible spectra, ^{1}H-NMR data, and MS measurements. For pea, apigenin-7-*O*-glucoside (5,7,4'-trihydroxyflavone-7-*O*-glucoside) and eriodictyol (5,7,3',4'-tetrahydroxyflavanone) were implicated as *nod*-gene inducers in *R.l.* bv. *viceae*,[3] while daidzein (7,4'-dihydroxyisoflavone) and genistein (5,7,4'-trihydroxyisoflavone) were identified from soybean as inducers in *B. japonicum*[4] and *R. fredii*[54] by UV/visible absorbance, ^{1}H-NMR, and MS data. Several other flavonoids, such as geraldone (7,4'-dihydroxy-3'-methoxyflavone) and 4'-hydroxy-7-methoxyflavone were identified as less active, natural *nod*-gene inducers.[2]

The concept that structural variations in flavonoids released from different legumes affect rhizobial *nod* gene transcription and determine their host plant specificity through interactions with NodD proteins is supported by at least

Table 3. Initial *nod*-gene inducing flavonoids identified from legumes.[1-4]

Plant Source	Compound(s)*
Alfalfa seed extract	Luteolin
Pea seed rinse	Apigenin-7-*O*-glucoside
	Eriodictyol
Clover seedling extract	7,4'-Dihydroxyflavone
	Geraldone
	4'-hydroxy-7-methoxyflavone
Soybean seedling extract	Genistein
	Daidzein

*Factors responsible for *nod*-gene-inducing activity detected originally in seed, root, or seedling exudate were identified by purifying biologically active compounds from the indicated plant sources.

two lines of evidence. First, when *nodD* genes from *R.l.* bv. *viceae*, *R.l.* bv.*trifolii*, and *R. meliloti* were cloned into a single *Rhizobium* strain lacking its normal symbiotic plasmid, they still induced the highest levels of *nodABC* transcription in the presence of flavonoids found in their normal host plant.[55] Second, the *nodD* gene product from a strain that nodulates siratro (*Macroptilium atropurpureum*), but not alfalfa, was activated by siratro seed flavonoids after being transferred to *R. meliloti*.[53]

Genetic analyses of *nodD* offer an additional indication that flavonoids act through the NodD gene product. Mutations in *nodD* of *R.l.* bv. *viceae* produced four different phenotypic classes, one of which induced *nod* gene transcription in the absence of flavonoids.[56] Other studies with a hybrid *nodD* gene containing 75% of the *nodD1* gene of *R. meliloti* and 27% of the *nodD* gene of *R.l.* bv. *trifolii* showed that the modified NodD protein activated inducible *nod* gene promoters constitutively without flavonoids.[57] Interestingly, the hybrid mutant strain produced higher rates of apparent N_2 fixation (acetylene reduction activity) than a control strain on white clover. Thus natural or induced changes in the *nodD* gene product can influence how flavonoids from different host plants induce rhizobial genes responsible for the initial stages of root nodule formation, and the NodD protein may be involved also in subsequent functioning of the root nodule.

As soon as it was established that some flavonoids are natural *nod*-gene inducers, the molecular specificity of rhizobia for flavonoids was tested by supplying numerous commercially available compounds.[1,3,58-61] The role of aromatic substitutions was emphasized from the very beginning by *nodC-lacZ* assays in *R. meliloti* which showed that luteolin was an active inducer but flavone was inactive.[1] Tests with *nodA-lacZ* fusions in rhizobia containing *nodD* from *R.l.* bv. *viceae* showed that flavone was inactive and 7-hydroxy-flavone produced about half the maximum induction obtained with luteolin, apigenin, naringenin (5,7,4'-trihydroxyflavanone), and eriodictyol.[52] Those types of studies showed that a C-7 hydroxylation (flavone numbering system) contributed to activity in most systems,[60] but the availability of different NodD proteins and varying levels of inducing activity permitted no global conclusions about structure-activity relationships in flavonoids. One concept from those studies which has been supported by later work with compounds actually released from legume plants is that a weak *nod*-gene inducer can attenuate the activity of a stronger inducer.[58] That phenomenon was demonstrated clearly for *R.l.* bv. *trifolii* with the strong *nod*-gene inducer 7,4'-dihydroxyflavone and chrysin (5,7-dihydroxyflavone), a weak inducer not found in white clover.[58] The *nod*-gene inhibitors formononetin (7-hydroxy-4'-methoxyisoflavone) and umbelliferone (7-hydroxycoumarin), which were indicated as being present in clover root exudates,[58] may serve an inhibitory function under natural conditions, but clear quantification of those compounds in the exudates is required to assess that possibility. The limitations of such studies were noted by workers who stated that "It is important to understand the synthesis and exudation of both inducing and inhibitory compounds in the developing root".[59]

NATURAL RELEASE OF FLAVONOIDS FROM SEEDS AND ROOTS

After it was established that some plant flavonoids are transcriptional signals to rhizobia (Table 3), further work was required to relate that concept to growing plants. It was not known, for example, whether flavonoids extracted from legume seeds with 2-butanone,[62] also are released into aqueous solution during germination. Studies that reported flavones and isoflavones in root exudates of lentil (*Lens culinaris*) and soybean (*Glycine max*)[14,16] supplied an early stimulus to this field. Those experiments, which were done under sterile conditions, identified 7,4'-dihydroxyflavone, 7,3',4'-trihydroxyflavone, and 7,4'-dihydroxy-3'-methoxyflavone in lentil root exudates, while coumestrol and daidzein were found in soybean root exudate.[16] D'Arcy-Lameta completed her work before it was known that flavonoids induce rhizobial *nod* genes, but other workers independently demonstrated that 7,3',4'-trihydroxyflavone is a moderate *nod*-gene inducer in *R.l.* bv. *viceae*,[3] the organism that nodulates lentil.

Studies with sterile alfalfa seedlings show clearly that transmission of *nod*-gene-inducing signals to *R. meliloti* involves more than the release of luteolin. Thus far, five flavonoids have been identified as natural *nod*-gene inducers in alfalfa exudates using UV/visible spectroscopy, ^{1}H-NMR, MS, and HPLC cochromatography with authentic standards (Table 4).[63,64] Although luteolin, a *nod*-gene inducer known from previous work,[1] was present in seed rinses,[64] it was not found in root exudate.[63] A *nodC-lacZ* fusion (Fig. 1) facilitated this work by detecting active molecules at far lower concentrations than would have been possible with normal analytical methods. The 4',4-dihydroxy-2-methoxychalcone identified in root exudate by this approach was 10-fold more powerful than luteolin as measured by half-maximum induction (I_{50}) of *nodC-lacZ* (Fig. 2).[63] The 7,4'-dihydroxyflavone and liquiritigenin (7,4'-dihydroxyflavanone) in root exudates had slightly higher I_{50} values than luteolin, and the maximum β–galactosidase activity (I_{max}) produced was considerably lower for both. Comparing I_{50} values for luteolin and chrysoeriol (3'-methylluteolin) indicates that chrysoeriol is three- to four-fold more active than luteolin.[64] The significance of different I_{max} values is difficult to interpret, but a lower I_{50} value presumably reflects a greater affinity of the flavonoid for binding to the NodD protein. As biochemical mechanisms underlying these genetic phenomena are described in more detail, the validity of this interpretation can be tested. A major difference between luteolin and 4',4-dihydroxy-2'-methoxychalcone is that the latter can activate protein products from both *nodD1* and *nodD2*, whereas luteolin activates only NodD1 protein.[65] A NodD2-activating molecule reported by other investigators in crude alfalfa seed rinse[66] has been detected in the seed rinse fraction with compounds listed in Table 4, and attempts to identify the molecule are progressing. A notable trait of this compound, which may be important in soil, is its high solubility in both water and methanol (C.M. Joseph, C.A. Maxwell and D.A. Phillips, unpublished data).

New studies of *nod*-gene-inducing flavonoids released by common bean under sterile conditions show this legume differs considerably from alfalfa (Table 5).[67,68] One of the few traits common to both plants is that seeds and roots release different flavonoids (Tables 4 and 5). The black-seeded bean used for these studies released much larger amounts of the active factors from both seeds

Table 4. Important flavonoids released into aqueous solution by sterile alfalfa (*Medicago sativa*) seeds and roots.[63-65,74,89]

Alfalfa Flavonoid*	nodD Product Activated	Amount pmol/plant/hr
Seed Compounds		
Luteolin	NodD1	70
Chrysoeriol	NodD1	10
5,3'-Dimethylluteolin	—	300
Luteolin-7-*O*-glucoside	—	800
5-Methylluteolin	—	825
Quercetin-3-*O*-galactoside	—	1,800
Root Compounds		
4',4-Dihydroxy-2'-methoxychalcone	NodD1 & NodD2	1
7,4'-Dihydroxyflavone	NodD1	2
Liquiritigenin	NodD1	1
Formononetin (Stressed Roots)	—	12

*Aglycones identified by UV/visible spectral shifts, ^{1}H-NMR studies, and MS data were confirmed by chemical and biological comparisons with authentic standards. The nod-gene-inducing activity was assayed in *R. meliloti* containing extra copies of nodD1 or nodD2 with a nodC-lacZ fusion.

(--) = no activity with either nodD1 or nodD2.

and roots than alfalfa, but most compounds from bean require higher concentrations to induce nod-gene transcription than those from alfalfa. For example, the lowest and highest I_{50} values measured with extra copies of nodD1 from the appropriate rhizobia were as follows: bean—genistein 200 nM, myricetin 1,400 nM; alfalfa—4',4-dihydroxy-2'-methoxychalcone 1 nM, liquiritigenin 20 nM. While luteolin activates NodD1, but not NodD2, protein in *R. meliloti*, the bean flavonoids activate all three nodD gene products in *R.l.* bv. *phaseoli*. Tests with aglycones from the bean flavonoids, showed that the mean

I_{50} value of active compounds in the root exudate was significantly lower with each of the three *nodD* genes in *R.l.* bv. *phaseoli* than the mean I_{50} value of active inducers in the seed rinse. The compound in bean root exudate that contributes most to that activity is genistein, which has the lowest I_{50} and highest I_{max} values of any *nod*-gene inducer identified in that system. Thus the strongest *nod*-gene-inducing flavonoids yet identified from both alfalfa and bean are released from roots, not seeds. Such quantitative comparisons must be qualified by the fact that they may not be representative of all rhizobia. For example, naturally occurring mutations in *nodD* genes could dramatically alter flavonoid-NodD interactions just as laboratory mutations do.[56,57]

Root exudates of vetch (*Vicia sativa* subsp. *nigra*) also have been studied under sterile conditions for the presence of *nod*-gene-inducing flavonoids.[52,61,69] Seven compounds released from this species activated a *nodD* gene from *R.l.* bv. *viceae*. On the basis of UV/visible spectra and ^{1}H-NMR studies, one compound was identified as 3,5,7,3'-tetrahydroxy-4-methoxy-flavanone, while another probably was 7,3'-dihydroxy-4'-methoxyflavanone (Table 6).[61] Unfortunately, no authentic standards were available for comparisons. Subsequent experiments using the same plants with an agar rooting medium rather

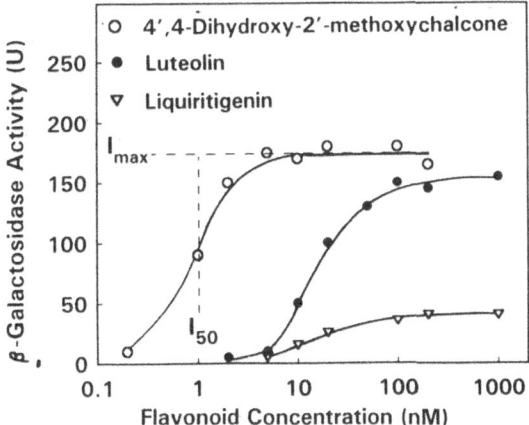

Fig. 2. Induction of *nodC-lacZ* fusion plasmid pRmM57 in *R. meliloti* by three flavonoids released naturally from alfalfa roots (4',4-dihydroxy-2'-methoxychalcone, liquiritigenin) or seeds (luteolin). Data show effects of flavonoid structure on the maximum *nod*-gene induction (I_{max}) and on the concentration required for half-maximum induction (I_{50}).[63]

Table 5. Major *nod*-gene-inducing flavonoids released into aqueous solution by sterile bean (*Phaseolus vulgaris*) seeds and roots.[67,68]

Aglycone*	Natural Form	*nodD* Product Activated	Amount nmol/plant/hr
Seed Compounds			
Delphinidin	3-*O*-glycoside	NodD1,D2,D3	8
Petunidin	3-*O*-glycosides	NodD1,D2,D3	146
Malvidin	3-*O*-glycosides	NodD1,D2,D3	229
Myricetin	3-*O*-glycoside	NodD1,D2,D3	28
Quercetin	3-*O*-glycosides	NodD1,D2,D3	26
Kaempferol	3-*O*-glycosides	NodD1,D2,D3	20
Root Compounds			
Genistein	7-*O*-glycoside	NodD1,D2,D3	2
Eriodictyol	aglycone	NodD1,D2,D3	12
Naringenin	aglycone	NodD1,D2,D3	14

*Aglycones identified by UV/visible spectral shifts, ^{1}H-NMR studies, and MS data were confirmed by chemical and biological comparisons with authentic standards. The *nod*-gene-inducing activity was assayed in *R.l.* bv. *phaseoli*. All aglycones and their natural glycoside(s) activated each of the three NodD proteins produced by *R.l.* bv. *phaseoli* (M. Hungria, unpublished data).

than a liquid system found only two *nod*-gene-inducing flavonoids released from roots under sterile conditions, and no chemical data for those two compounds were reported.[70]

Seed rinses from *Vicia sativa* subsp. *nigra* have not been studied for *nod*-gene-inducers, but *V. faba* reportedly releases 7,4'-dihydroxyflavone, 7,3',4'-trihydroxyflavone, a 7-*O*-glycoside of quercetin (3,5,7,3',4'-pentahydroxy-flavone), and a 7-*O*-glycoside of kaempferol (3,5,7,4'-tetrahydroxyflavone) on the basis of UV/visible shift reagent studies and chromatographic comparisons with authentic standards.[71] Recent chromatographic studies of seed coat extracts from *V. faba* seeds of eight different colors concluded that a number of flavonoid aglycones, including myricetin (3,5,7,3',4',5'-hexahydroxyflavone), kaempferol, quercetin, apigenin, cyanidin (3,5,7,3',4'-pentahydroxyflavylium), delphinidin (3,5,7,3',4',5'-hexahydroxyflavylium), petunidin (3,5,7,4',5'-pentahydroxy-3'-methoxyflavylium), and malvidin (3,5,7,4'-pentahydroxy-3',5'-dimethoxyflavylium) were present.[72] However, the *nod*-gene-inducing activity of these compounds has not been assessed with *R.l.* bv. *viceae*.

PLANT CONTROLS OVER FLAVONOID SIGNALS

Because release of any compound from plants involves at least an energy cost associated with synthesis and possibly a cost for storage and/or exudation, it is reasonable to expect that plants regulate production of *nod*-gene-inducing flavonoids. Such controls relating to flavonoid synthesis, storage, exudation, or modification after exudation may operate at the genetic and/or biochemical level. Clear differences in structure (5-hydroxy vs. 5-deoxy; Table 4) and time of release for seed and root *nod*-gene-inducing flavonoids in alfalfa suggest the release of these compounds from the two organs may be controlled by dissimilar factors.

Apparently *nod*-gene-inducing flavonoids released from alfalfa seeds exist before germination as preformed molecules which can be modified by both plant and bacterial enzymes. Surgical experiments showed that those compounds are released from the seed coat, not from the cotyledons, and that rinsing seeds briefly with 50% methanol removed the same flavonoids that were released naturally into the aqueous solution during germination.[73] The simplest interpretation of those results is that *nod*-gene-inducing flavonoids released by alfalfa seeds are deposited on the seed coat while the seed is maturing and that synthesis during germination is not required for their release. One of the major flavonoids on the seed coat, luteolin-7-*O*-glucoside, is not itself an active *nod*-gene inducer (Table 4), but it can be hydrolyzed to luteolin by extracellular β-glucosidase activity exuded from the seed and by *R. meliloti* cells.[73] Thus the luteolin-7-*O*-glucoside apparently serves as a reservoir of *nod*-gene-inducing activity on the seed coat.

Alfalfa roots, by contrast, link synthesis closely to the release of *nod*-gene-inducing flavonoids. When roots of intact seedlings were incubated with ^{14}C-phenylalanine in the presence of the phenylalanine ammonia-lyase (PAL) inhibitor L-2-aminooxy-3-phenylpropionic acid (AOPP), synthesis and release of 4',4-dihydroxy-2'-methoxychalcone were decreased by more than 90% relative to AOPP-free controls.[74] Similar results were obtained for 7,4'-dihydroxyflavone, but internal pools of liquiritigenin glycosides allowed AOPP-treated roots to release this flavanone at about 50% of the normal rate.

The site of synthesis and/or release of *nod*-gene-inducing flavonoids from legume roots is poorly understood at the cellular level. Recent studies, however, have related these processes to particular root zones or tissues. When 3-day-old alfalfa seedlings were placed on agar plates covered by *R. meliloti* containing pRmM57 with its *nodC-lacZ* fusion (Fig. 1), spraying the root with 4-methylumbelliferyl-β-D-galactoside (MUG) revealed β-galactosidase activity was present only in bacteria surrounding the root tip and in the region of young root hairs.[59] Appropriate controls supported the interpretation that the fluorescent breakdown product of MUG was produced by induction of the bacterial *nodC-lacZ* fusion. Detailed studies with soybean roots showed that cotton string placed directly on the root tip collected more daidzein than string touching tissue 1 cm behind the tip.[75] Those observations suggest that some *nod*-gene-inducing flavonoids may be excreted from cells that are released from the rootcap;[76] additional studies should clarify that possibility.

Metabolic studies are consistent with the concept that *nod*-gene-inducing flavonoids released from alfalfa roots are synthesized in that organ. When excised roots, hypocotyls, and cotyledons from alfalfa seedlings were exposed to ^{14}C-phenylalanine, all three tissues released ^{14}C-labeled molecules of the normal *nod*-gene-inducing flavonoids exuded from intact seedlings.[74] The relative proportions of the three ^{14}C-labeled compounds released from excised alfalfa roots, however, were similar to those measured from intact roots. Although excised hypocotyls synthesized flavonoids at much higher rates than excised roots, the above rates for roots indicate that transport of flavonoids from hypocotyls to the roots is not required to account for the ^{14}C-phenylalanine incorporation into *nod*-gene inducers released from roots of intact seedlings.[74]

New data indicate that the presence of rhizobial cells around the root can alter how plants synthesize and/or release *nod*-gene-inducing flavonoids. Studies with soybeans showed that roots inoculated with *B. japonicum* contained additional daidzein and genistein.[77] Experiments with vetch indicate the presence of *R.l.* bv. *viceae* bacteria strongly affects the release of *nod*-gene-inducing flavonoids into root exudate.[70,78] In the presence of rhizobia, vetch roots released six flavanones and two chalcones that induce *nod* genes in *R.l.* bv. *viceae* (Table 6). Only two unidentified *nod*-gene-inducing flavonoids were released from sterile vetch roots in these studies, and thus the relationship of these results to the seven *nod*-gene inducers reported previously for this same species under sterile conditions[61] is unclear.

Whether the new *nod*-gene inducers released in the presence of rhizobia also regulate other, presently unknown, bacterial genes remains to be determined.

Another point that should be established in these types of experiments is whether additional flavonoids that do not induce *nod* genes are released by the plant in response to rhizobia. If such flavonoids are released from legume roots in response to rhizobia, they could be part of the well-known plant response to microbes and other elicitors which results in the release of isoflavonoid phytoalexins.[9] Initial experiments with alfalfa roots suggest that *R. meliloti* stimulates a release of both *nod*-gene inducers and other flavonoids, which do not activate *nodD1* (F.D. Dakora and D.A. Phillips, unpublished data). Whether flavonoids released by vetch or alfalfa roots in response to rhizobia are produced by new synthesis or by hydrolysis of stored conjugates remains to be determined.

The concept that plants may control the level of *nod*-gene-inducing activity in the rhizosphere by releasing inhibitors[3,58] was further supported by the demonstration that compounds released naturally from alfalfa roots interact to affect *nodC-lacZ* transcription in vitro.[79] Those data indicate that the weak inducers, 7,4'-dihydroxyflavone and liquiritigenin, present in root exudate can decrease the effectiveness of the very strong *nod*-gene inducer 4',4-dihydroxy-2'-methoxychalcone. Thus, it is conceivable that legumes can alter the effectiveness of a strong inducer in the rhizosphere by releasing weaker inducers which would compete for binding to NodD proteins.

Table 6. Major *nod*-gene-inducing flavonoids identified from vetch (*Vicia sativa* subsp. *nigra*) root exudates in the absence[61] (-) or presence[70] (+) of *R.l.* bv. *viceae*.

Compound	Rhizobium
3,5,7,3'-Tetrahydroxy-4'-methoxyflavanone	—
7,3'-Dihydroxy-4'-methoxyflavanone	±
2',4',4-Trihydroxychalcone	+
4',4-Dihydroxy-2'-methoxychalcone	+
Naringenin	+
Liquiritigenin	+
7,4'-Dihydroxy-3'-methoxyflavanone	+
5,7,4'-Trihydroxy-3'-methoxyflavanone	+
5,7,3'-Trihydroxy-4'-methoxyflavanone	+

OTHER MICROBIAL PROCESSES AFFECTED BY FLAVONOIDS

Reports indicate that low concentrations of flavonoids can affect microbial processes other than root nodule formation. The small quantities of flavonoids released by roots (Tables 4 and 5) occur in the exudate together with higher concentrations of sugars and amino acids. While the latter compounds serve as easily metabolized carbon substrates,[80] flavonoids in root exudates act as plant signals to microbes. The microbial responses to these nanomolar or micromolar concentrations of flavonoids represent reactions that generally increase the potential for subsequent plant-microbe contact. The major reactions to flavonoids demonstrated thus far include effects on chemotaxis, growth, and development of bacteria and fungi.

Chemotaxis

R. meliloti,[81] *R.l.* bv. *viceae*,[82] and *R.l.* bv. *phaseoli*,[83] all show positive chemotaxis to some flavonoids. In each case, the responses were evident at concentrations much lower than those required for chemotactic effects of carbohydrates or amino acids, and only certain flavonoids were active. In *R.l.* bv. *phaseoli*, 10 μM apigenin and 10 μM luteolin were strong chemoattractants, but naringenin had no effect.[83] *R.l.* bv. *viceae* moved toward 5 μM naringenin, 10 μM kaempferol, and 50 μM apigenin, but hesperitin (5,7,3'-trihydroxy-4'-methoxyflavanone) produced no chemotaxis.[82] *R. meliloti* appeared considerably more sensitive to flavonoids because it gave a positive chemotactic response to 10 nM luteolin, but it did not respond to either apigenin or naringenin.[81] Although chromosomal genes in rhizobia are involved in chemotactic responses to carbon substrates,[84] the symbiotic plasmid, pSym,[82] or more specifically *nodA*, *nodC*, and *nodD*,[81] which are carried on pSym, contribute to the flavonoid-induced chemotaxis. Conversely, *R. meliloti* strains which were nonchemotactic to amino acids and sugars because of chromosomal gene mutations still responded to the presence of roots and root tip extracts.[85] Recent results show that *R. meliloti* also is chemotactic to 7,4'-dihydroxyflavone (W.D. Bauer, personal communication). That observation suggests rhizobia capable of nodulating alfalfa move toward both the seed zone (in response to luteolin) and up the gradient of 7,4'-dihydroxyflavone exuded from the root (Table 4). Chemotaxis studies with flavonoids thus far have involved rhizobia because of the proven effects of flavonoids on *nod* gene transcription. Although *Agrobacterium tumefaciens* is attracted to 10 nM luteolin,[81] there are few data on how flavonoids affect other soil microbes. It seems probable that published reports showing positive chemotactic effects of total, unfractionated root exudates[86,87] reflect the presence of flavonoids in those solutions, and as new studies are conducted with known concentrations of particular compounds, a general role for flavonoids as chemoattractants in soil may become evident.

Growth Rate

Low micromolar concentrations of flavonoids have been reported to stimulate growth rate of rhizobia in two laboratories.[88,89] In the case of *B. japonicum*, one of two strains studied grew faster with 1 μM daidzein, but not 10 μM.[88] Results with *R. meliloti* showed that 2 to 5 μM luteolin and quercetin increased growth rate when added to a defined minimal medium.[89] The measured increases in optical density, viable cell counts, and total protein for *R. meliloti* were reproducible when cells were growing slowly on the defined medium, but no flavonoid effect on growth rate was detected when cells were growing more rapidly on a complex medium. The flavonoid promotion of growth rate was not affected by the concentrations (0.25 to 10 $g \cdot L^{-1}$) or types (mannitol or glucose) of C substrates tested, and it occurred in *R. meliloti, R.l.* bv. *trifolii*, and *Pseudomonas putida*, but not *Bacillus subtilis*, or *Agrobacterium tumefaciens*. Apparently a 5,7-dihydroxy substitution pattern is required for the observed effect, because chrysin promoted growth as much as luteolin or quercetin, but neither flavone, 5-hydroxyflavone nor 7-hydroxyflavone had any impact on bacterial growth.[89] These structural requirements were supported by tests that showed none of the *nod*-gene-inducing flavonoids in alfalfa root exudate (Table 4) promoted microbial growth. Quantification of these low concentrations of luteolin in the medium over time showed no evidence that the compound was metabolized, and thus it was concluded that flavonoids probably enhance growth through a regulatory event in *R. meliloti*. The process apparently does not involve *nodD* genes because a strain of R. *meliloti* with mutations in the three *nodD* genes still grew faster with quercetin, which does not induce *nodABC* transcription in this bacterium. One importance of these data lies in the fact that most microbes are generally believed to be either resting or growing at less than optimum rates in soil. Thus growth factors from plants could favor development of certain bacterial populations.

Mycorrhizal Fungi

Some of the most striking evidence that flavonoids may be important signals from plants to soil microbes other than symbiotic rhizobia comes from recent work with mycorrhizal fungi. Root exudates and extracts from pine (*Pinus sylvestris*) enhance spore germination of several *Suillus* species that form ectomycorrhizal associations with that tree, and the effect has been attributed to a diterpene resin, abietic acid.[90] Other workers have documented the promotive effects of root exudates on in vitro spore germination and/or hyphal growth of vesicular-arbuscular mycorrhizal (VAM) fungi,[91-93] and recently it was shown that commercially available flavonoids promote spore germination of the VAM fungus *Gigaspora margarita*.[94] It was not surprising, therefore, when tests showed that some flavonoids released naturally from alfalfa seeds and roots promoted *in vitro* spore germination of two *Glomus* species that form mycorrhizae on alfalfa.[95] The notable point was that quercetin-3-*O*-galactoside and 7,4'-dihydroxyflavone, which are the most abundant flavonoids in seed and root exudates of healthy, sterile alfalfa seedlings (Table 4), were highly active inducers of *Glomus* spore germination. Luteolin-7-*O*-glucoside and liquiritigenin also enhanced spore germination significantly at 2.5 µM concentrations, but 4',4-dihydroxy-2'-methoxychalcone had no effect, and formononetin significantly inhibited germination. Thus the phenomenon was not a general response to all flavonoid structures. Although 2.5 µM quercetin also produced significant increases in hyphal growth and branching in vitro after 21 days, those responses may have resulted from the four-fold increase in spore germination after 3 days rather than a separate effect of quercetin on hyphal development.[95] Whether alfalfa plants exposed to *Glomus* produce additional flavonoids, as they do when *R. meliloti* cells are present, is being investigated. As various flavonoid factors are identified, some combination of them may allow development of the pure *in vitro* culture methodologies for *Glomus* which are crucial to advance fundamental studies of VAM fungi.

AGRICULTURAL SIGNIFICANCE OF FLAVONOID SIGNALS

The discovery that flavonoids are plant signals to microbes has important implications for understanding and managing soil ecosystems. The induction of bacterial genes in rhizobia by plant flavonoids may be only one of many cases in which plant compounds influence activities of soil microbes. Current evidence suggests a similar phenomenon occurs in VAM fungi.[92-95] Because degradation of organic matter by microbes and their activities in mineral nutrient cycles are required for soil formation and plant nutrition, significant effects of flavonoids on growth of bacteria or fungi could be crucial factors for maintaining soil ecosystems on which agriculture is based. A broad view of flavonoids as plant signals to soil microbes recognizes that because plant genes are responsible for producing flavonoids, all the agronomic tools available for altering those genes and for managing plant species that release certain flavonoids from living or dead roots are available for modifying the flavonoid profile of soils.

Two research groups have reported that adding flavonoids to the rhizosphere of alfalfa increased nodulation and N_2 fixation by *R. meliloti* under controlled conditions. In one example, 10 μM luteolin had a positive effect on Hairy Peruvian alfalfa;[96] in the other case 10 μM naringenin produced a similar response in Rijka T9 alfalfa.[97] Presumably those results indicate that, under the conditions of the experiments, the natural flavonoids released from seeds and roots of those alfalfa cultivars limited nodulation. Thus adding more flavonoids may have enhanced nodulation by promoting growth of *R. meliloti* and by inducing *nod* genes in a larger proportion of the rhizobial population. However, attempts to extend these observations to other alfalfa cultivars by adding luteolin or 4',4-dihydroxy-2'-methoxychalcone to rhizospheres have not increased nodulation (Y. Kapulnik, personal communication). In addition to our present understanding as to which flavonoids are released by alfalfa and how those compounds affect *R. meliloti*, more detailed studies to determine when flavonoids limit nodulation are needed to clarify the basis of these apparently conflicting results.

A major limitation to the use of genetically improved rhizobial strains is the presence of highly competitive, but poorly effective, indigenous rhizobia.[98] Data available for *B. japonicum* show that not all strains nodulating soybean respond identically to flavonoids. The *nod* genes in *B. japonicum* USDA 110, for example, were induced by daidzein in the presence of chrysin and naringenin, while the daidzein effect on other strains was greatly reduced by those compounds.[99] The possibility that nodulation by the outstanding strain USDA 110 would be favored by adding chrysin and naringenin to the rhizosphere, however, was not supported by competition experiments involving indigenous and inoculated rhizobia.[99] Other workers, who tested approximately 1,000 compounds for inhibition of genistein-induced transcription of *nodY-lacZ* in *B. japonicum*, identified 7-hydroxy-5-methylflavone as a very strong inhibitor.[100] Complete inhibition of *nod*-gene induction by 0.3 μM genistein was produced with 0.2 μM 7-hydroxy-5-methylflavone in the bacterial strain used as a screening tool, but subsequent experiments showed that, regardless of the inhibitor tested, some less desirable soil rhizobia always transcribed *nod* genes. Thus both studies concluded that natural variation among *B. japonicum* probably will prevent chemical inhibitors of *nod*-gene induction from being commercially useful.[99,100] Additional problems could evolve as rhizobia in the soil develop resistance to the inhibitor by mutating *nodD*. An alternative method for favoring nodulation of soybean by certain strains of *B. japonicum* may involve the isoflavonoid-inducible *nolA* gene, which allows rhizobia to nodulate certain soybean genotypes.[101] Whether that genotype-specific nodulation is based on the presence of unique flavonoids has not been reported.

FUTURE RESEARCH DIRECTIONS

Future research on plant flavonoid signals to microbes undoubtedly will advance in several directions. First, molecular genetic studies will establish mechanisms by which flavonoids interact with NodD proteins and possibly other gene products. Whether *nodD* genes are part of a larger group of regulatory proteins[102] that respond to small molecules such as flavonoids is a critically important question for soil microbiology and agriculture. Second, new studies will clarify how particular flavonoids are synthesized and released from living and dead plant roots. Third, data collected on flavonoids in soil will determine whether these compounds persist at physiologically active (i.e. µM) concentrations without inducing catabolic genes in microbes. One example of a flavonoid that has been identified by HPLC, MS and ^{1}H-NMR in soil extracts is medicarpin (3-hydroxy-9-methoxypterocarpan).[103] Ten mg of medicarpin was extracted from each kg of soil, but percent recovery was not estimated, and because of its hydrophobic nature, the actual concentration of medicarpin in soil solution could not be calculated. The availability of reporter genes (Fig. 1) should help measure low concentrations of certain flavonoids in soil. Determining K_m values for potential inactivating enzymes, such as peroxidases,[104] may help clarify whether biological processes contribute significantly to the degradation of these molecules.

As the presence of particular flavonoids is demonstrated in soil, microbiologists undoubtedly will test those compounds for effects on microbes responsible for processes important in natural and agricultural ecosystems, including mineralization and biological reactions affecting the cycling of important nutrients such as nitrogen, phosphorus, and sulfur. Another major area of investigation will examine how these compounds affect pathogenic microbes or their biocontrol agents. In such studies, it seems probable that flavonoid differences between seed and root zones (Tables 4 and 5) may produce dissimilar microbial responses in these two ecochemical zones.

One especially exciting direction for future research will be to use the new understanding of flavonoid-NodD interactions to develop rhizobia that form nodules on non-legumes.[105] It is well known that *Rhizobium* forms functional nodules on the non-legume *Parasponia*,[20,21] and preliminary data now show that some rhizobia can form a few effective (acetylene-reducing) root nodules on rice.[106] The concept that many non-legumes are potential hosts for rhizobia, when initial barriers to bacterial infection are removed, is supported by the fact that treating seedling roots of oilseed rape (*Brassica napus*) with hydrolytic enzymes permitted several rhizobial species to form nodules.[107] Following that reasoning, one line of investigation will be to determine how the *nodD* gene from *Rhizobium* strain NGR234, which permits nodulation of *Parasponia*, interacts with flavonoids from roots of various plants. Thus far, vanillin (4-hydroxy-3-methoxybenzaldehyde) and isovanillin, which were claimed to be present in wheat root extracts on the basis of HPLC retention time and UV spectral analysis, have been identified as *nod*-gene inducers in this rhizobial strain.[106] A number of other phenolics, not present in the wheat root extract, also induced *nod* gene transcription in the presence of this *nodD* gene.[108] Although this work may not produce effective root nodules on wheat in the immediate future, a rhizobial strain ANU536, which carries the NGR234 *nodD* gene, formed a few ineffective root nodules on rice seedlings.[105] Advancing our understanding of how flavonoids control host specificity of rhizobia will explain many of the observations in Tables 1 and 2, and perhaps will help overcome agricultural problems related to rhizobial specificity.

CONCLUSION

Certain flavonoids released from plants regulate activities of soil microbes at micromolar, or even nanomolar, concentrations. Processes affected by these compounds have been examined most thoroughly in symbiotic, N_2-fixing rhizobial bacteria where they induce transcription of nodulation genes and promote both chemotaxis and growth. Important effects on other soil microbes are evident in the capacity of some flavonoids in seed and root exudates to promote spore germination in mycorrhizal fungi. Data on the amount and identity of flavonoids released from several crop plants now offer a basis for molecular genetic and ecological studies of how these compounds may control rhizosphere biology and soil formation. Because flavonoids are biologically active at such low concentrations, it will be important to determine whether their presence in soil is controlled primarily by their release from living and dead plants or by their inactivation through biological and chemical processes.

ACKNOWLEDGMENTS

Funding from the USDA Competitive Research Grants Program and the U.S.-Israel Binational Agricultural Research and Development Fund (BARD), supported the author's research on alfalfa and bean *nod*-gene-inducing flavonoids described in this review. I thank past and present scientific collaborators for their creative contributions to this work and value especially suggestions from C.M Joseph, M. Hungria, F.D. Dakora and M. Bongue on this manuscript.

REFERENCES

1.   PETERS, N.K., FROST, J.W., LONG, S.R. 1986. A plant flavone, luteolin, induces expression of *Rhizobium meliloti* nodulation genes. Science 233:977-980.

2.   REDMOND, J.W., BATLEY, M., DJORDJEVIC, M.A., INNES, R.W., KUEMPEL, P.L., ROLFE, B.G. 1986. Flavones induce expression of nodulation genes in *Rhizobium*. Nature 323:632-635.

3.   FIRMIN, J.L., WILSON, K.E., ROSSEN, L., JOHNSTON, A.W.B. 1986. Flavonoid activation of nodulation genes in *Rhizobium* reversed by other compounds present in plants. Nature 324:90-92.

4.   KOSSLAK, R.M., BOOKLAND, R., BARKEI, J., PAAREN, H.E., APPELBAUM, E.R. 1987. Induction of *Bradyrhizobium japonicum* common *nod* genes by isoflavones isolated from *Glycine max*. Proc. Natl. Acad. Sci. USA 84:7428-7432.

5.   STACHEL, S.E., NESTER, E.W., ZAMBRYSKI, P.C. 1986. A plant cell factor induces *Agrobacterium tumefaciens vir* gene expression. Proc. Natl. Acad. Sci., USA 83:379-383.

6.   STACHEL, S.E., MESSENS, E., VAN MONTAGU, M., ZAMBRYSKI, P. 1985. Identification of the signal molecules produced by wounded plant cells that activate T-DNA transfer in *Agrobacterium tumefaciens*. Nature 318:624-629.

7.   PETERS, N.K., VERMA, D.P.S. 1990. Phenolic compounds as regulators of gene expression in plant-microbe interactions. Molec. Plant-Microbe Inter. 3:4-8.

8.   VanETTEN, H.D., MATTHEWS, D.E., MATTHEWS, P.S. 1989.

Phytoalexin detoxification: Importance for pathogenicity and practical implications. Annu. Rev. Phytopathol. 27:143-164.

9.   DIXON, R.A., LAMB, C.J. 1990. Molecular communication in interactions between plants and microbial pathogens. Annu. Rev. Plant Physiol. Plant Mol. Biol. 41:339-367.

10. GAJENDIRAN, N., MAHADEVAN, A. 1990. Growth of *Rhizobium* sp. in the presence of catechol. Plant Soil 125:207-211.

11. LUNDEGARDH, H., STENLID, G. 1944. On the exudation of nucleotides and flavanone from living roots. Arkiv Bot. 31A:1-27.

12. WHIPPS, J.M., LYNCH, J.M. 1986. The influence of the rhizosphere on crop productivity. In: Advances in Microbial Ecology, Vol. 9. (K.C. Marshall, ed.) Plenum Publ. Corp., New York, pp. 187-244.

13. KANDASAMY, D., PRASAD, N.N. 1979. Colonization by rhizobia of the seed and roots of legumes in relation to exudation of phenolics. Soil Biol. Biochem. 11:73-75.

14. D'ARCY, A.L. 1982. Etude des exsudats racinaires de soja et de lentille I. Cinetique d'exsudation des composés phénoliques, des amino acides et des sucres, au cours des premiers jours de la vie des plantules. Plant Soil 68:399-403.

15. RAO, A.S. 1990. Root flavonoids. Bot. Rev. 56:1-84.

16. D'ARCY-LAMETA, A. 1986. Study of soybean and lentil root exudates II. Identification of some polyphenolic compounds, relation with plantlet physiology. Plant Soil 92:113-123.

17. JORDAN, D.C. 1984. Genus I. Rhizobium. In: Bergey's Manual of Systematic Bacteriology, Vol. 1. (N.R. Krieg, J.G. Holt, eds.) Williams and Wilkins, Baltimore, pp. 235-242.

18. JORDAN, D.C. 1984. Genus II. Bradyrhizobium. In: Bergey's Manual of Systematic Bacteriology, Vol. 1. (N.R. Krieg, J.G. Holt, eds.) Williams and Wilkins, Baltimore, pp. 242-244.

19. YOUNG, J.P.W., JOHNSTON, A.W.B. 1989. The evolution of specificity in the legume-*Rhizobium* symbiosis. Trends Ecol. Evolut. 4:341-349.

20. TRINICK, M.J. 1973. Symbiosis between *Rhizobium* and the non-legume, *Trema aspera*. Nature 244:459-460.

21. AKKERMANS, A.D.L., ABDULKADIR, S., TRINICK, M.J. 1978. N_2-fixing root nodules in Ulmaceae: *Parasponia* or (and) *Trema* spp.? Plant Soil 49:711-715.

22. KEYSER, H.H., BOHLOOL, B.B., HU, T.S., WEBER, D.F. 1982. Fast-growing rhizobia isolated from root nodules of soybean. Science

215:1631-1632.

23. SCHOLLA, M.H., ELKAN, G.H. 1984. *Rhizobium fredii* sp. nov., a fast-growing species that effectively nodulates soybeans. Inter. J. Syst. Bacteriol. 34:484-486.

24. CHEN, W.X., YAN, G.H., LI, J.L. 1988. Numerical taxonomic study of fast-growing soybean rhizobia and a proposal that *Rhizobium fredii* be assigned to *Sinorhizobium* gen. nov. Inter. J. Syst. Bacteriol. 38:392-397.

25. DREYFUS, B., GARCIA, J.L., GILLIS, M. 1988. Characterization of *Azorhizobium caulinodans* gen. nov., sp. nov., a stem-nodulating nitrogen-fixing bacterium isolated from *Sesbania rostrata*. Inter. J. Syst. Bacteriol. 38:89-98.

26. LINDSTRÖM, K., LEHTOMÄKI, S. 1988. Metabolic properties, maximum growth temperature and phage sensitivity of *Rhizobium* sp. (*Galega*) compared with other fast-growing rhizobia. FEMS Microbiol. Lett. 50:277-287.

27. EAGLESHAM, A.R.J., ELLIS, J.M., EVANS, W.R., FLEISCHMAN, D.E., HUNGRIA, M., HARDY, R.W.F. 1990. The first photosynthetic *Rhizobium*: Characteristics. In: Nitrogen Fixation: Achievements and Objectives. (P.M. Gresshoff, L.E. Roth, G. Stacey, W.E. Newton, eds.) Chapman and Hall, New York, pp. 805-811.

28. BROCKWELL, J., HELY, F.W. 1966. Symbiotic characteristics of *Rhizobium meliloti*:an appraisal of the systematic treatment of nodulation and nitrogen fixation interactions between hosts and rhizobia of diverse origins. Aust. J. Agric. Res. 17:885-899.

29. DOWNIE, J.A., ECONOMOU, A., SCHEU, A.K., JOHNSTON, A.W.B., FIRMIN, J.L., WILSON, K.E., CUBO, M.T., MAVRIDOU, A., MARIE, C., DAVIES, A., SURIN, B.P. 1990. The *Rhizobium leguminosarum* bv. *viciae* NodO protein compensates for the exported signal made by the host-specific nodulation genes. In: Nitrogen Fixation: Achievements and Objectives. (P.M. Gresshoff, L.E. Roth, G. Stacey, W.E. Newton, eds.) Chapman and Hall, New York, pp. 201-206.

30. ROLFE, B.G., GRESSHOFF, P.M. 1988. Genetic analysis of legume nodule initiation. Annu. Rev. Plant Physiol. Plant Mol. Biol. 39:297-319.

31. LONG, S.R. 1989. *Rhizobium* genetics. Annu. Rev. Genet. 23:483-506.

32. LONG, S.R. 1989. Rhizobium-legume nodulation: Life together in the underground. Cell 56:203-214.

33. KONDOROSI, E., BANFALVI, Z., KONDOROSI, A. 1984. Physical and genetic analysis of a symbiotic region of *Rhizobium meliloti*:. Identification of nodulation genes. Mol. Gen. Genet. 193:445-452.

34. DJORDJEVIC, M.A., SCHOFIELD, P.R., ROLFE, B.G. 1985. Tn*5* mutagenesis of *Rhizobium trifolii* host-specific nodulation genes result in mutants with altered host-range ability. Mol. Gen. Genet. 200:463-471.

35. MARVEL, D.J., KULDAU, G., HIRSCH, A., RICHARDS, E., TORREY, J.G., AUSUBEL, F.M. 1985. Conservation of nodulation genes between *Rhizobium meliloti* and a slow-growing *Rhizobium* strain that nodulates a nonlegume host. Proc. Natl. Acad. Sci., USA 82:5841-5845.

36. FISHER, R.F., TU, J.K., LONG, S.R. 1985. Conserved nodulation genes in *Rhizobium meliloti* and *Rhizobium trifolii*. Appl. Environ. Microbiol. 49:1432-1435.

37. LEROUGE, P., ROCHE, P., FAUCHER, C., MAILLET, F., TRUCHET, G., PROMÉ, J.C., DÉNARIÉ, J. 1990. Symbiotic host-specificity of *Rhizobium meliloti* is determined by a sulphated and acylated glucosamine oligosaccharide signal. Nature 344:781-784.

38. LEROUGE, P., ROCHE, P., PROMÉ, J.C., FAUCHER, C., VASSE, J., MAILLET, F., CAMUT, S., DE BILLY, F., BARKER, D.G., DÉNARIÉ, J., TRUCHET, G. 1990. *Rhizobium meliloti* nodulation genes specify the production of an alfalfa-specific sulfated lipo-oligosaccharide signal. In: Nitrogen Fixation: Achievements and Objectives. (P.M. Gresshoff, L.E. Roth, G. Stacey, W.E. Newton, eds.) Chapman and Hall, New York, pp. 177-186.

39. DIAZ, C.L., MELCHERS, L.S., HOOYKAAS, P.J.J., LUGTENBERG, B.J.J., KIJNE, J.W. 1989. Root lectin as a determinant of host-plant specificity in *Rhizobium*-legume symbiosis. Nature 338:579-581.

40. GÖTTFERT, M., HORVATH, B., KONDOROSI, E., PUTNOKY, P., RODRIGUEZ-QUIÑONES, F., KONDOROSI, A. 1986. At least two *nodD* genes are necessary for efficient nodulation of alfalfa by *Rhizobium meliloti*. J. Mol. Biol. 191:411-420.

41. HONMA, M.A., AUSUBEL, F.M. 1987. *Rhizobium meliloti* has three functional copies of the *nodD* symbiotic regulatory gene. Proc. Natl. Acad. Sci., USA 84:8558-8562.

42. APPELBAUM, E.R., THOMPSON, D.V., IDLER, K., CHARTRAIN, N. 1988. *Rhizobium japonicum* USDA 191 has two *nodD* genes that differ in primary structure and function. J. Bacteriol. 170:12-20.

43.   MULLIGAN, J.T., LONG, S.R. 1989. A family of activator genes regulate expression of *Rhizobium meliloti* nodulation genes. Genetics 122:7-18.

44.   DAVIS, E.O., JOHNSTON, A.W.B. 1990. Analysis of three *nodD* genes in *Rhizobium leguminosarum* biovar *phaseoli; nodD1* is preceded by *nolE*, a gene whose product is secreted from the cytoplasm. Molec. Microbiol. 4:921-932.

45.   RODRIGUEZ-QUIÑONES, F., BANFALVI, Z., MURPHY, P., KONDOROSI, A. 1987. Interspecies homology of nodulation genes in *Rhizobium*. Plant Mol. Biol. 8:61-75.

46.   HONG, G.F., BURN, J.E., JOHNSTON, A.W.B. 1987. Evidence that DNA involved in the expression of nodulation (*nod*) genes in *Rhizobium* binds to the product of the regulatory gene *nodD*. Nucl. Acids Res. 15:9677-9690.

47.   BANFALVI, Z., NIEUWKOOP, A., SCHELL, M., BESL, L., STACEY, G. 1988. Regulation of *nod* gene expression in *Bradyrhizobium japonicum*. Mol. Gen. Genet. 214:420-424.

48.   DAVIS, E.O., JOHNSTON, A.W.B. 1990. Regulatory functions of the three *nodD* genes of *Rhizobium leguminosarum* biovar. *phaseoli*. Molec. Microbiol. 4:933-941.

49.   INNES, R.W., KUEMPEL, P.L., PLAZINSKI, J., CANTER-CREMERS, H., ROLFE, B.G., DJORDJEVIC, M.A. 1985. Plant factors induce expression of nodulation and host-range genes in *Rhizobium trifolii*. Mol. Gen. Genet. 201:426-432.

50.   MULLIGAN, J.T., LONG, S.R. 1985. Induction of *Rhizobium meliloti nodC* expression by plant exudate requires *nodD*. Proc. Natl. Acad. Sci., USA 82:6609-6613.

51.   ROSSEN, L., SHEARMAN, C.A., JOHNSTON, A.W.B., DOWNIE, J.A. 1985. The *nodD* gene of *Rhizobium leguminosarum* is auto-regulatory and in the presence of plant exudate induces the *nodA,B,C* genes. EMBO J. 4:3369-3373.

52.   ZAAT, S.A.J., WIJFFELMAN, C.A., SPAINK, H.P., VAN BRUSSEL, A.A.N., OKKER, R.J.H., LUGTENBERG, B.J.J. 1987. Induction of the *nodA* promoter of *Rhizobium leguminosarum* sym plasmid pRL1JI by plant flavanones and flavones. J. Bacteriol. 169:198-204.

53.   HORVATH, B., BACHEM, C.W.B., SCHELL, J., KONDOROSI, A. 1987. Host-specific regulation of nodulation genes in *Rhizobium* is mediated by a plant-signal, interacting with the *nodD* gene product. EMBO J. 6:841-848.

54. SADOWSKY, M.J., OLSON, E.R., FOSTER, V.E., KOSSLAK, R.M., VERMA, D.P.S. 1988. Two host-inducible genes of *Rhizobium fredii* and characterization of the inducing compound. J. Bacteriol. 170:171-178.

55. SPAINK, H.P., WIJFFELMAN, C.A., PEES, E., OKKER, R.J.H., LUGTENBERG, B.J.J. 1987. *Rhizobium* nodulation gene *nodD* as a determinant of host specificity. Nature 328:337-340.

56. BURN, J., ROSSEN, L., JOHNSTON, A.W.B. 1987. Four classes of mutations in the *nodD* gene of *Rhizobium leguminosarum* biovar. *viciae* that affect its ability to autoregulate and/or activate other *nod* genes in the presence of flavonoid inducers. Genes & Dev.1:456-464.

57. SPAINK, H.P., OKKER, R.J.H., WIJFFELMAN, C.A., TAK, T., GOOSEN-DE ROO, L., PEES, E., VAN BRUSSEL, A.A.N., LUGTENBERG, B.J.J. 1989. Symbiotic properties of rhizobia containing a flavonoid-independent hybrid *nodD* product. J. Bacteriol. 171:4045-4053.

58. DJORDJEVIC, M.A., REDMOND, J.W., BATLEY, M., ROLFE, B.G. 1987. Clovers secrete specific phenolic compounds which either stimulate or repress *nod* gene expression in *Rhizobium trifolii*. EMBO J. 6:1173-1179.

59. PETERS, N.K., LONG, S.R. 1988. Alfalfa root exudates and compounds which promote or inhibit induction of *Rhizobium meliloti* nodulation genes. Plant Physiol. 88:396-400.

60. ROLFE, B.G. 1988. Flavones and isoflavones as inducing substances of legume nodulation. Biofactors 1:3-10.

61. ZAAT, S.A.J., SCHRIPSEMA, J., WIJFFELMAN, C.A., VAN BRUSSEL, A.A.N., LUGTENBERG, B.J.J. 1989. Analysis of the major inducers of the *Rhizobium nodA* promoter from *Vicia sativa* root exudate and their activity with different *nodD* genes. Plant Molec. Biol. 13:175-188.

62. GEHRING, E., GEIGER, H. 1980. Die Flavonoide der Samen von *Medicago* x *varia* Martyn c.v. Cardinal (Fabaceae). Z. Naturforsch 35C:380-383.

63. MAXWELL, C.A., HARTWIG, U.A., JOSEPH, C.M., PHILLIPS, D.A. 1989. A chalcone and two related flavonoids released from alfalfa roots induce *nod* genes of *Rhizobium meliloti*. Plant Physiol 91:842-847.

64. HARTWIG, U.A., MAXWELL, C.A., JOSEPH, C.M., PHILLIPS, D.A. 1990. Chrysoeriol and luteolin released from alfalfa seeds induce

nod genes in *Rhizobium meliloti*. Plant Physiol 92:116-122.

65. HARTWIG, U.A., MAXWELL, C.A., JOSEPH, C.M., PHILLIPS, D.A. 1990. Effects of alfalfa *nod* gene-inducing flavonoids on *nodABC* transcription in *Rhizobium meliloti* strains containing different *nodD* genes. J. Bacteriol. 172:2769-2773.

66. GYÖRGYPAL, Z., IYER, N., KONDOROSI, A. 1988. Three regulatory *nodD* alleles of diverged flavonoid-specificity are involved in host-dependent nodulation by *Rhizobium meliloti*. Mol. Gen. Genet. 212:85-92.

67. HUNGRIA, M., JOSEPH, C.M., PHILLIPS, D.A. 1991. Antho-cyanidins and flavonols, major *nod*-gene inducers from seeds of a black-seeded common bean (*Phaseolus vulgaris* L.). Plant Physiol. (in press).

68. HUNGRIA, M., JOSEPH, C.M., PHILLIPS, D.A. 1991. *Rhizobium nod*-gene inducers exuded naturally from roots of common bean (*Phaseolus vulgaris* L.). Plant Physiol. (in press).

69. ZAAT, S.A.J., WIJFFELMAN, C.A., MULDERS, I.H.M., VAN BRUSSEL, A.A.N., LUGTENBERG, B.J.J. 1988. Root exudates of various host plants of *Rhizobium leguminosarum* contain different sets of inducers of *Rhizobium* nodulation genes. Plant Physiol. 86:1298-1303.

70. RECOURT, K., SCHRIPSEMA, J., KIJNE, J.W., VAN BRUSSEL, A.A.N., LUGTENBERG, B.J.J. 1991. Inoculation of *Vicia sativa* subsp. *nigra* roots with *Rhizobium leguminosarum* biovar *viciae* results in release of *nod* gene activating flavanones and chalcones. Plant Mol. Biol. Plant Mol. Biol. 16:841-852.

71. TOMAS-LORENTE, F., GARCIA-GRAU, M.M., TOMAS-BARBERAN, F.A. 1990. Flavonoids from *Vicia faba* seed exudates. Z. Naturforch. 45C:1070-1072.

72. NOZZOLILLO, C., RICCIARDI, L., LATTANZIO, V. 1988. Flavonoid constituents of seed coats of *Vicia faba* (Fabaceae) in relation to genetic control of their color. Can. J. Bot. 67:1600-1604.

73. HARTWIG, U.A., PHILLIPS, D.A. 1991. Release and modification of *nod*-gene-inducing flavonoids from alfalfa seeds. Plant Physiol. 95:804-807.

74. MAXWELL, C.A., PHILLIPS, D.A. 1990. Concurrent synthesis and release of *nod*-gene-inducing flavonoids from alfalfa roots. Plant Physiol. 93:1552-1558.

75. GRAHAM, T.L. 1991. Flavonoid and isoflavonoid distribution in

developing soybean seedling tissues and in seed and root exudates. Plant Physiol. 95:594-603.

76.   HAWES, M.C. 1990. Living plant cells released from the root cap: A regulator of microbial populations in the rhizosphere? Plant Soil 129:19-27.

77.   CHO, M.J., HARPER, J.E. 1991. Effect of inoculation and nitrogen on isoflavonoid concentration in wild-type and nodulation-mutant soybean roots. Plant Physiol. 95:435-442.

78.   VAN BRUSSEL, A.A.N., RECOURT, K., PEES, E., SPAINK, H.P., TAK, T., WIJFFELMAN, C.A., KIJNE, J.W., LUGTENBERG, B.J.J. 1990. A biovar-specific signal of *Rhizobium leguminosarum* bv. *viciae* induces increased nodulation gene-inducing activity in root exudate of *Vicia sativa* subsp. *nigra*. J. Bacteriol. 172:5394-5401.

79.   HARTWIG, U.A., MAXWELL, C.A., JOSEPH, C.M., PHILLIPS, D.A. 1989. Interactions among flavonoid *nod* gene inducers released from alfalfa seeds and roots. Plant Physiol. 91:1138-1142.

80.   ROVIRA, A.D. 1969. Plant root exudates. Bot. Rev. 35:35-57.

81.   CAETANO-ANOLLÉS, G., CRIST-ESTES, D.K., BAUER, W.D. 1988. Chemotaxis of *Rhizobium meliloti* to the plant flavone luteolin requires functional nodulation genes. J. Bacteriol. 170:3164-3169.

82.   ARMITAGE, J.P., GALLAGHER, A., JOHNSTON, A.W.B. 1988. Comparison of the chemotactic behaviour of *Rhizobium leguminosarum* with and without the nodulation plasmid. Molec. Microbiol. 2:743-748.

83.   AGUILAR, J.M.M., ASHBY, A.M., RICHARDS, A.J.M., LOAKE, G.J., WATSON, M.D., SHAW, C.H. 1988. Chemotaxis of *Rhizobium leguminosarum* biovar *phaseoli* towards flavonoid inducers of the symbiotic nodulation genes. J. Gen. Microbiol. 134:2741-2746.

84.   ZIEGLER, R.J., PIERCE, C., BERGMAN, K. 1986. Mapping and cloning of a *fla-che* region of the *Rhizobium meliloti* chromosome. J Bacteriol. 168:785-790.

85.   BERGMAN, K., GULASH-HOFFEE, M., HOVESTADT, R.E., LAROSILIERE, R.C., RONCO, P.G., SU, L. 1988. Physiology of behavioral mutants of *Rhizobium meliloti*:. Evidence for a dual chemotaxis pathway. J Bacteriol. 170:3249-3254.

86.   GAWORZEWSKA, E.T., CARLILE, M.J. 1982. Positive chemotaxis of *Rhizobium leguminosarum* and other bacteria towards root exudates from legumes and other plants. J. Gen. Microbiol 128:1179-1188.

87.   MALEK, W. 1989. Chemotaxis in *Rhizobium meliloti* strain L5.30. Arch. Microbiol. 152:611-612.

88. D'ARCY-LAMETA, A., JAY, M. 1987. Study of soybean and lentil root exudates III. Influence of soybean isoflavonoids on growth of rhizobia and some rhizospheric microorganisms. Plant Soil 101:267-272.

89. HARTWIG, U.A., JOSEPH, C.M., PHILLIPS, D.A. 1991. Flavonoids released naturally from alfalfa seeds enhance growth rate of *Rhizobium meliloti*. Plant Physiol. 95:796-803.

90. FRIES, N., SERCK-HANSSEN, K., DIMBERG, L.H., THEANDER, O. 1987. Abietic acid, an activator of basidiospore germination in ectomycorrhizal species of the genus *Suillus* (Boletaceae). Experimental Mycology 11:360-363.

91. GRAHAM, J.H. 1982. Effect of citrus root exudates on germination of chlamydospores of the vesicular-arbuscular mycorrhizal fungus, *Glomus epigaeum*. Mycologia 74:831-835.

92. ELIAS, K.S., SAFIR, G.R. 1987. Hyphal elongation of *Glomus fasciculatus* in response to root exudates. Appl. Environ. Microbiol. 53:1928-1933.

93. BÉCARD, G., PICHÉ, Y. 1989. Fungal growth stimulation by CO_2 and root exudates in vesicular-arbuscular mycorrhizal symbiosis. Appl. Environ. Microbiol. 55:2320-2325.

94. GIANINAZZI-PEARSON, V., BRANZANTI, B., GIANINAZZI, S. 1989. *In vitro* enhancement of spore germination and early hyphal growth of a vesicular-arbuscular mycorrhizal fungus by host root exudates and plant flavonoids. Symbiosis 7:243-255.

95. TSAI, S.M., PHILLIPS, D.A. 1991. Flavonoids released naturally from alfalfa promote development of symbiotic *Glomus* spores *in vitro*. Appl. Environ. Microbiol. 57:1485-1488.

96. KAPULNIK, Y., JOSEPH, C.M., PHILLIPS, D.A. 1987. Flavone limitations to root nodulation and symbiotic nitrogen fixation in alfalfa. Plant Physiol. 84:1193-1196.

97. JAIN, V., GARG, N., NAINAWATEE, H.S. 1990. Naringenin enhanced efficiency of *Rhizobium meliloti*-alfalfa symbiosis. World J. Microbiol. 6:434-436.

98. TRIPLETT, E.W. 1990. The molecular genetics of nodulation competitiveness in *Rhizobium* and *Bradyrhizobium*. Mol. Plant-Microbe Inter. 3:199-206.

99. KOSSLAK, R.M., JOSHI, R.S., BOWEN, B.A., PAAREN, H.E., APPELBAUM, E.R. 1990. Strain-specific inhibition of *nod* gene induction in *Bradyrhizobium japonicum* by flavonoid compounds. Appl. Environ. Microbiol. 56:1333-1341.

100. CUNNINGHAM, S., KOLLMEYER, W.D., STACEY, G. 1991. Chemical control of interstrain competition for soybean nodulation by *Bradyrhizobium japonicum*. Appl. Environ. Microbiol. 57:1886-1892.

101. SADOWSKY, M.J., CREGAN, P.B., GOTTFERT, M., SHARMA, A., GERHOLD, D., RODRIGUEZ-QUINONES, F., KEYSER, H.H., HENNECKE, H., STACEY, G. 1991. The *Bradyrhizobium japonicum nolA* gene and its involvement in the genotype-specific nodulation of soybeans. Proc. Natl. Acad. Sci., USA 88:637-641.

102. HENIKOFF, S., HAUGHN, G.W., CALVO, J.M., WALLACE, J.C. 1988. A large family of bacterial activator proteins. Proc. Natl. Acad. Sci., USA 85:6602-6606.

103. DORNBOS, D.L., SPENCER, G.F., MILLER, R.W. 1990. Medicarpin delays alfalfa seed germination and seedling growth. Crop Sci. 30:162-166.

104. BARZ, W., BLESS, W., BÖRGER-PAPENDORF, G., GUNIA, W., MACKENBROCK, U., MEIER, D., OTTO, C., SÜPER, E. 1990. Phytoalexins as part of induced defence reactions in plants: their elicitation, function and metabolism. In: Bioactive Compounds from Plants. (D.J. Chadwick, J. Marsh, eds.) Wiley, Chichester, pp. 140-153.

105. ROLFE, B.G., BENDER, G.L. 1990. Evolving a *Rhizobium* for non-legume nodulation. In: Nitrogen Fixation: Achievements and Objectives. (P.M. Gresshoff, L.E. Roth, G. Stacey, W.E. Newton, eds.) Chapman and Hall, New York, pp. 779-780.

106. JING, Y., LI, G., JIN, G., SHAN, X., ZHANG, B., GUAN, C., LI, J. 1990. Rice root nodules with acetylene reduction activity. In: Nitrogen Fixation: Achievements and Objectives. (P.M. Gresshoff, L.E. Roth, G. Stacey, W.E. Newton, eds.) Chapman and Hall, New York, pp. 829.

107. AL-MALLAH, M.K., DAVEY, M.R., COCKING, E.C. 1990. Nodulation of oilseed rape (*Brassica napus*) by rhizobia. J. Expt. Bot. 41:1567-1572.

108. LE STRANGE, K.K., BENDER, G.L., DJORDJEVIC, M.A., ROLFE, B.G., REDMOND, J.W. 1990. The *Rhizobium* strain NGR234 *nodD1* gene product responds to activation by the simple phenolic compounds vanillin and isovanillin present in wheat seedling extracts. Molec. Plant-Microb. Inter. 3:214-220.

Chapter Eight

FLAVONOID SULFATION: PHYTOCHEMISTRY, ENZYMOLOGY AND MOLECULAR BIOLOGY

Luc Varin

Department of Biochemistry
University of Montréal
Montréal, Québec, Canada H3C 3J7

INTRODUCTION

Sulfate ester formation is the most recently recognized conjugation reaction of flavonoid compounds. Although the first sulfated compound was reported as early as 1937,[1] it was not until 1975 that flavonoid sulfate esters have been considered to be of common occurrence in a number of plant families.[2-5] In a recent review,[6] however, more than 100 sulfated flavonoids

Phenolic Metabolism in Plants, Edited by H.A. Stafford
and R.K. Ibrahim, Plenum Press, New York, 1992

have been reported in 250 species belonging to 17 dicotyledon and 16 monocotyledon families. Most of these compounds are mono- to tetrasulfate esters of common hydroxyflavones and hydroxyflavonols or their methyl ethers, and less commonly of their glucosylated derivatives. In some flavonoid glycosides, the sulfate group may be linked to the 3- or 6-position of the sugar moiety. Because of instability of the sulfate ester bond, especially in acid media, many sulfated flavonoids may have escaped detection, and the sulfate esters of flavonoid classes other than flavones and flavonols have yet to be described.

In contrast with the known physiological roles of sulfated conjugates (phenols, steroids, bile acids, xenobiotics) in animal tissues,[7] the functional significance of flavonoid sulfates in plant tissues is not clear. Apart from their involvement in detoxification of reactive hydroxyl groups, their accumulation in plants growing in saline or marshy habitats suggests a possible role in sequestering of sulfate ions.[2] In addition, their solubility in aqueous solutions may facilitate their storage in hydrophilic cellular compartments.

Recently, sulfated metabolites have been shown to be responsible for the seismonastic and gravitropic movements of plants[8] (turgorins) and for the cortical cell division during early nodule initiation (NodRM1) in the interaction between *Rhizobium meliloti* and alfalfa.[9] In both cases, the presence of the sulfate group was required for biological activity suggesting that the sulfation reaction plays an important role in the recognition and signalling processes.

This chapter summarizes our current knowledge of the structural variation and natural distribution of flavonoid sulfates, as well as the phytochemical and spectroscopic methods used for their identification, including the use of HCl and arylsulfatase shifts in UV spectral analysis, ion pairing coupled with ^{1}H NMR and of sulfation shifts in ^{13}C NMR and FABMS in structural analysis.[10] Recently developed methods for the chemical synthesis of specifically sulfated flavonoids[10] made it possible to synthesize the substrates necessary to purify and characterize a number of position-specific sulfotransferases (STs). Furthermore, cDNA cloning of two flavonol sulfotransferases allowed for the identification of sequences involved in substrate and position specificity of these enzymes.

PHYTOCHEMICAL ASPECTS OF SULFATED FLAVONOIDS

Structural Variation

Most simple flavonoid sulfate esters are fairly represented in plants with sulfation at positions 7>>3'>4'>6>8 and 3>>7>4'>3', for flavones and flavonols respectively.[6] Of the common flavonol aglycones, quercetin is invariably sulfated at all positions except the 5-hydroxyl group, presumably because of H-bonding to the 4-carbonyl group. Among the 12 polysulfated flavonoids identified to date, six were purified from the genus *Flaveria* (Asteraceae), a very rich source of flavonoid sulfates.[6] *F. bidentis* is the only known species to contain a variety of mono- to tetrasulfate esters of quercetin (Table 1).[11] In addition, a large number (18 of 22) of flavonol sulfates are *O*-methylated, so that there seems to be an association between *O*-methylation and sulfate ester formation.[6] On the other hand, 18 of 56 sulfated flavonols contain both sulfate and sugar (or other) substitution. In the latter group, the sugar may be separately attached to a hydroxyl group different from that linked to sulfate (e.g. kaempferol 3-glucuronide-7-sulfate), or the sulfate is linked to the flavonol through the sugar moiety (e.g. kaempferol 3-β-(6"-sulfatoglucoside).[6] 6-Methoxykaempferol (eupafolin)[12] and 6-methoxyquercetin (patuletin)[13] 3,3'-disulfates have only been characterized from *F. chloraefolia* (Table 1). In addition, sulfate esters of 8-hydroxyquercetin (gossypetin), the 3-sulfate, 3-sulfate-8-glucoside and 3-sulfate-8-glucuronide, have so far been described from the Malvaceae.[14]

Table 1. The flavonol sulfates of *Flaveria bidentis* and *F. chloraefolia*.

Flaveria bidentis	*Flaveria chloraefolia*
Quercetin 3-sulfate	Quercetin 3-sulfate
Isohamnetin 3-sulfate	Patuletin 3-sulfate
Quercetin 3,4'-disulfate	Ombuin 3-sulfate
Quercetin 3,7-disulfate	Eupaletin 3-sulfate
Quercetin 3,7,3'-trisulfate	Spinacetin 3-sulfate
Quercetin 3,7,4'-trisulfate	Eupatolitin 3-sulfate
Quercetin 3-acetyl-7,3',4'-trisulfate	Eupatin 3-sulfate
Quercetin 3,7,3',4'-tetrasulfate	6-Methoxykaempferol 3-sulfate
	Quercetin 3,3'-disulfate
	Patuletin 3,3'-disulfate
	Quercetin 3,4'-disulfate

Most sulfated flavonoids have been isolated as their K-salts, although other cations have been reported in a few instances, such as Na-salts in those of *Lippia*,[15] and Ca-salts in *Brickellia*.[16] A novel equimolar mixture of quercetin 3-sulfate and patuletin 3-sulfate has been isolated from *F. chloraefolia* containing three inorganic ions: Na, K and/or Ca.[17]

Detection and Purification

Sulfated flavonoids are easily detected on cellulose TLC by their characteristic arrow-shaped spots. They display high mobility in water as a solvent, but with little distinction among the polysulfate esters. In butanol solvents (e.g. BAW, 3:1:1), on the other hand, their mobility is reversed with a separation pattern inversely proportional to the number of sulfate groups.[6] The fact that flavonoid sulfates are more polar than their corresponding glycosides, and the latter more so than their aglycones, allows the separation of the three flavonoid types using the appropriate solvent systems.[2,5] Flavonoid sulfates can also be resolved from glycosides by HPLC on RP C_{18} column, after ion pairing with tetrabutylammonium dihydrogen phosphate (TBADP),[10] where the polarity of the flavonoid sulfates is decreased with the increasing number of sulfate groups, whereas that of glycosides is not affected. Screening of plant extracts for flavonoid sulfates can be accomplished by electrophoresis on Whatman No. 3 paper using HOAc-HCOOH buffer of pH 2.2,[18] where they move toward the anode with increasing mobilities relative to the number of sulfate groups. In addition, flavonol sulfates move farther than flavone sulfates, and substitution in the 3-position increases electrophoretic mobility over those in the 7- or 3'-positions.[4] However, the use of electrophoretic mobilities in screening for flavonoid conjugates should be interpreted with caution, since flavonoid glycosides migrate at the level of monosulfate esters and their migration increases with increasing number of sugar residues.[6]

Flavonoid sulfates can be precipitated from concentrated aqueous extracts by the addition of MeOH/EtOH, followed by column chromatography on Polyclar, Sephadex LH-20 or G-10 using H_2O, gradients of aq. MeOH, or water-methanol-methyl ethyl ketone-acetone (13:3:3:1, v/v/v/v). Sulfated flavonoids may be selectively separated from glycosides by ion pairing with tetra-butyl-ammonium hydrogen sulfate (TBAHS) where they separate as TBA-salts.

Identification and Structural Determination

Proper identification of flavonoid sulfates involves determination of the parent aglycone, as well as the number and position of sulfate groups, using different hydrolytic and spectroscopic methods.

Acid hydrolysis. Total acid hydrolysis (2N HCl, 15 min, 95°) yields the flavonoid aglycone and sulfate, as well as the sugar in case of sulfated flavonoid glycosides. After extraction in EtOAc, the aglycone can be identified by chromatographic, UV and mass spectroscopic methods.[19] The sulfate ion may be precipitated with $BaCl_2$, the associated cation determined by atomic absorption and FABMS (see below), and the sugar by TLC on cellulose. Partial hydrolysis with organic acids (e.g. 10% HOAc) allows the detection of intermediate products of the parent molecule, e.g. sulfated kaempferol glucoside gives kaempferol 3-glucoside, as well as kaempferol, glucose and sulfate. Furthermore, controlled hydrolysis of polysulfated flavonoids yields intermediates of lower sulfation levels.

Enzymatic hydrolysis. Selective cleavage of sulfated glucosides and uronides can be achieved using specific hydrolases, such as β-glucosidase, β-glucuronidase and aryl sulfatase.[6] The position of attachment of the sugar moiety can be determined by comparison of UV absorption spectra of the intact flavonoid and its hydrolysis product. However, when the sulfate group is attached to the sugar moiety, enzymatic cleavage releases the aglycone and the sulfated sugar. Purified aryl sulfatase acts primarily on the phenolic sulfate esters, leaving the alcoholic ester bond (as in flavonoid 3-sulfates) intact.[10] Furthermore, controlled enzymic hydrolysis allows the identification of partially sulfated intermediates in polysulfated compounds.

IR and UV spectroscopy. The appearance in IR spectra of two strong bands at *ca* 1200 (S=O) and 1040 (C-O-S) cm^{-1} is indicative of the presence of sulfate groups. The position of their attachment can be deduced from determining the UV spectra of the intact molecule in aq. MeOH and after the addition of spectral shift reagents.[20] While flavonoid sulfates exhibit the same UV spectral characteristics as those of the corresponding methyl ethers, sulfation at positions 3 and/or 4' induce an appreciably more important hypsochromic shift in Band I than those reported for methylation or glucosylation, due to the electron withdrawing effect of the sulfate group.[10] Furthermore, unlike methyl groups, the ease of hydrolysis of the sulfate groups with HCl and/or aryl sulfatase, except the 3-sulfate, causes a bathochromic shift in Band I due to formation of the corresponding aglycone or its 3-sulfate ester.[21] The HCl reagent is of diagnostic value for the detection of 3'/4'-sulfated flavones and of 3-

sulfated flavonols.[22]

Methylation. Total methylation of the intact sulfated flavonoid, followed by hydrolysis and UV spectral and/or EIMS analysis, reveals the OH groups that were originally sulfated. The same procedure also applies to the sulfate group attached to the sugar moiety of a sulfated flavonoid glycoside; the methylated sugar formed may be identified by TLC using reference compounds.[6]

^{13}C NMR spectroscopy. Because of its electron withdrawing nature, O-sulfation causes a decreased electron density of the carbons *ortho* and *para*, and an increased electron density of the carbon carrying the sulfate group; thus resulting in a downfield shift for the former carbons and an upfield shift for the latter. Calculated shifts of 7, 4' and/or 3'-sulfated naturally occurring and synthetic flavonoids are similar to those reported for simple phenol sulfate esters. Flavonoids with 3',4'-dihydroxy groups exhibit downfield shifts induced by 3'-/4'-sulfation, which are more pronounced for the *ortho* carbon lacking the hydroxyl than that carrying the phenolic group.[6,10] On the other hand, individual effects of 3'- and 4'-sulfation are not cumulative in the case of 3',4'-*o*-disulfated compounds, such as quercetin 3,7,3',4'-tetrasulfate, since the downfield shifts for the carbons *ortho* and *para* to the sulfate groups are lower than expected, whereas the upfield shifts for carbons carrying the sulfate groups tend to be higher.[6,23] Finally, flavonoid 3-sulfation exhibits a more pronounced downfield shift for *ortho* C-2, similar to those reported for 3-glycosylation and 3-methylation, which is characteristic of an olefinic rather than an aromatic system.[6,23]

Fast atom bombardment mass spectrometry. FABMS of flavonoid sulfates can be carried out in the positive or negative mode. Flavonoid monosulfate esters give two major, [M] and [M-SO$_3$] peaks in their negative FABMS spectra.[17] When K or Na are the counter ions, the respective [M+K-H] and [M+Na-H] fragments are observed as well. However, when a mixture of cations such as K$^+$, Na$^+$ and Ca^{2+} are present in the sample, only the former can be visible in the spectrum because of exchange phenomena.[17] Recently, negative FABMS spectra of several natural and synthetic flavonoid disulfate esters have revealed the presence of [M+cation], [M-SO$_3$+H] and [M-2SO$_3$+H] peaks.[15,21,23,24] Those of trisulfated compounds exhibit ions at [M^{+2} cations], [M-SO$_3$+H$^+$ cation].[21,24] The fragmentation patterns characteristic of flavonoid mono-, di- and trisulfates have recently been summarized.[6]

Synthesis of Flavonoid Sulfate Esters

Chemical synthesis. Few reagents, such as sulfamic acid, sulfur trioxide adducts or N,N'-dicyclohexylcarbodiimide (DCC) in the presence of sulfuric acid, have been utilized for the synthesis of sulfated flavonoids. However, these reagents proved to be of little use due to the formation of complex mixtures of sulfate ester derivatives, or the destruction of DCC before any sulfate ester was formed.[6] On the other hand, using TBAHS+DCC as sulfating reagent allows the stepwise sulfation of positions 3,7 and 4' of flavonoids in good yield, following the order 7>4'>3.[24] Since the 3-OH is usually chelated with the adjacent carbonyl group, it can only be sulfated in the presence of large excess of reagent, and over an extended reaction time to give the flavonol 3,7,4'-trisulfate ester. In this manner, a number of flavone and flavonol mono- to tetrasulfate esters have been synthesized.[24] Moreover, due to its weak acidic nature the 3'-OH group exhibits high reactivity towards the sulfating reagent only if the 4'-OH is methylated, but not sulfated, possibly because of the steric hindrance of the bulky 4'-sulfate. On the other hand, 4'-sulfation in presence of 3'-sulfate group can be achieved by using sulfur trioxide-trimethylamine complex in aqueous potassium carbonate.[23]

Enzymatic synthesis. Highly purified aryl sulfatase mediates the stepwise hydrolysis of sulfate groups esterified to flavonoids, except that at position 3.[12,13,17] The stepwise desulfation of polysulfated flavonoids follows the order 7/4'>3'>>>3.[21,23] This property allows for the preparation of 3-sulfated intermediates, as well as flavonol 3-sulfates which are known for their common occurrence in plants.[6,21,23]

BIOSYNTHESIS AND SPATIAL DISTRIBUTION OF SULFATED FLAVONOIDS

Biosynthetic studies of sulfated flavonoids in *F. bidentis* from [3H]cinnamate and [35S]sulfate indicated that the terminal bud and the first pair of expanded leaves are the most active organs in sulfated flavonoid biosynthesis.[11] The incorporation of label from sulfate was proportional to the number of sulfate ester linkages of the individual flavonoids (Table 1). The fact that double labeling experiments resulted in constant incorporation ratios of both [3H] and [35S] labels into flavonoid sulfates, indicates that sulfation is a later step in their biosynthesis.[11]

HPLC studies of the distribution of flavonoid sulfates in different parts of *F. bidentis* indicates their highest accumulation (as μmol/g fresh tissue) in the terminal bud and the first pair of expanded leaves, which drops dramatically in older leaves.[25] In addition, sulfotransferase (ST) activity corresponds with the amount of sulfated flavonoids in these organs. ST activity decreases from the upper to the lower internode segments, and from the base to the apex of leaves. Western blot analysis of protein extracts shows that variations in ST activities in different tissues correlate well with the amounts of immunodetected enzyme protein. It seems, therefore, that both sulfated flavonoids and ST activity are associated with actively growing tissues, although they are absent in the root system.[25]

ENZYMATIC SULFATION OF FLAVONOIDS

In contrast with the extensive studies on the sulfation of endogenous metabolites and xenobiotics in animal tissue,[26,27] it was only recently that enzymatic sulfation of flavonoids has been demonstrated for the first time in plants.[28,29] This was rendered possible by the acquisition of an extensive library of specifically substituted flavonoid compounds,[10] as well as the development of a reliable assay for ST activity.[30] This assay makes use of TBADP which generates an ion pair with the enzymatically formed flavonoid sulfate ester. The latter can then be extracted in an organic solvent, such as ethyl acetate, whereas the sulfate donor (3'-phosphoadenosine 5'-phosphosulfate, PAPS) remains in the aqueous reaction medium.[30] Using cell-free extracts of *F. bidentis* as the enzyme source and quercetin as substrate, resulted in the formation of its mono- to tetrasulfated derivatives. These results suggested the existence of a family of ST activities involved in the biosynthesis of polysulfated flavonols in this species (Table 1).[28]

FLAVONOL SULFOTRANSFERASES

Three distinct, flavonol-specific sulfotransferases (3, 3' and 4'-STs) were highly purified from the shoot tips of *F. chloraefolia* by fractional precipitation with ammonium sulfate, followed by chromatography on Sephacryl S-200, 3'-phosphoadenosine 5'-phosphate (PAP)-Agarose affinity column and chromato-focusing on Mono P.[31] Using the same protocol, another enzyme (7-ST) was partially purified from the shoot tips of *F. bidentis*.[32] These enzymes exhibit expressed specificity for positions 3 of various flavonol acceptors (3-ST), 3' and

Table 2. Properties of *Flaveria* sulfotransferases.

Property	3-ST	3'-ST	4'-ST	7-ST
K_m Flavonoid (μM)[1]	0.2	0.29	0.36	0.24
K_m PAPS (μM)	0.2	0.35	0.38	0.33
pH optimum	6.0, 8.5	7.5	7.5	7.5
Apparent pI	5.4	6.0	5.1	6.5
Apparent mol wt (D)	35,000	35,000	35,000	35,000

[1] With quercetin as substrate for the 3-ST, quercetin 3-sulfate for the 3' and 4'-STs and quercetin 3,3'-disulfate for the 7-ST.

4' of flavonol 3-sulfate (3' and 4'-STs) and 7 of quercetin 3,3'- and -3,4'-disulfate (7-ST). The four enzymes can only be resolved by chromatofocusing and exhibit distinct pI values between pH 5.1 and 6.5 (Table 2). They have similar molecular weights, exhibit no requirement for divalent cations, and are not inhibited by SH group reagents. Their K_m values for both the sulfate donor, PAPS and the flavonoid acceptors are in the micromolar range (Table 2).[31,32]

Sequential Order of Enzymatic Sulfation

The natural occurrence of flavonol disulfate esters in *F. chloraefolia*,[13,17] and of a variety of flavonol mono- to tetrasulfates in *F. bidentis*,[11] raises the question as to the sequential order of enzymatic sulfation in both species. It is proposed that the first two steps of enzymatic sulfation (Scheme 1) are probably identical in both species, although quercetin 3,3'-disulfate does not accumulate in detectable amounts in *F. bidentis*. However, further sulfation of quercetin 3,3'- and -3,4'-disulfates by the 7-ST gives rise to 3,7,3'- and 3,7,4'-trisulfate ester derivatives; both of which are natural constituents of this species (Table 1).[11] The fact that *F. bidentis* is the only species known to accumulate quercetin 3,7,3',4'-tetrasulfate, implies the existence in this plant of one or two ring B-specific STs which catalyze further sulfation of quercetin 3,7,3'- and -3,7,4'-trisulfates to their tetrasulfate ester derivatives.[32]

Another question concerns the biosynthetic origin of quercetin 3,7-

disulfate which accounts for about 70% of the flavonoids in *F. bidentis*.[11] Since the 7-ST of this species accepts only quercetin 3,3'- and -3,4'-disulfates, but not the 3-sulfate, as substrates, its involvement in the formation of the 3,7-disulfate should be excluded. Although there is no evidence, as yet, for enzymatic desulfation in plants, it is conceivable that flavonoid-specific aryl sulfatases may exist in *Flaveria* in a manner similar to those of animal tissues[33] and microorganisms.[34] The action of such sulfatases on quercetin tri- or tetrasulfate esters may explain the accumulation of quercetin 3,7-disulfate in *F. bidentis* (Scheme 1).[30]

Purification and Properties of Flavonol 3-Sulfotransferase

The fact that sulfation at position 3 is the first step in the enzymatic synthesis of polysulfated flavonoids,[31] and that flavonol 3-sulfates are of common occurrence in plants,[6] calls for a detailed study of this novel enzyme.

The flavonol 3-ST has been purified to apparent homogeneity from *F. chloraefolia* by ammonium sulfate precipitation and successive chromatography

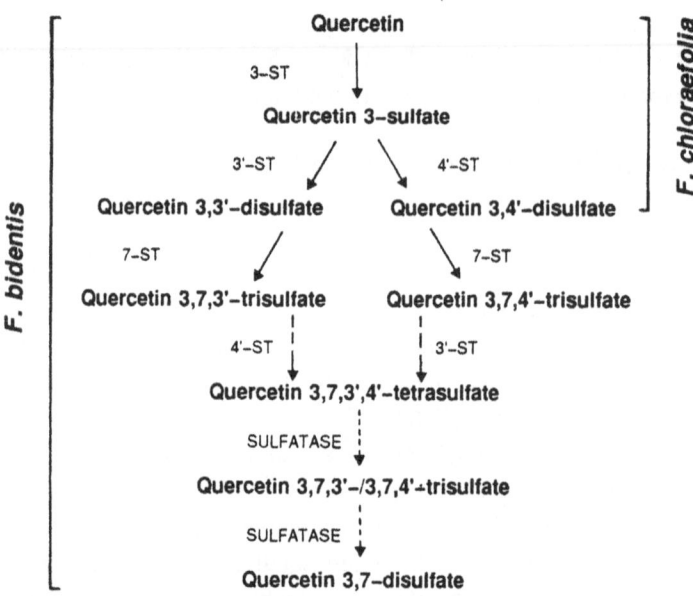

Scheme 1. Proposed model for the biosynthesis of polysulfated flavonols in *Flaveria* species.

Table 3. Purification table for flavonol 3-sulfotransferase.

Purification step	Total Protein (mg)	Specific activity (pkat/mg)	Total activity (pkat)	Purification fold	% Recovery
Dowex	1087	0.68	739	—	100
Sephadex G-100	106	7.19	766	10.6	106
DEAE-Sephacel	25	26.86	671	39.5	91
Hydroxyapatite	3.7	56.91	208	83.7	28
PAP-agarose	0.425	413.2	176	607.6	24
Mono Q	0.050	1356.8	68	1994	9.2

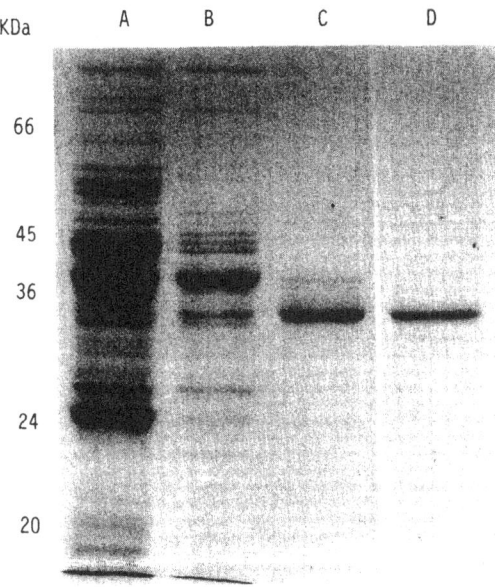

Fig. 1. SDS-PAGE of fractions collected during purification of the 3-ST. Lanes (A) DEAE-sephacel (ca. 10 µg), (B) Hydroxyapatite (ca. 10 µg), (C) PAP-agarose (ca. 4 µg), (D) Mono Q (ca. 1 µg). Proteins were visualized by Coomassie staining. Numbers on the left correspond to molecular weight markers.

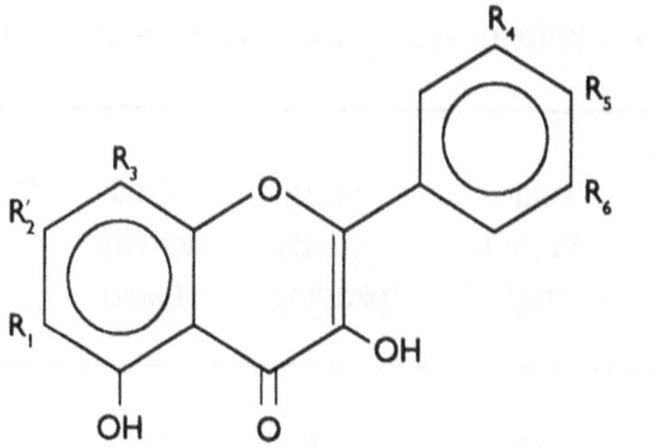

Flavonoid	R_1	R_2	R_3	R_4	R_5	R_6
Quercetin	H	OH	H	OH	OH	H
Kaempferol	H	OH	H	H	OH	H
Rhamnetin	H	OMe	H	OH	OH	H
Isorhamnetin	H	OH	H	OMe	OH	H
Patuletin	OMe	OH	H	OH	OH	H
Quercetagetin	OH	OH	H	OH	OH	H
Gossypetin	H	OH	OH	OH	OH	H
Myricetin	H	OH	H	OH	OH	OH
Galangin	H	OH	H	H	H	H

Scheme 2. Structural formulae of the flavonoids described in this paper.

on Sephadex G-100, DEAE-Sephacel, hydroxyapatite, affinity chromatography on PAP-agarose and finally on a Mono Q column (Table 3).[35] The Mono Q-purified protein migrated as a single band on SDS-PAGE with an M_r of 34,500 (Fig. 1), which was consistent with the M_r of the native enzyme (35,000) as determined by gel filtration on a Superose 12 column.[35]

The 3-ST exhibited strict specificity for position 3 of several flavonols in the following order of decreasing activity, with rhamnetin> isorhamnetin> quercetin> patuletin> kaempferol, as sulfate acceptors (Scheme 2). Substrate inhibition was observed when either of the above compounds was assayed at concentrations higher than 2 μM. The 3-ST accepted neither flavonols with additional hydroxyl groups at position 6 (quercetagetin), 8 (gossypetin) or 5' (myricetin), nor flavonols lacking ring-B hydroxylation (e.g. galangin) (Scheme 2).[35] The high affinity of the 3-ST for both the sulfate and the flavonol acceptors indicates that both metabolites are present in low concentration inside the cells at the site of synthesis. Furthermore, the inhibition of enzyme activity at flavonol concentration above K_m suggests a very tight control of flavonol sulfate biosynthesis. Similar substrate inhibition patterns were obtained with the hydroxysteroid ST from human liver[36] and for the flavonoid O-methyl-transferases of *Chrysosplenium americanum*.[37]

Substrate interaction kinetics of the flavonol substrate and the sulfate donor, PAPS gave converging lines that are consistent with a sequential bisubstrate binding mechanism. Product inhibition studies showed competitive inhibition between PAPS and PAP, and noncompetitive patterns between the flavonol substrate and either of the reaction products, PAP or the flavonol 3-sulfate. The kinetic analysis is consistent with an ordered Bi Bi mechanism where the sulfate donor is the first substrate to bind to the enzyme, and PAP is the final product released.[35]

The purified 3-ST was cleaved with trypsin and the peptides fractionated by reverse-phase HPLC on a Vydac C_{18} microbore column. Two peptides were selected for amino acid sequencing and yielded the following residues: Tyr-Lys-Asp-Ala-Trp-Asn-His-Gln-Glu-Phe-Leu-Glu-Gly-Arg, and Leu-Ser-Val-Glu-Glu-Ala-Phe-Asp-Glu-Phe-Cys-Gln-Gly-Ile-Ser-Ser-Cys-Gly-Pro-Tyr-X-Glu-His-Ile-Lys-Gly-Tyr-Lys. The amino acid released at cycle 26 of the second peptide could not be identified, although a tryptophan residue was tentatively assigned based on the absorption of this peptide at 292 nm.[35] It is interesting to note that comparison of the two peptide sequences with those reported for other mammalian STs[38-40] did not reveal any significant amino acid similarities, nor did antibodies raised against the phenol ST from human liver[41] recognize *Flaveria* 3-ST on Western blotting (C.N. Falany, personal communication).

Polyclonal antibodies raised against *F. chloraefolia* flavonol 3-ST[35] were found to cross-react with the 3'- and 4'-STs of the same plant,[31] suggesting epitope similarities among the three enzymes (Fig. 2). This result was not unexpected in view of the similar catalytic functions and chromatographic behavior of the three enzymes. The anti-3-ST antibody was used for the isolation of cDNA clones coding for *F. chloraefolia* flavonol STs.

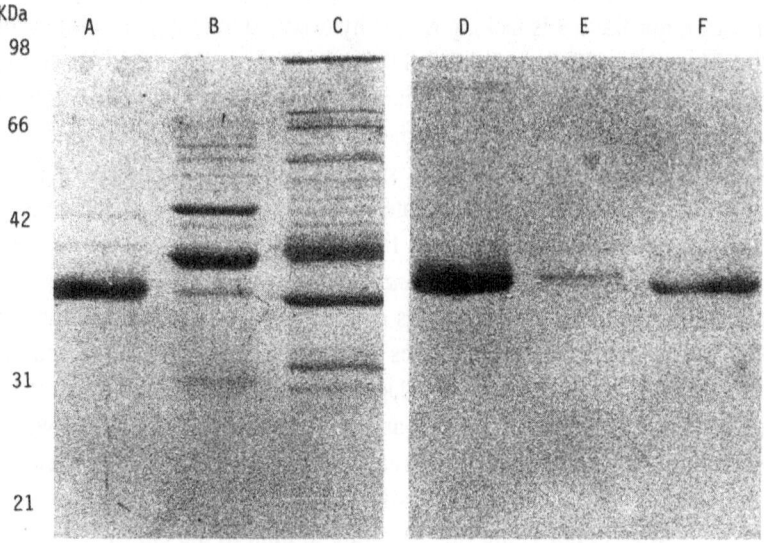

Fig. 2. Cross reactivity of the anti-3-ST antibodies with the 3-(A,D),
3'-(B,E), and the 4'-STs (C,F). Proteins were electrophoresed on 12%
SDS-PAGE and electrotransferred to nitrocellulose. Lanes A, B and C
were stained with amido black for protein detection. Lanes D, E and F
were immunodetected with the anti-3-ST antibodies. Numbers on the
left correspond to molecular weight markers.

MOLECULAR CLONING OF *F. chloraefolia* FLAVONOL STs

cDNA clones coding for the flavonol 3- (pFST3) and 4'-STs (pFST4')
were isolated from a cDNA library produced from poly(A)[+] mRNAs isolated
from shoot tips of *F. chloraefolia*.[42]

Several direct and indirect evidence confirm the authenticity of pFST3
and pFST4' as coding for the 3- and 4'-ST respectively. First, the matches
between the deduced and directly determined amino acid sequences of the 3-ST
(Fig.3). Second, the recognition by the anti-3-ST antibodies of a protein
expressed in the bacteria expressing flavonol ST activities but not in bacteria
harboring a control plasmid. Finally, the most significant evidence is the
position and substrate specificity of the ST activities expressed in *E. coli*
harboring the pFST3 and pFST4' plasmids.

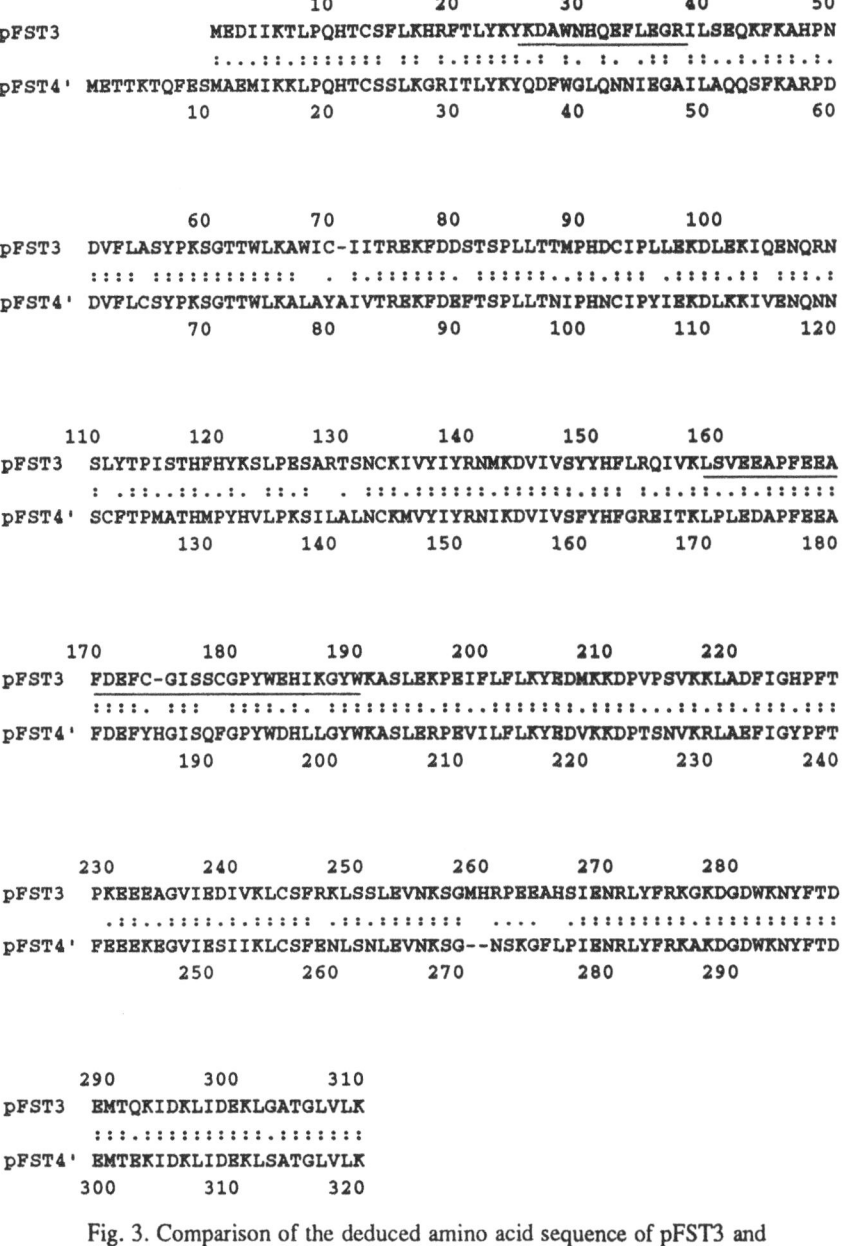

Fig. 3. Comparison of the deduced amino acid sequence of pFST3 and pFST4'. (:) indicates the presence of the same residue; (.) represents a conserved residue; dashes represent gaps introduced for maximal alignment. The underlined amino acids of pFST3 correspond to the sequence of the two tryptic peptides.

The low frequency of cDNA clones coding for the 3-ST (6 clones) compared to the 4'-ST (42 clones) is quite surprising when we consider that the 3-ST is catalysing the first step in the sequential formation of polysulfated flavonoids. Furthermore, no cDNA clone coding for the 3'-ST was isolated. This low representation of the 3- and 3'-ST might be due to the higher instability of their mRNAs, or to the low expression of their respective genes at the time of mRNA extraction.

The complete nucleotide sequence of pFST3 and pFST4' has been determined. Comparison of the deduced amino acid sequence of pFST3 and pFST4' revealed an overall positional identity of 69% in 311 amino acid residues between the two proteins (Fig. 3). The homology between the two STs at both the nucleotide and amino acid level suggests that early in the evolutionary process, gene duplication followed by sequence divergence between the two loci has occurred to the point where they no longer function as duplicate loci but rather as complementary genes. The high amino acid sequence similarities provide an explanation for the difficulty observed in trying to separate the two proteins by chromatographic methods and the cross-reactivity of the anti 3-ST antibodies with the 4'-ST. The even distribution of the amino acid identities all over the protein sequences renders difficult the identification of domains which might be necessary for catalysis or might act as determinants for the specificity of the two enzymes.

The two STs of *F. chloraefolia* also share sequence similarities with other STs characterized from animal tissues (Fig. 4). About 28.9% and 30.7% identity is observed with the hydroysteroid ST cDNA clone isolated from rat liver[38] and the estrogen ST cDNA clone isolated from bovine placenta.[40] Furthermore, 30.7% identity is observed between the 3-ST cDNA clone and the senescence marker protein 2 (SMP2) isolated from rat liver.[43] The common evolutionary origin of plant and animal STs becomes more evident when we analyze their aligned sequences (Fig. 4). In addition to their similar length, the presence of identical amino acid residues is distributed all over the sequences with four especially well conserved regions. The stability of the sequence in these four regions through divergent evolution suggests that they play an important role in the function of these proteins. We can speculate on the role of these regions for the appropriate folding of the proteins to create the catalytic domain or in the binding of the co-substrate PAPS.

The catalytic function of SMP-2 was unknown when the cloning of its cDNA was reported.[43] Recently, however, Agura *et al.* reported the isolation of a cDNA clone coding for an hydroxysteroid ST which has 74% homology with the SMP-2 deduced amino acid sequence.[39] This result, in addition to the

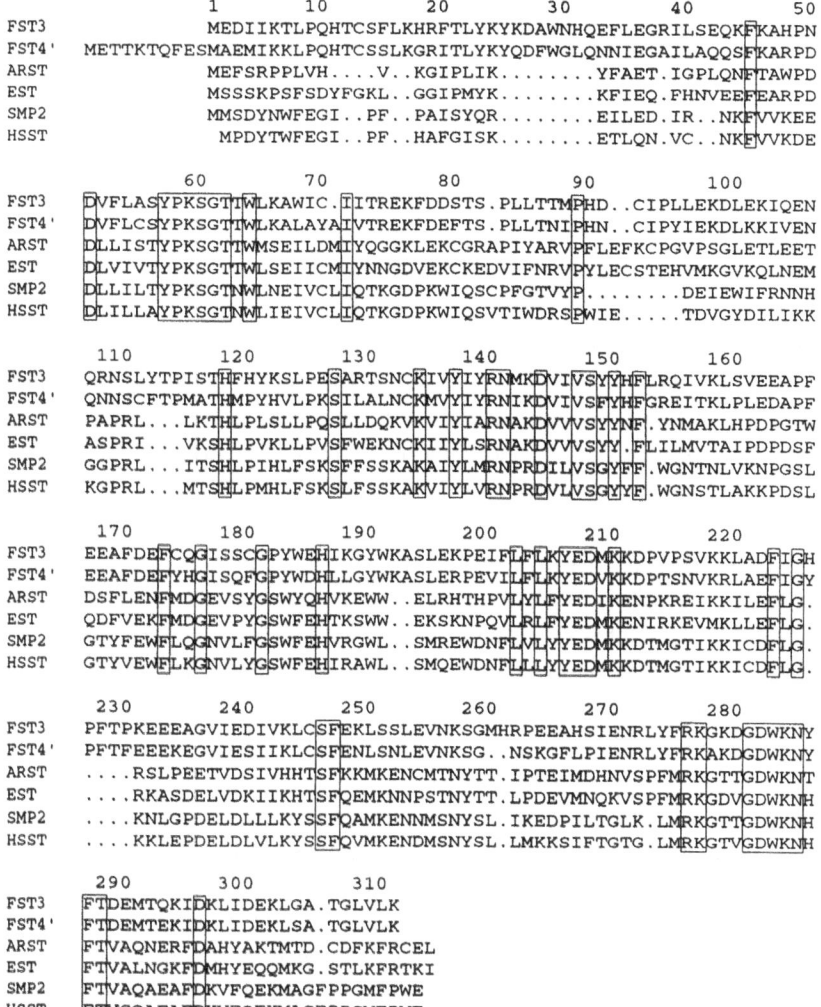

```
             1        10        20        30        40        50
FST3         MEDIIKTLPQHTCSFLKHRFTLYKYKDAWNHQEFLEGRILSEQKFKAHPN
FST4'   METTKTQFESMAEMIKKLPQHTCSSLKGRITLYKYQDFWGLQNNIEGAILAQQSFKARPD
ARST         MEFSRPPLVH....V..KGIPLIK.......YFAET.IGPLQNFTAWPD
EST          MSSSKPSFSDYFGKL..GGIPMYK.......KFIEQ.FHNVEEFEARPD
SMP2         MMSDYNWFEGI..PF..PAISYQR.......EILED.IR..NKFVVKEE
HSST         MPDYTWFEGI..PF..HAFGISK.......ETLQN.VC..NKFVVKDE

             60        70        80        90        100
FST3         DVFLASYPKSGTTIWLKAWIC.ILITREKFDDSTS.PLLTTMPHD..CIPLLEKDLEKIQEN
FST4'        DVFLCSYPKSGTTIWLKALAYAIVTREKFDEFTS.PLLTNIPHN..CIPYIEKDLKKIVEN
ARST         DLLISTYPKSGTTWMSEILDMIYQGGKLEKCGRAPIYARVPFLEFKCPGVPSGLETLEET
EST          DLVIVTYPKSGTTWLSEIICMIYNNGDVEKCKEDVIFNRVPYLECSTEHVMKGVKQLNEM
SMP2         DLLILTYPKSGTNWLNEIVCLIQTKGDPKWIQSCPFGTVYP........DEIEWIFRNNH
HSST         DLILLAYPKSGTNWLIEIVCLIQTKGDPKWIQSVTIWDRSPWIE.....TDVGYDILIKK

             110        120        130        140        150        160
FST3         QRNSLYTPISTHFHYKSLPESARTSNCKIVYIYRNMKDVIVSYYHFLRQIVKLSVEEAPF
FST4'        QNNSCFTPMATHMPYHVLPKSILALNCKMVYIYRNIKDVIVSFYHFGREITKLPLEDAPF
ARST         PAPRL...LKTHLPLSLLPQSLLDQKVKVIYIARNAKDVVVSYYNF.YNMAKLHPDPGTW
EST          ASPRI...VKSHLPVKLLPVSFWEKNCKIIYLSRNAKDVVVSYY.FLILMVTAIPDPDSF
SMP2         GGPRL...ITSHLPIHLFSKSFFSSKAKAIYLMRNPRDILVSGYFF.WGNTNLVKNPGSL
HSST         KGPRL...MTSHLPMHLFSKSLFSSKAKVIYLVRNPRDVLVSGYYF.WGNSTLAKKPDSL

             170        180        190        200        210        220
FST3         EEAFDEFCQGISSCGPYWEHIKGYWKASLEKPEIFLFLKYEDMKKDPVPSVKKLADFIGH
FST4'        EEAFDEFYHGISQFGPYWDHLLGYWKASLERPEVILFLKYEDVKKDPTSNVKRLAEFIGY
ARST         DSFLENFMDGEVSYGSWYQHVKEWW..ELRHTHPVLYLFYEDIKENPKREIKKILEFLG
EST          QDFVEKFMDGEVPYGSWFEHTKSWW..EKSKNPQVLRLFYEDMKENIRKEVMKLLEFLG
SMP2         GTYFEWFLQGNVLFGSWFEHVRGWL..SMREWDNFLVLYYEDMKKDTMGTIKKICDFLG
HSST         GTYVEWFLKGNVLYGSWFEHIRAWL..SMQEWDNFLLLYYEDMKKDTMGTIKKICDFLG

             230        240        250        260        270        280
FST3         PFTPKEEEAGVIEDIVKLCSFEKLSSLEVNKSGMHRPEEAHSIENRLYFRKGKDGDWKNY
FST4'        PFTFEEEKEGVIESIIKLCSFENLSNLEVNKSG..NSKGFLPIENRLYFRKAKDGDWKNY
ARST         ....RSLPEETVDSIVHHTSFKKMKENCMTNYTT.IPTEIMDHNVSPFMRKGTTGDWKNT
EST          ....RKASDELVDKIIKHTSFQEMKNNPSTNYTT.LPDEVMNQKVSPFMRKGDVGDWKNH
SMP2         ....KNLGPDELDLLLKYSSFQAMKENNMSNYSL.IKEDPILTGLK.LMRKGTTGDWKNH
HSST         ....KKLEPDELDLVLKYSSFQVMKENDMSNYSL.LMKKSIFTGTG.LMRKGTVGDWKNH

             290        300        310
FST3         FTDEMTQKIDKLIDEKLGA.TGLVLK
FST4'        FTDEMTEKIDKLIDEKLSA.TGLVLK
ARST         FTVAQNERFDAHYAKTMTD.CDFKFRCEL
EST          FTVALNGKFDMHYEQQMKG.STLKFRTKI
SMP2         FTVAQAEAFDKVFQEKMAGFPPGMFPWE
HSST         FTVSQAEAFDKVFQEKMAGFPPGMFPWE
```

Fig. 4. Amino acid alignment of *F. chloraefolia* pFST3 and pFST4', rat liver hydroxysteroid ST (HSST), bovine placenta oestrogen ST (EST), rat liver aryl ST (ARST) and rat liver senescence marker protein 2 (SMP2). Boxes show residues common to the six sequences.

homology observed with the deduced amino acid sequences of the flavonol STs, strongly suggest that SMP-2 codes for a ST.

PERSPECTIVE FOR FUTURE WORK

The availability of antibodies against both the sulfated flavonols and the ST proteins, as well as two ST cDNA clones, provide the basic tools for new experimental strategies in order to define the role of sulfate conjugation with flavonoids in plants. These include the localization of the ST proteins at the tissue and subcellular level using immunocytochemistry; the study of environmental and developmental factors governing ST gene expression; the isolation of genomic clones and analysis of *cis*-acting gene regulatory signals such as promoters, enhancers and silencers by the introduction of dissected gene constructs in transgenic plants; as well as the mapping of the amino acids which are critical for ST activity by site directed mutagenesis.

Another strategy involves the *in vivo* manipulation of plants to create a new sulfate metabolic sink through the expression of the 3-ST gene into flavonol-producing plants which accumulate sulfur-containing deleterious metabolites, such as glucosinolates in the Brassicacae family. This implies that the newly formed sulfated metabolites do not have a deleterious effect on plant growth, or any effect on seed quality of the forage produced by such a transgenic plant. It also implies that overexpression of the flavonol ST enzymes would compete with the ST involved in glucosinolate biosynthesis for the sulfate donor PAPS. The recent work of D.D. Lefebvre demonstrates the feasibility of such an approach in creating an artificial sulfur sink.[44] This was achieved by introducing a foreign gene coding for a product rich in the amino acid cysteine which reduced the sulfur pool of the plant available for glucosinolate biosynthesis and, consequently, reduced metabolite accumulation. The production of transgenic *Brassica napus* harboring the flavonol 3-ST gene with the aim of creating a new variety with low glucosinolate content is presently in progress.

ACKNOWLEDGMENTS

I would like to thank Dr. Denis Barron, Dr. Normand Brisson, Dr. Vincenzo DeLuca and Dr. Ragai Ibrahim who shared their data with me for this publication.

REFERENCES

1. KAWAGUCHI, R., KIM, K.W. 1937. The constituents of *Persicaria hydropiper*. J. Pharm. Sci. Japan 57:767-769.

2. HARBORNE, J.B. 1975. Flavonoid sulphates: a new class of sulphur compounds in higher plants. Phytochemistry 14:1147-1155.

3. HARBORNE, J.B. 1977. Flavonoid sulphates: a new class of natural products of ecological significance in plants. In: Progress in Phytochemistry, Vol. 4. (L. Reinhold, J.B. Harborne, T. Swain, eds.) Pergamon Press, New York, pp. 189-208.

4. HARBORNE, J.B., WILLIAMS, C.A. 1982. Flavone and flavonol glycosides. In: The Flavonoids: Advances in Research. (J.B. Harborne, T.J. Mabry, eds.) Chapman and Hall, pp. 261-312.

5. TIMMERMANN, B.N. 1980. Phytochemical investigations of the genus *Brickellia* (Compositae) emphasizing flavonoids. Ph.D. Thesis, University of Texas, Austin, TX.

6. BARRON, D., VARIN, L., IBRAHIM, R.K., HARBORNE, J.B., WILLIAMS, A.C. 1988. Sulphated flavonoids—an update. Phytochemistry 27:2375-2395.

7. MULDER, G.J. 1981. Sulfation of Drugs and Related Compounds. CRC Press, Boca Raton, FL.

8. SCHILDKNECHT, H., MEIER-AUGENSTEIN, W. 1990. The Pulvinus: Motor Organ For Leaf Movement. (R.L. Satter, H.L. Gorton and T.C. Vogelmann, eds.) American Society of Plant Physiologists, Rockville, MD. pp. 205-213.

9. TRUCHET, G., ROCHE, P., LEROUGE, P., VASEE, J., CAMUT, S., DEBILLY, F., PROME, J.C., DENARIÉ, J. 1991. Sulphated lipo-oligosaccharide signals of *Rhizobium meliloti* elicit root nodule organogenesis in alfalfa. Nature 351:670-673.

10. BARRON, D. 1987. Advances in phytochemistry, organic synthesis, spectral analysis and enzymatic synthesis of sulfated flavonoids. Ph.D. Thesis, Concordia University, Montréal.

11. VARIN, L., BARRON, D., IBRAHIM, R.K. 1986. Identification and biosynthesis of sulfated and glucosylated flavonoids in *Flaveria bidentis*. Z. Naturforsch. 41c:813-819.

12. BARRON, D., IBRAHIM, R.K. 1987. 6-Methoxykaempferol 3-monosulfates from *Flaveria chloraefolia*. Phytochemistry 26:2085-2088.

13. BARRON, D., IBRAHIM, R.K. 1987. Quercetin and patuletin 3,3'-disulfates from *Flaveria chloraefolia*. Phytochemistry 26:1181-1184.

14. NAWWAR, M.A.M., BUDDRUS, J. 1981. A gossypetin glucuronide sulfate from the leaves of *Malva sylvestris*. Phytochemistry 20:2446-2448.

15. TOMAS-BARBERAN, F.A., HARBORNE, J.B., SELF, R. 1987. Twelve 6-oxygenated flavone sulphates from *Lippia nodiflora* and *L. canescens*. Phytochemistry 26:2281-2285.

16. MUES, R., TIMMERMANN, B.N., OHNO, M., MABRY, T.J. 1979. 6-Methoxyflavonoids from *Brickellia californica*. Phytochemistry 18:1379-1383.

17. BARRON, D., COLEBROOK, L., IBRAHIM, R.K. 1986. An equimolar mixture of quercetin 3-sulfate and patuletin 3-sulfate from *Flaveria chloraefolia*. Phytochemistry 25:1719-1721.

18. WILLIAMS, C.A., HARBORNE, J.B., CLIFFORD, H.T. 1973. Negatively charged flavones and tricin as chemosystematic markers in the Palmae. Phytochemistry 12:2417-2430.

19. HARBORNE, J.B., ed. 1989. Methods in Plant Biochemistry. Vol. 1. Academic Press, New York.

20. MARKHAM, K.R. 1982. Techniques of Flavonoid Identification. Academic Press, London.

21. BARRON, D., IBRAHIM, R.K. 1988. Synthesis of flavonoid sulfates. II. The use of aryl sulfatase in the synthesis of flavonol 3-sulfates. Z. Naturforsch. 43c:625-630.

22. BARRON, D., IBRAHIM, R.K. 1988. Hydrochloric acid and aryl sulfatase as reagents for UV spectral detection of 3- and 4'-sulfated flavonoids. Phytochemistry 27:2335-2338.

23. BARRON, D., IBRAHIM, R.K. 1988. Synthesis of flavonoid sulfates. III. Synthesis of 3',4'-*o*-disulfates using sulfur trioxide-trimethyl-amine complex, and of 3-sulfates using aryl sulfatase. Z. Naturforsch. 43c:631-635.

24. BARRON, D., IBRAHIM, R.K. 1987. Synthesis of flavonoid sulfates. I. Stepwise sulfation of positions 3, 7, and 4' using *N,N'*-dicyclohexyl-carbodiimide and tetrabutylammonium hydrogen sulfate. Tetrahedron 43:5197-5202.

25. HANNOUFA, A., VARIN, L., IBRAHIM, R.K. 1991. Spatial distribu-tion of flavonoid conjugates in relation to glucosyltransferase and sulfotransferase activities in *Flaveria bidentis*. Plant Phys. 97:259-263.

26. ROY, A.B. 1981. Sulfotransferases. In: Sulfation of Drugs and Related

Compounds. (G.J. Mulder, ed.) CRC Press, Boca Raton, FL, pp. 83-130.

27. HOBKIRK, R. 1985. Steroid sulfotransferases and steroid sulfate sulfatases: characteristics and biological roles. Can. J. Biochem. Cell Biol. 63:1127-1144.

28. VARIN, L., BARRON, D., IBRAHIM, R.K. 1987. Enzymatic synthesis of sulfated flavonols in *Flaveria spp.* Phytochemistry 26:135-138.

29. VARIN, L. 1990. Enzymology of flavonoid sulfation: purification, characterization and molecular cloning of a number of flavonol sulfotransferases from *Flaveria spp.* Ph.D. Thesis, Concordia University, Montréal.

30. VARIN, L., BARRON, D., IBRAHIM, R.K. 1987. Enzymatic assay for flavonol sulfotransferase. Anal. Biochem. 161:176-180.

31. VARIN, L., IBRAHIM, R.K. 1989. Partial purification and characterization of three flavonol-specific sulfotransferases from *Flaveria chloraefolia*. Plant Physiol. 90:977-981.

32. VARIN, L., IBRAHIM, R.K. 1991. Partial purification and some properties of flavonol 7-sulfotransferase from *Flaveria bidentis*. Plant Physiol. 95:1254-1258.

33. POWELL, G.M., OLAVESEN, A.H. 1981. The fate of sulfate esters *in vitro*. In: Sulfation of Drugs and Related Compounds (G.J. Mulder, ed.) CRC Press, Boca Raton, FL, pp. 187-212.

34. DODGSON, K.S., WHITE, G.F., FITZGERALD, J.W. 1982. The Arylsulfatases. In: Sulfatases of Microbial Origin. (K.S. Dodgson, ed.) CRC Press, Boca Raton, FL, pp. 103-168.

35. VARIN, L., IBRAHIM, R.K. 1992. Novel flavonol 3-sulfotransferase: purification, kinetic properties and partial amino acid sequence. J. Biol. Chem. 267:1858-1863.

36. FALANY, C.N., VAZQUEZ, M.E., 1989. Purification and characterization of human liver dehydroepiandrosterone sulfotransferase. Biochem. J. 260:641-646.

37. IBRAHIM, R.K., DELUCA, V., KHOURI, H., LATCHINIAN, L., BRISSON, L., BARRON, D. CHAREST, P.M. 1987. Enzymology and compartmentation of polymethylated flavonol glucosides in *Chrysosplenium americanum*. Phytochemistry 26:1237-1245.

38. OGURA, K., KAJITA, J., NARIHATA, H., WATABE, T., OZAWA, S., NAGATA, K., YAMAZOE, Y. KATO, R. 1989. Cloning and sequence analysis of a rat liver cDNA encoding hydroxysteroid sulfotransferase. Biochem. Biophys. Res. Commun. 165:168-174.

S., NAGATA, K., YAMAZOE, Y. KATO, R. 1990. cDNA cloning
of the hydroxysteroid sulfotransferase ST sharing a strong homology
in amino acid sequence with the senescence marker protein SMP-2 in
rat livers. Biochem. Biophys. Res. Commun. 166:1494-1500.

40.   NASH, A., GLENN, W.K., MOORE, S.S., KERR, J., THOMPSON,
A.R., THOMPSON, E.O.P. 1988. Oestrogen sulfotransferase:
Molecular cloning and sequencing of cDNA for the bovine placental
enzyme. Aust. J. Biol. Sci. 41:507-516.

41.   FALANY, C.N., VAZQUEZ, M.E., HEROUX, J.A. ROTH, J.A. 1990.
Purification and characterization of human liver phenol-sulfating
phenol sulfotransferase. Arch. Biochem. Biophys. 278:312-318.

42.   VARIN, L., DELUCA, V., IBRAHIM, R.K., BRISSON, N. 1992.
Molecular characterization of two plant flavonol sulfotransferases
Proc. Natl. Acad. Sci. *(USA)* (in press).

43.   CHATTERJEE, B., MAJUMDAR, D., OZBILEN, O., MURTY, R.,
ROY, A.K. 1987. Molecular cloning and characterization of cDNA
for androgen-repressible rat liver protein SMP-2' J. Biol. Chem.
262:822-825.

44.   LEFEBVRE, D.D. 1990. Expression of mammalian metallothionein
suppresses glucosinolate synthesis in *Brassica campestris*. Plant
Physiol. 93:522-524.

Chapter Nine

SYNTHESIS AND BASE-CATALYZED TRANSFORMATIONS OF PROANTHOCYANIDINS

Daneel Ferreira, Jan P. Steynberg, Johann F.W. Burger
and Barend C.B. Bezuidenhoudt

FRD Research Unit for Polyphenol and Synthetic Chemistry
Department of Chemistry, University of the Orange Free State
P.O. Box 339, Bloemfontein, 9300 South Africa

INTRODUCTION

The proanthocyanidins (oligoflavanoids or condensed tannins) are major phenolic constituents of woody and some herbaceous plants.[1] Their exceptionally high concentrations in the barks and heartwoods of a variety of tree

Phenolic Metabolism in Plants, Edited by H.A. Stafford
and R.K. Ibrahim, Plenum Press, New York, 1992

species have resulted in their commercial extraction with the initial objective of applying the extracts in leather manufacture.[2] The chemistry of these exceptionally reactive natural products represents an intricate but fascinating area of research, creating exciting challenges to those involved in their chemical manipulation. Factors limiting[3] this area of research have been the complexity of condensed tannin composition, the lack of a universal method for both the synthesis and assessment of the absolute stereochemistry of the interflavan linkage, the phenomenon of dynamic rotational isomerism about interflavanoid bonds thus complicating NMR spectral investigations, and the lack of knowledge regarding the points of bonding at nucleophilic centres of the flavan-3-ol chain extender units. The more significant initial work on oligoflavanoids was, therefore, limited mainly to an analytical approach involving biflavanoids.[4-7] Notable exceptions were the earlier efforts by Geissman and Yoshimura[8] and by Weinges and his co-workers[9] at synthesizing procyanidin biflavanoid derivatives. However, by their very nature, these synthetic methods would not permit their extension to higher oligomers.

Against this background we and others embarked on exploring methods of synthesis with a view to comparing the physical data of synthetic and naturally occurring compounds, hence eliminating most of the complicating factors described above. In additon, availability of the synthetic oligomers offered the opportunity to study their base-catalyzed transformations; such an approach being pursued in view of the fact that the majority of the industrial applications of condensed tannins involve their dissolution and/or reaction at alkaline pH.[10,11] Principles which have emanated from these studies constitute the subject of this paper.

SYNTHESIS OF OLIGOFLAVANOIDS

Selection of Precursors to Synthetic Oligoflavanoids

In those flavanoid metabolic pools that possess the potential for proanthocyanidin formation, flavan-3,4-diols, e.g. (1), when considered as p-hydroxybenzyl alcohols, represent precursors to C-4 carbocations (2) under mild acidic conditons.[12] These may be trapped by interaction with the nucleophilic centres of the ubiquitous flavan-3-ols,[12] e.g. (3), exhibiting *m*-substituted A-rings (Scheme 1) to form predominantly the (4α,8)- and (4β,8)-biflavanoids (4) and (5), and to a lesser extent also the (4,6)-regiomers (6) and (7).[13] Substitution at the remaining and more potent nucleophilic site of the D-ring

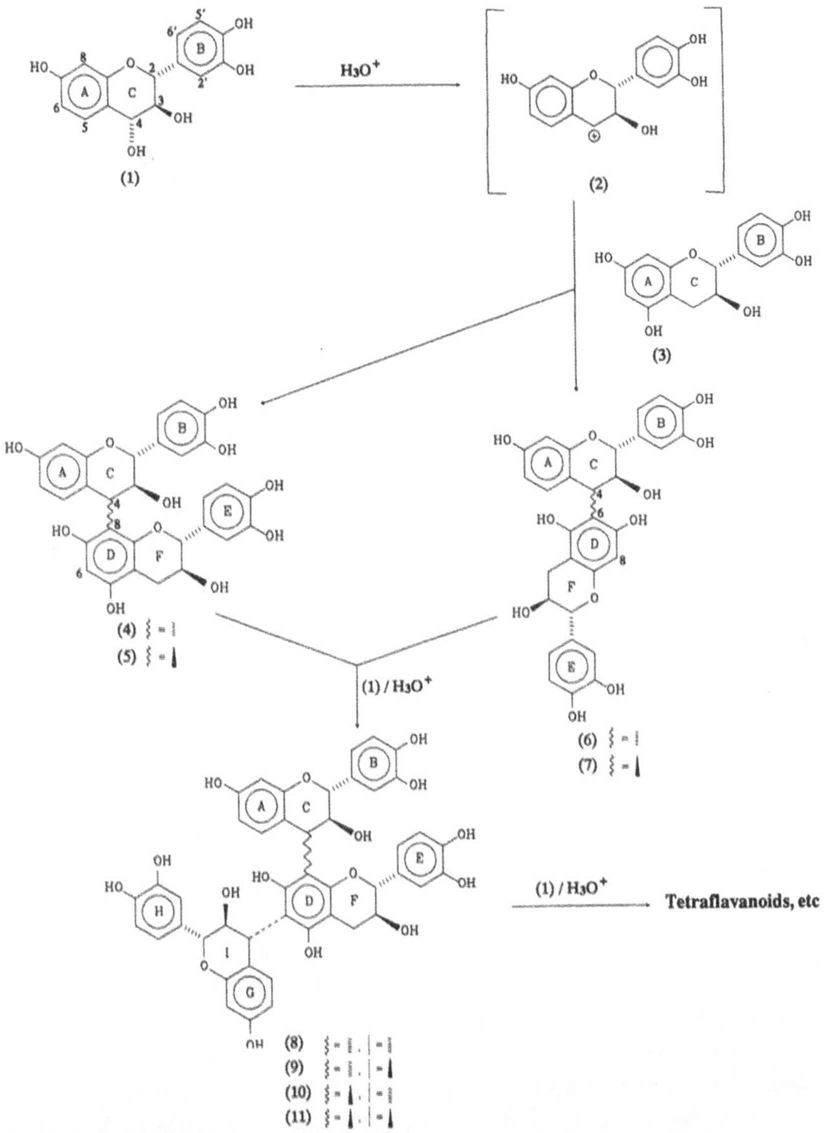

Scheme 1. Flavan-3,4-diols and flavan-3-ols as precursors to oligoflavonoids under mild acidic conditions.

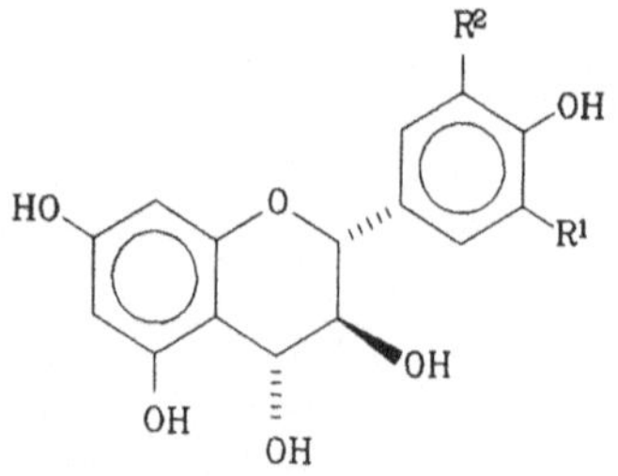

(12) $R^1 = H, R^2 = OH$
(13) $R^1 = R^2 = OH$
(14) $R^1 = R^2 = H$

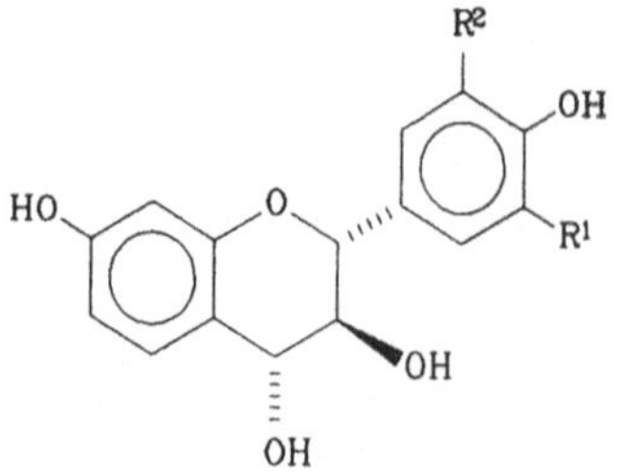

(1) $R^1 = H, R^2 = OH$
(15) $R^1 = R^2 = OH$
(16) $R^1 = R^2 = H$

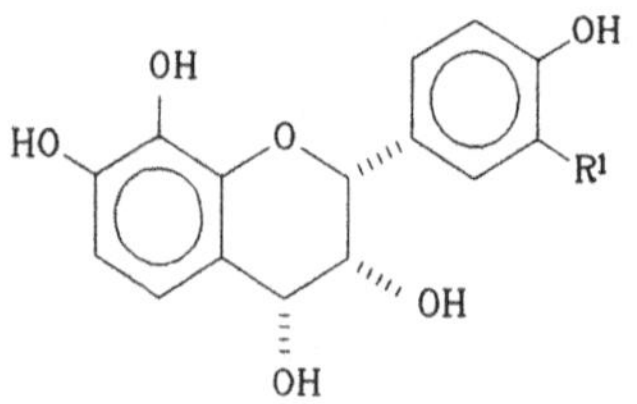

(17) $R^1 = OH$
(18) $R^1 = H$

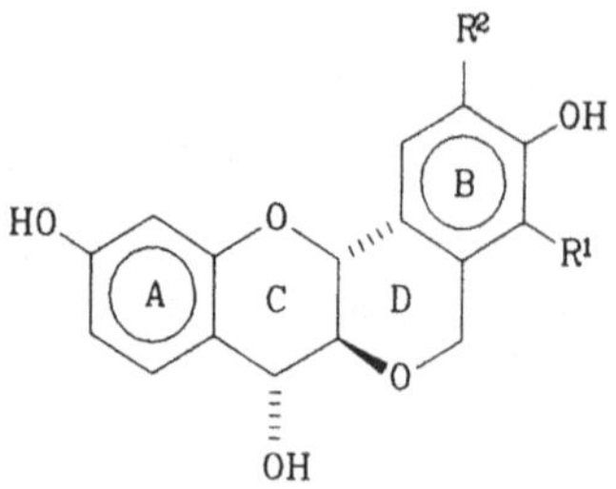

(19) $R^1 = H, R^2 = OH$
(20) $R^1 = OH, R^2 = H$

compared to that of the A-ring of these biflavanoids by carbocation (2) would then lead to the 'branched' or 'angular' triflavanoids (8)-(11)[14,15] which may serve as the precursors to tetraflavanoids,[16,17] and eventually also to the higher oligomers.

Flavan-3,4-diols (leucoanthocyanidins) as Incipient Electrophiles

The flavan-3,4-diols, via their C-4 carbocations e.g. (2), may serve as a source of the chain extender units in a semi-synthetic approach to oligo-flavanoids. The stability of these carbocations is partially dependent on the degree of delocalization of the positive charge over the A-ring. It may be predicted that such delocalization will be most effective for flavan-3,4-diols with phloroglucinol-type A-rings (12)-(14), intermediate in efficiency for resorcinol-type compounds (1), (15), and (16), and still less effective for pyrogallol-type (-)-melacacidins (17) and (-)-teracacidins (18). These concepts provide a simple

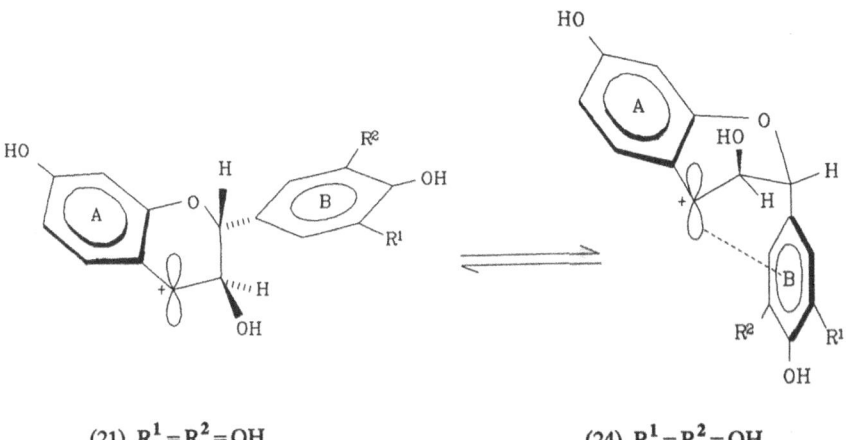

(21) $R^1 = R^2 = OH$
(22) $R^1 = H, R^2 = OH$
(23) $R^1 = R^2 = H$

(24) $R^1 = R^2 = OH$
(25) $R^1 = H, R^2 = OH$
(26) $R^1 = R^2 = H$

rationale for the striking instability of leucocyanidins (12), leucodelphinidins (13) and leucopelargonidins (14), and hence their absence from natural sources containing oligomers derived from them. Such a phenomenon contrasts with the stability and wide distribution of the natural compounds (1), (15)-(18).

The potential of the B-ring to contribute towards stabilizing C-4 carbocations of type (2) via an A-conformation (25) has been overlooked for a long time. First proposed by Brown,[18] recognized by us and others on several occasions,[19-22] and formally designated the A-conformer by Porter,[23] this represents a half-chair/sofa conformation for the pyran ring in which the B-ring occupies an axial-(25) as opposed to the 'customary' equatorial orientation in the E-conformer (22). This ability of the B-ring to additionally stabilize C-4 carbocations via an A-conformation was strikingly demonstrated by the different rates of condensation observed for (+)-leucorobinetinidin (15),[24] (+)-mollisacacidin (1)[24] and (+)-guibourtacacidin (16).[25] Benzylic carbocations of type (2) may thus be additionally stabilized by charge donation from the B-ring. The more electron-rich pyrogallol function in the (+)-leucorobinetinidin carbocations [(21)↔(24)] is more effective than the pyrocatechol functionality in (+)-mollisacacidin analogues [(22)↔(25)] and the mono-oxygenated moiety in the (+)-guibourtacacidin ions [(23)↔(26)], hence leading to condensation rates decreasing in the order (15) > (1) > (16).

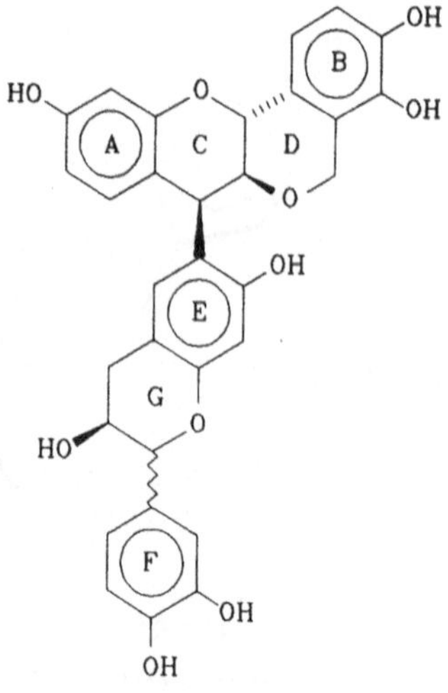

(27) { = }
(28) { = |

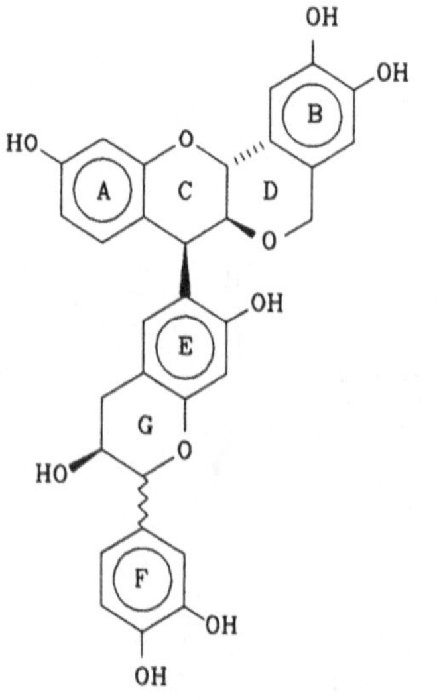

(29) { = }
(30) { = |

(+)-Peltogynol (19) and (+)-mcpanol (20), which also possess the potential for C-4 carbocation formation, are non-reactive under conditions that readily promote coupling reactions involving flavan-3,4-diols and flavan-3-ols.[26] Forcing conditions are required for promoting condensations of (19) and (20) with nucleophilic phenols and the reactions are characterized by low yields.[26,27] The increased energy requirements for these condensation reactions similarly result from the C-rings of compounds (19) and (20), being restricted to an (E) C-3 sofa conformation of type (21) by the D-ring, hence eliminating contributions by an A-conformer of type (24) towards a decrease in the activation energy.[27] Despite the adverse influence of these effects on the *in vitro* formation of propeltogynidins and promopanidins, we have recently demonstrated[27] the natural occurrence of a variety of these metabolites, e.g. the promopanidins (27) and (28), and the propeltogynidins (29) and (30), in the heartwood of *Colophospermum mopane*.

Nucleophilic Flavanoids

The majority of oligoflavanoids with $C4(sp^3)$-$C6/8(sp^2)$ interflavanyl linkages consist of a flavan-3-ol as chain extender unit that originates as an electrophilic unit from a flavan-3,4-diol, and a chain terminating unit comprised of a nucleophilic C-4 deoxy flavan-3-ol with a phloroglucinol A-ring. Most prominent amongst these are the procyanidins[6,28,29] with their (+)-catechin (3) and (-)-epicatechin (33) constituent units, the profisetinidins[5,15] based on (-)-fisetinidol (35), its enantiomer (+)-fisetinidol, (+)-epifisetinidol (37a), (+)-catechin (3) and (-)-epicatechin (33),[30] and the prorobinetinidins[24,31] composed of (-)-robinetinidol (36), (+)-catechin (3) and (+)-gallocatechin (31). In addition, there are several other groups with limited distribution but also possessing a lower flavan-3-ol unit with C-4 deoxy heterocycle. Flavan-3-ols with phloroglucinol A-rings, e.g. (+)-catechin (3), would be expected to represent stronger nucleophilic substrates than their resorcinol counterparts, e.g. (-)-fisetinidol (35).

Flavanoids possessing C-4 carbonyl functions exhibit reduced nucleophilicities of their aromatic A-rings.[2] The inductive effect of the 4-hydroxyl function of flavan-3,4-diols or of its C-4 carbocation equivalent similarly reduces their tendency for self-condensation.[26] Examples where the heterocycles of the terminal 'lower' units are oxygenated at C-4 are restricted to four dimeric (38)-(41) and a single trimeric profisetinidin (42) with terminal 3,4-diol function,[4,32] and the profisetinidins (43)-(45) with constituent dihydroflavonol

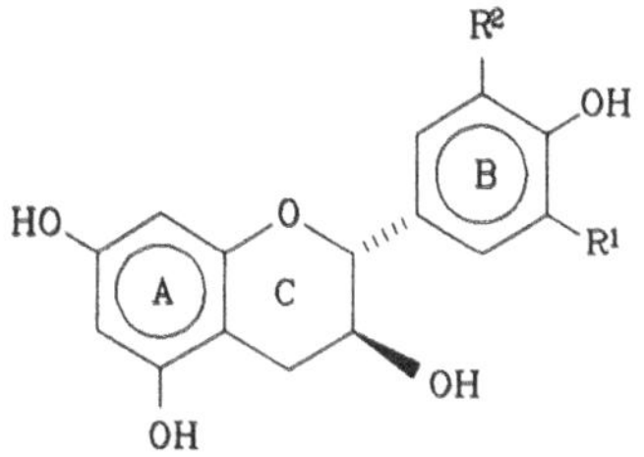

(3) $R^1 = OH, R^2 = H$
(31) $R^1 = R^2 = OH$
(32) $R^1 = R^2 = H$

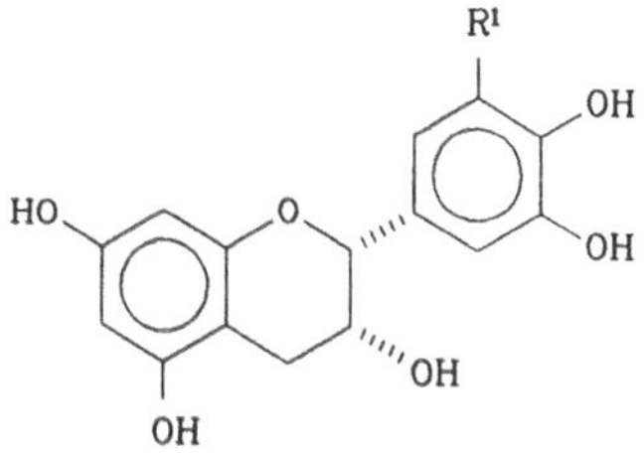

(33) $R^1 = H$
(34) $R^1 = OH$

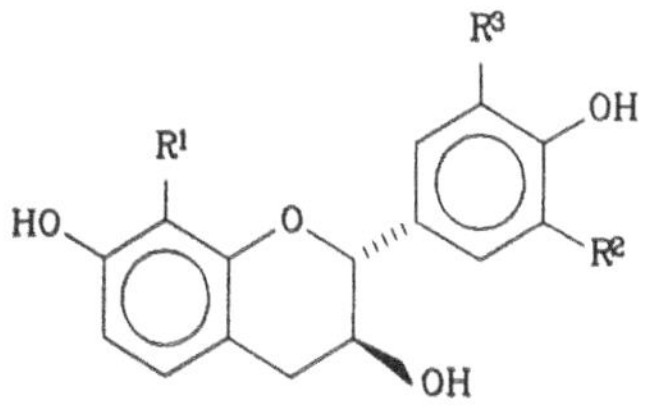

(35) $R^1 = R^3 = H, R^2 = OH$
(36) $R^1 = H, R^2 = R^3 = OH$
(37) $R^1 = R^2 = OH, R^3 = H$

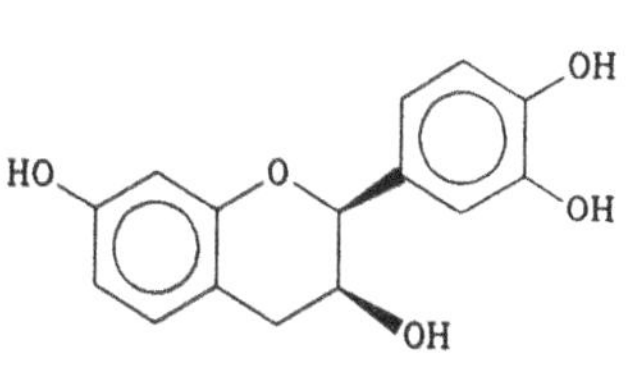

(37a)

(38) ⌇ = ⋮
(39) ⌇ = │

(40) ⌇ = ⋮
(41) ⌇ = │

(42)

(43) ⌇ = ⋮ , R^1 = H
(44) ⌇ = ⋮ , R^1 = OH
(45) ⌇ = │ , R^1 = OH

DEF-units.[33] Acid-induced self-condensation of (+)-mollisacacidin (1) afforded biflavanoids (39) and (41) and the triflavanoid (42) in exceptionally low yields together with high molecular condensates. The conditions for self-condensation of the flavan-3,4-diol are generally more drastic[32] than those permitting the facile condensation of the flavan-3,4-diol with its flavan-3-ol analog, (-)-fisetinidol (35).[13,34] The more prolonged or drastic conditions required for initial dimerization of flavan-3,4-diol (1) to biflavanoids (38)-(41) should subsequently result in the preferential and accelerated condensation with the 'upper' ABC units to form higher condensates. Such a conjecture is supported by the uncontrollable nature of self-condensation of the flavan-3,4-diol which leads to high condensates rather than to oligomers of intermediate mass.

Owing to the absence of a flavan-3,4-diol- or dihydroflavonol analog of (-)-epicatechin (33) in natural sources, the semi-synthetic approach towards the structural elucidation of procyanidins had to be adapted for analogues with $(2R,3R)$-2,3-cis constituent units. Acid-catalyzed thiolysis (toluene-α-thiol[28,29,35] or benzenethiol[18]) of oligomeric procyanidins, e.g. (46), with (-)-epi-catechin chain extender units, and the subsequent utilization of the C-4 thio ethers (47) and (48) as electrophiles in condensation with appropriate flavan-3-ol nucleophiles were successfully implemented by Haslam and his collaborators [28,29,35] in the synthesis of these types of procyanidins. Conversion of thio-ether (48) into the A-ring quinone-methide (49) under mild basic conditions by Hemingway and Foo[36,37] and subsequent trapping by the (-)-epicatechin-(4β,6)-(+)-catechin (50), led to synthesis of the first 'branched' procyanidin trimer (51) in higher yield than the linear analogue (52). This observation suggested that naturally occurring procyanidin polymers may be highly branched[37] despite the fact that such analogues have not yet been encountered in natural sources. The synthesis of the first procyanidins with 3,4-cis-configuration, e.g. (53),[38-40] has, however, similarly preceded the first recognition[41] of the (4β,8)-bis-(+)-catechin (54) in Nature.

Acid-catalyzed reactions to produce flavan-4-carbocations, e.g. (2), or A-ring quinone methides, e.g. (49), either from flavan-3,4-diols or metabolites capable of yielding diols *via* reduction, e.g. dihydroflavonols (55), (56), or from interflavanoid bond cleavage of oligomeric/polymeric proanthocyanidins, that react with the A-rings of flavan-3-ols to produce oligomeric proanthocyanidins, have been so successfully employed that they have been called biomimetic syntheses.[13,42] It should also be emphasized that the controversy regarding the intermediacy of a C-4 carbocation (2), or an A-ring quinone methide (49)[43] in the acid mediated condensation of flavan-3,4-diols with nucleophiles, is actually irrelevant since C-4 in either species is sp[2] hybridized with similar heterocyclic

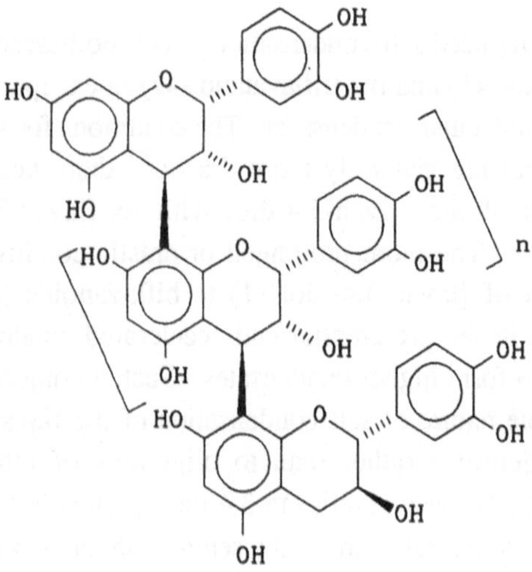

(46)

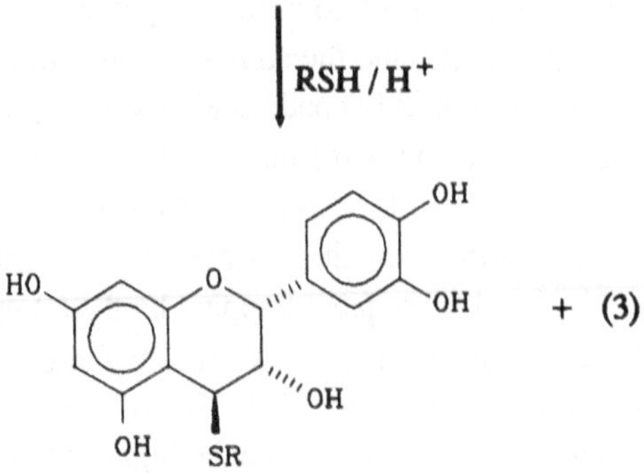

$$RSH / H^+$$

+ (3)

(47) R = CH₂Ph

(48) R = Ph

ring geometry, thus resulting in the same stereochemical course of the coupling step (*vide infra*). The formation of A-ring quinone methide intermediates, nevertheless, constitutes a viable mechanism for the condensation of 4-substituted flavans over a wide range of pH values.

Bonding Positions at Nucleophilic Centres

One of the more important problems that hampered progress of the chemistry of condensed tannins for a long time, was the differentiation between the alternative of C4-C8 and C4-C6 [cf. structures (4) and (6), Scheme 1] interflavanoid links in those instances where the 'lower' terminal flavan-3-ol unit is composed of a substituted (+)-catechin or (-)-epicatechin. Differentiation of (4,6)- and (4,8)-regiomers, e.g. the heptamethyl ether diacetates of (4) and (6), based on the absolute values of chemical shifts of 6- and 8-H(D),[44] is both temperature- and solvent-dependent, thus limiting its general utility. Nuclear Overhauser effect (NOE) difference spectroscopy (^{1}H NMR) has recently been elegantly applied[45] to distinguish between the C-8 (57) and C-6 (58) substituted (+)-catechin moieties of oligoflavanoids; thus, NOE association of the residual proton with either two methoxy groups (6-H→ 5,7-OMe) confirming 8-C (57) or with one methoxy group (8-H→ 7-OMe) indicative of 6-C-linked units (58).

(53)   (54)

(55) $R^1 = OH$
(56) $R^1 = H$

Despite the remarkable preference of flavan-3-ols with resorcinol-type A-rings to be electrophilically substituted at C-6,[13] analogues where C-8 (A-ring) and C-6 (B-ring) of (-)-fisetinidol (35) and (+)-epifisetinidol (37a) served as nucleophilic centres have recently been identified. Amongst these are the (4α,8)-linked profisetinidins (59) and (60) and the (+)-guibourtinidol-(4α,8)-(-)-fisetinidol (61),[46] and the C→ E-ring profisetinidins (62) and (63)[47] and proguibourtinidin (64)[46] from the heartwood of *Colophospermum mopane*.

The oligoflavanoids which exhibit 'abnormal' bonding patterns, e.g. (59)-(64), are usually encountered in natural sources which do not possess significant concentrations of flavan-3-ols with phloroglucinol-type A-rings or which are devoid of flavan-3-ols. The latter situation is prevalent in the heartwood of *Guibourtia coleosperma* where oxygenated stilbenes replaced flavan-3-ols as nucleophiles in the biosynthetic sequence leading to a series of (+)-guibourtinidol-stilbene bi- and tri-flavanoids, e.g. (65) and (66).[48]

(57)

(58)

(59) $\xi = \vdots$, $R^1 = OH$
(60) $\xi = \big|$, $R^1 = OH$
(61) $\xi = \vdots$, $R^1 = H$

(62) $\xi = \vdots$, $R^1 = OH$
(63) $\xi = \big|$, $R^1 = OH$
(64) $\xi = \vdots$, $R^1 = H$

(65)

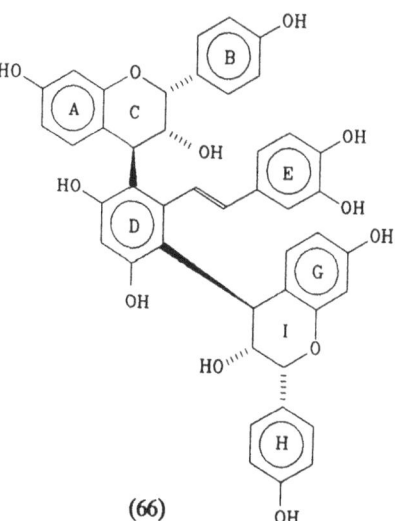

(66)

Stereochemical Course of Condensation and Methods for Determining the Absolute Configuration at C-4 (C-ring)

The generation of carbocations, e.g. (2), or A-ring quinone methides,[43] e.g. (49), from typical flavan-3,4-diols and their trapping by nucleophilic phenolic nuclei, were found to proceed under mild acidic conditions, at ambient temperatures, and over short periods with a minimum of side reactions.[49] Under these conditions coupling reactions of 5,7-dihydroxyflavan-3,4-diols, e.g. (12), are presumably under thermodynamic control[28] in contrast to the kinetic regulation[49] of coupling involving 5-deoxy (A-ring) analogs, e.g. (1).

Assuming that carbocationic intermediates possess sofa conformations, nucleophilic attack on the ion with (2R,3R)-2,3-*cis* configuration (67) proceeds from the less hindered 'upper' side, presumably with neighbouring group participation of the 3-axial hydroxyl in an E-conformation and by the 2-axial B-ring in an A-conformation, whereas reaction with a 2,3-*trans* carbocation (68) is directed preferentially from the less hindered 'lower' side, i.e., the reaction proceeds with a moderate degree of stereoselectivity.

Since flavan-3,4-diols (leucoanthocyanidins) and nucleophilic flavan-3-ols with resorcinol- or phloroglucinol-type A-rings, as well as analogous bi- and tri-flavanoid condensation products accompany condensed tannins, the aforementioned mechanism should and in fact does accurately predict the configuration at C-4 of all constituent 2,3-*trans* and 2,3-*cis* flavanoid units in natural condensed tannins. The reactions of (2R,3S,4R)-(+)-leucocyanidin (12) are more highly directed to the 3,4-*trans* isomers of type (4)[13,50] owing to the greater steric restraint due to the 5-hydroxyl group. Coupling of (+)-leucocyanidin (12) and (+)-catechin (3) gave the 4α-linked dimers with no evidence for any 3,4-cis dimers.[51] This latter result was consistent with the common occurrence of (4α,8)-bis-(+)-catechin, (+)-catechin-(4α,8)-(-)-epicate-chin, and their (4α,6)-regiomers in lower yields from plants containing procyanidins. Recently, however, the synthesis and natural occurrence in low amounts of procyanidins exhibiting 2,3-*trans*-3,4-*cis* linkages have been demonstrated.[38-41]

Apart from defining the ideal conditions for biomimetic condensations, the derivatives of the optically pure 4-arylflavan-3-ols offered the opportunity to formulate a chiroptical rule that defines the absolute configuration at C-4 of flavanoid units of this type and hence in biflavanoids and higher oligomers. Thus, the absolute configuration of the interflavanoid bond could be correlated with the sign of the CD band near 230 nm (probably an ^1L_a transition), a positive sign being correlated with a 4β (69), and a negative with a 4α (70),

(67) : Complete stereoselectivity

(68) : Moderate stereoselectivity

configuration, regardless of the configuration of the rest of the molecule. In much the same way, Haslam and his co-workers[52] demonstrated that the CD spectra of procyanidin oligomers possess a short wavelength couplet centred near 200 nm. A positive couplet was correlated with a 4β and a negative couplet with a 4α interflavanoid bond configuration. The CD method thus supplemented their previous indirect method based on ^{13}C NMR chemical-shift differences.[29]

However, 4-arylflavan-3-ol derivatives with 2,3-*cis*-3,4-*cis* configuration, e.g. (71), and also some with all-trans configurations, do not obey the aromatic quadrant rule,[53] hence leading to exceptions[54] to the afore-mentioned observations. Analogs which do not conform to this otherwise simple rule also exhibit 'abnormal' ^{1}H NMR coupling constants which have incorrectly been ascribed to heterocyclic boat conformations.[54] Deviations in coupling constants and also the exceptions to the aromatic quadrant rule are at present more satisfactorily explained in terms of an equilibrium between E- and A-conformers[23] of the heterocycle of 4-arylflavan-3-ols and related compounds.

(69)

(70)   (71)

Conformation of Oligoflavanoids

Conformational analysis of oligoflavanoids is in principle concerned with the pyran heterocycle conformation and with the phenomenon of conformational isomerism owing to restricted rotation about the interflavanyl linkage. The advent of ^{1}H NMR spectroscopy enabled Clark-Lewis and collaborators[55] to propose C-ring conformations approximating a half-chair, with the B-ring in an equatorial position, for a series of flavans with phenolic groups protected by methylation and with various heterocyclic ring substituents. Numerous ^{1}H NMR investigations have since borne out these findings. More recently, X-ray crystallographic studies of (-)-epicatechin (33),[56] 8-bromo-3',4',5,7-tetramethyl ether derivatives of (+)-catechin[57] and (-)-epicatechin,[58] and (+)-leuco-cyanidin (12),[59] have been published which generally support the NMR conclusions.

Porter *et al.*[23] has considered the factors that contribute to the equilibrium constant of the C-ring of flavan-3-ols and demonstrated that it may be described by the equilibrium shown in Figure 1 (top) where E- and A-conformers[23] are those with the B-ring equatorial or axial respectively. Figure 1 (bottom) depicts the ground-state energy conformations which may be adopted by the flavan heterocycle. Figure 2[23] gives the relative stereochemistry of groups at C-2 and C-3 for the E- and A-conformations of (+)-catechin (3) [(72) and (73)] and (-)-epicatechin (33) [(74) and (75)].

The boat conformation represents the high-energy transition state for the interconversion of E- and A-conformers. An unequal conformational energy for these conformers is manifested by an unequal population of the two states, the one with lower energy being populated to a greater extent. ^{1}H NMR measurements in conjunction with theoretical calculations[22] demonstrated that the E:A ratio for (+)-catechin (3) and (-)-epicatechin (33) were 62:38 and 86:14 respectively. Acetylation of the 3-hydroxy group stabilized the A-conformation and altered the ratio to 48:52.

Substitution at C-4 by a hydroxy or aryl substituent strongly favours the E-conformation due to minimization of 1,3-diaxial interactions and the *pseudo*-allylic or A(1,3)-strain effect.[23] It was also concluded that the favoured E-conformers of (+)-catechin-4 and (-)-epicatechin-4 units would adopt C(2)-sofa and half-chair conformations, respectively as minimum energy species.

The profound effect of the dynamic equilibrium between E- and A-conformers on the dihedral angles of heterocyclic protons and hence their ^{1}H NMR coupling constants is obvious. Prior to the pioneering work of Porter and Mattice,[23] the observation of 'abnormal' $^3J_{H,H}$ -values for C-4 substituted flavan-3-ols with 2,4-*cis* arrangement of substituents has often led to the erroneous assumption of a preferred boat or twisted boat conformation for these and other analogs, thus exhibiting atypical coupling constants and Cotton effects in their CD spectra (cf. refs. 2, 26, and 54).

(4,8)-Linked procyanidin dimers possess detectable conformational isomers resulting from steric interactions in the vicinity of the interflavanoid bond[29,60] which exhibit two sets of ^{1}H NMR signals[61] and heterogeneous fluorescence decay.[62] Conformations (76) and (77) of the 2,3-*cis*- and *trans* dimers respectively, correspond to that in which the C-4 proton eclipses the aromatic A-ring of the lower flavanoid unit, with the bulky C-2 and C-4a substituents occupying positions of least steric interaction with the lower flavan unit. In $CDCl_3$, conformation (79) predominates in 2,3-*trans* dimer peracetates whereas conformers (76) and (78) occur in nearly equal proportions for 2,3-*cis* dimer deca-acetates.[29,61]

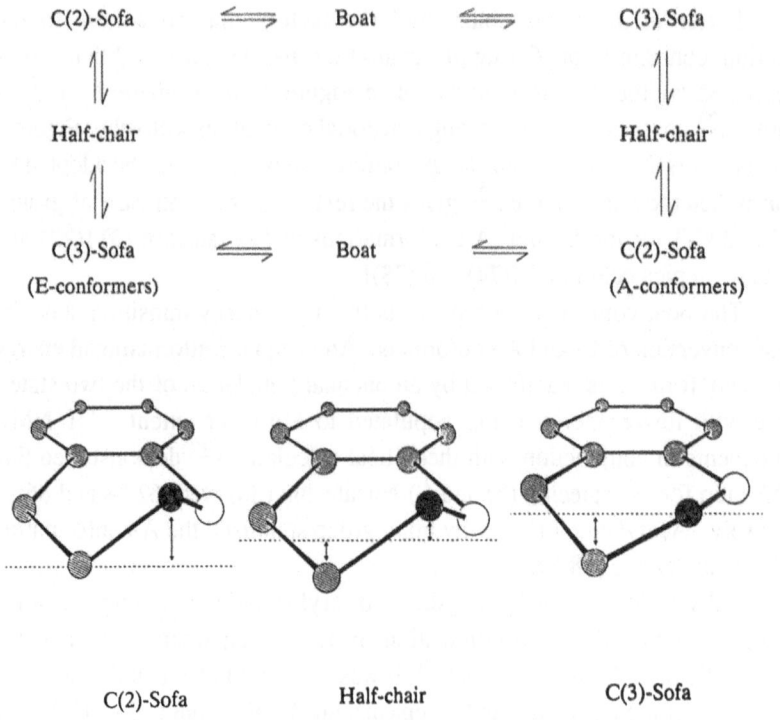

C(2)-Sofa   ⇌   Boat   ⇌   C(3)-Sofa

⇅   ⇅

Half-chair   Half-chair

⇅   ⇅

C(3)-Sofa   ⇌   Boat   ⇌   C(2)-Sofa
(E-conformers)   (A-conformers)

C(2)-Sofa   Half-chair   C(3)-Sofa

Fig. 1. (Top) Factors that contribute to the equilibrium constant of the C-ring of flavan-3-ols according to Porter *et al.*.[28] (Bottom) Ground-state energy conformations possible for the flavan heterocyte. The hatched areas indicate the projection of the A-ring.

These models have also led to proposal[63] of a transition polarization diagram to account for the sign of the short-wavelength CD couplet of dimeric procyanidins.

Interest in the proanthocyanidin polymers is increasing due to their potential as a renewable source of useful chemicals,[64] their probable use by plants as a defense mechanism,[65] and their formation of complexes with a variety of naturally occurring and synthetic polymers.[66] Owing to the purported importance of the conformation about the interflavanoid bond towards these phenomena, Mattice and his co-workers have recently launched an intensive investigation aimed at an understanding of the conformations of dimeric procyanidins and higher polymers.[67]

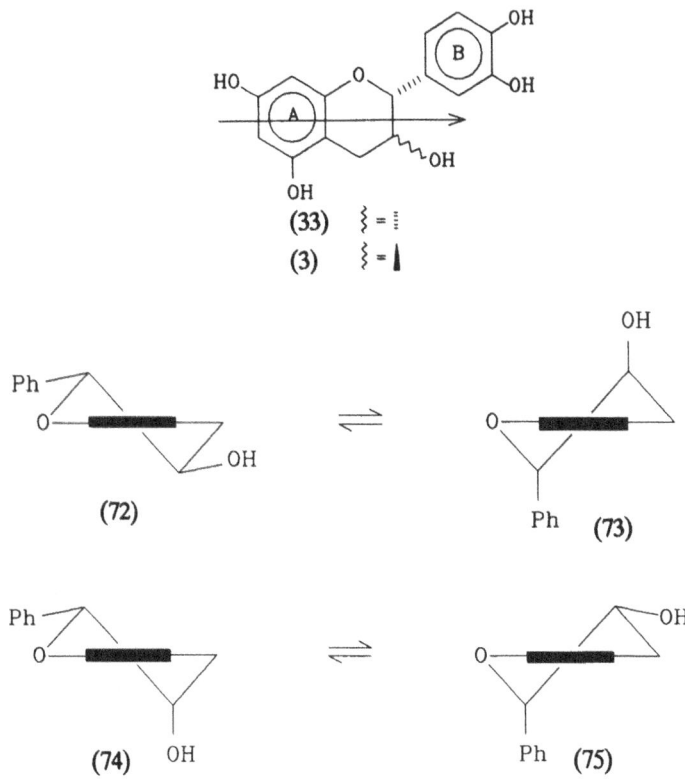

Fig. 2. Relative stereochemistry of groups at C-2 and C-3 for the E- and A- conformations of (+)-catechin (3) [(72) (73)] and (-)-epicatechin (33) [(74) (75)]. The conformations are viewed in the sense indicated by the arrow in structures (3) and (33), and the solid line in (72)-(75) represents the A-ring plane.

BASE-CATALYZED REARRANGEMENTS OF OLIGOFLAVANOIDS

Introduction

Condensed tannins are often extracted and/or allowed to react at alkaline pH in the course of manufacture of specialty polymers such as tannin-based adhesives. These preparations invariably exhibit increased acidity and lower

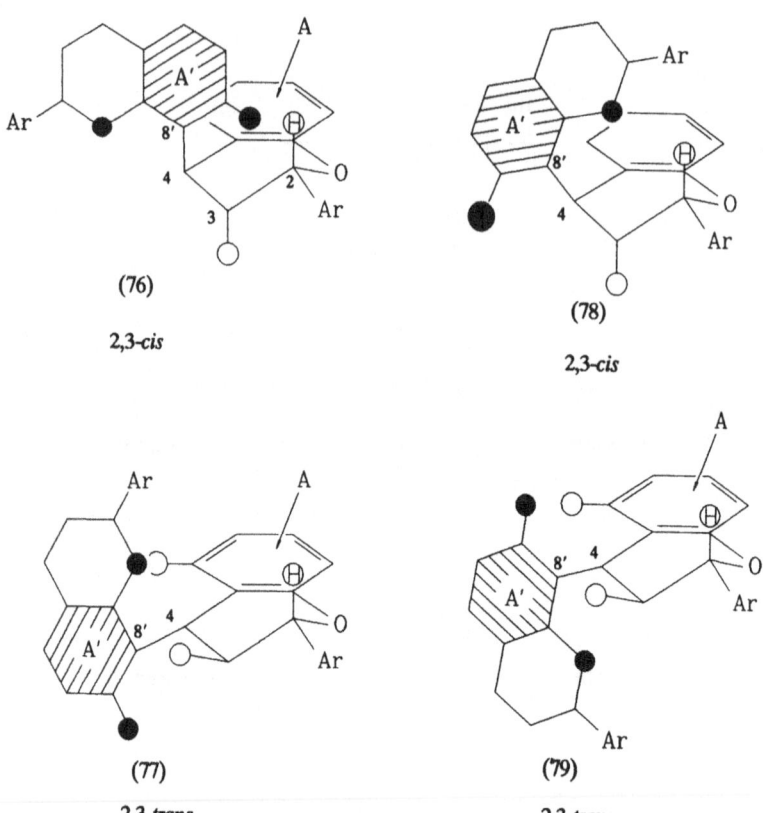

(76)

2,3-cis

(78)

2,3-cis

(77)

2,3-trans

(79)

2,3-trans

reactivity towards aldehydes than those obtained by neutral-solvent extraction;[68,69] such phenomena have been attributed to the presence of catechinic acid-type rearrangement products.[68]

The facile epimerization of (+)-catechin (3) to (+)-epicatechin (83) and of (-)-epicatechin (33) to (-)-catechin (84) in basic or neutral solution is well established.[70,71] The mechanism proposed by Whalley[72] (Scheme 2) proceeding through ionization of the 4'-hydroxyl group and B-ring quinone-methide intermediates (80) and (81), via a reversible Michael addition, is supported by the fact that (+)-catechin tetra-O-methyl ether remains unchanged after prolonged heating in alkaline solution. Quinone-methide (80) presumably also serves as precursor to the formation of (+)-catechinic acid (82)[68] through interaction of C-8 (A-ring) and the si-face at C-2.

More recently, however, Powell and his co-workers[73] found that opening of the pyran ring of (+)-catechin (3) for epimerization or nucleophilic addition, is greatly retarded by the total exclusion of oxygen. On these premises, it was suggested that the formation of epimerization and rearrangement products at alkaline pH may proceed through a one-electron (radical) mechanism (Scheme 3) rather than the ionic mechanism proposed earlier.[68,72] Thus, the B-ring radical anion (85) derived from autoxidation[74,75] of (+)-catechin (3) could rearrange to the intermediate radicals (86) and (87). Pyran recyclization of radical (86) would then lead to the formation of (+)-epicatechin (83) via (89) while radical (87) may be susceptible to irreversible rearrangement to (+)-catechinic acid (82) via (88). A similar sequence of transformations is also applicable to (-)-epicatechin (33).

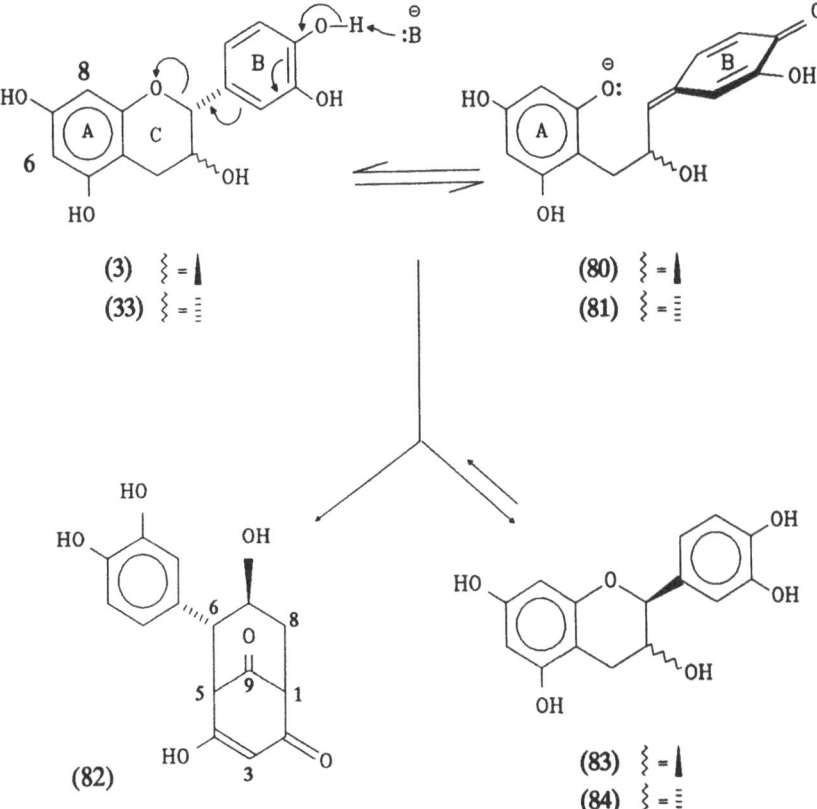

Scheme 2. Epimerization of C-2 of flavan-3-ols and route to the formation of (+)-catechinic acid (82) via a two-electron mechanism.

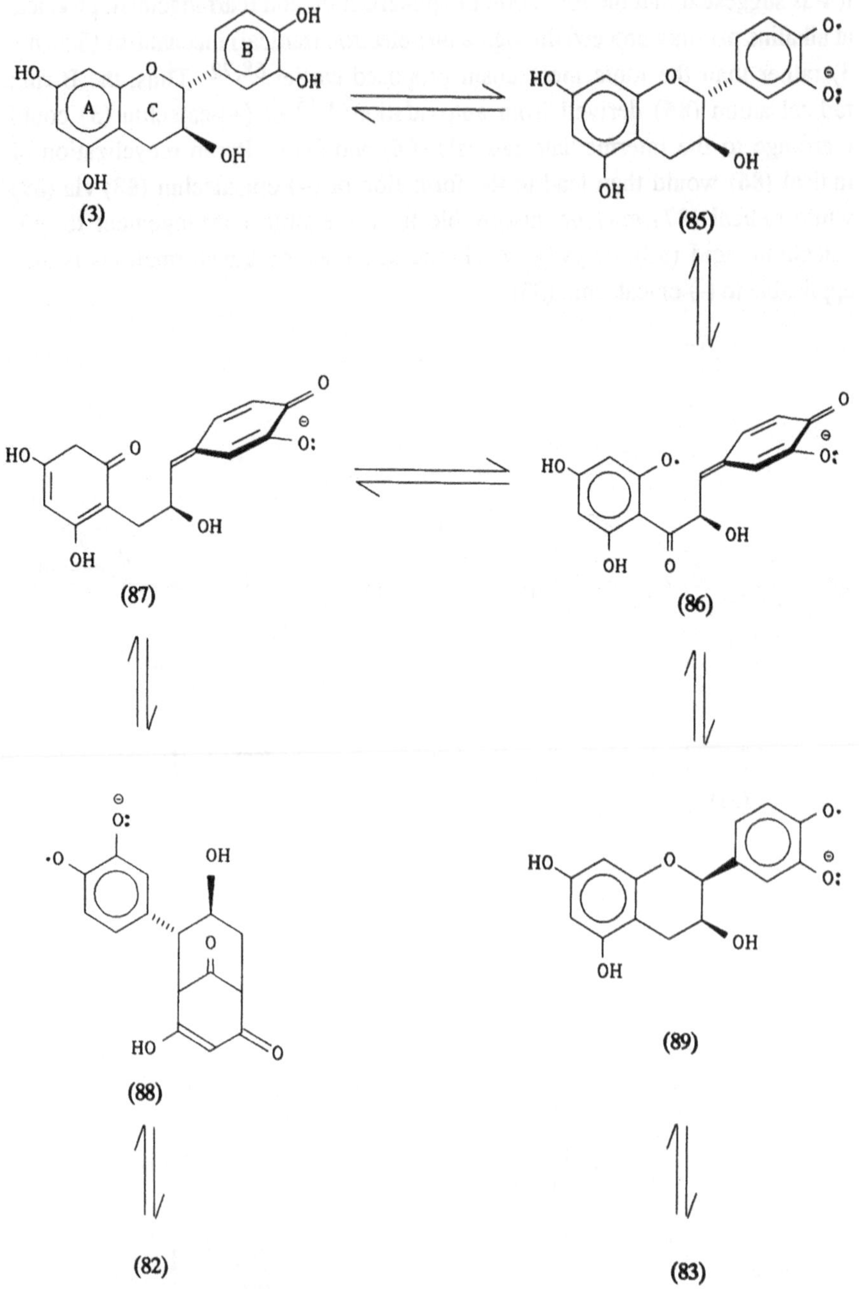

Scheme 3. One-electron mechanism for the epimerization of (+)-catechin and rearrangement to (+)-catechinic acid (82)

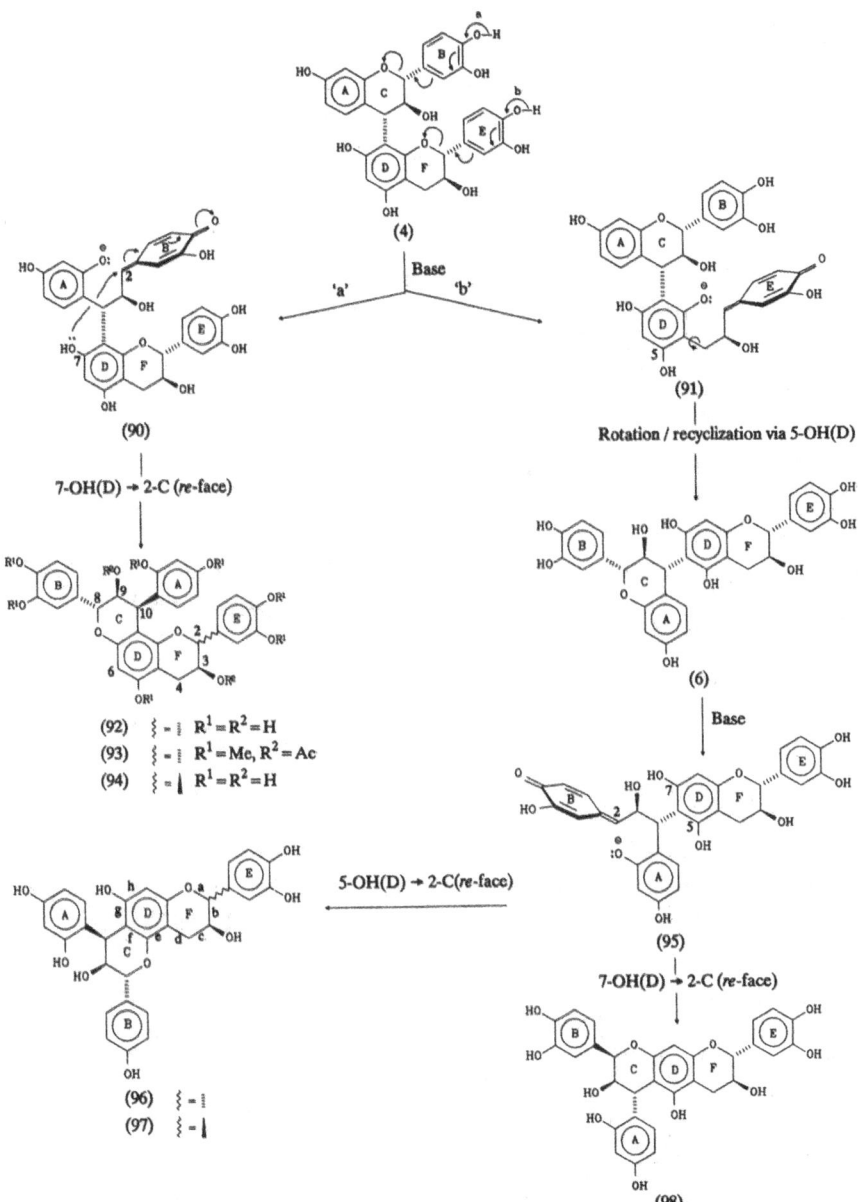

Scheme 4. Base-catalyzed formation of tetrahydropyrano-chromenes (92), (94), (96), and (98) from (-)-fisetinidol-(4α,8)-(+)-catechin (4). Quinone-methides (90), (91), and (95) are postulated and have not been isolated.

The research efforts on base-catalyzed transformations of oligoflavanoids were triggered by a few authorative reports by Hemingway and his collaborators. These studies have largely focused on flavan derivatives with 'good' leaving groups at C-4,[36,37] the effects of external nucleophiles on intramolecular rearrangements, and the lability of the interflavanyl bond and pyran ring at high pH values.[76,77] The observed rearrangements have proved to serve as useful models for interpretation of the reactions of polymeric procyanidins in alkaline solution without the addition of an external nucleophile. In addition, these results help to explain the low aldehyde reactivity and acidity of polymeric procyanidins that have been extracted from plant tissue or reacted at alkaline pH.

The natural occurrence of a novel class of C-ring isomerized condensed tannins, termed phlobatannins, has recently been demonstrated.[78] The structures of these functionalized tetrahydropyranochromenes, e.g. (92), were established by application of ^{1}H NMR NOE difference spectroscopy to their phenolic methyl ether acetates, e.g. (93). In each instance, NOE associations of 2-OMe (A-ring) with 3-H(A) and of 4-OMe(A), with both 3-H(A) and 5-H(A) indicate the 'liberation' of resorcinol moieties from pyran heterocycles, compared with involvement in the C-ring of the presumed (-)-fisetinidol-(4,8)-(+)-catechin precursor (4) (*vide infra*). In addition, the ^{1}H NMR spectra of the derivatives are characterized by the typical absence of the effects of dynamic rotational isomerism at ambient temperatures. Such a phenomenon not only permits their facile investigation by NMR at these temperatures but also allows easy differentiation from their oligoflavanoid precursors.

Initial identification of the pyran-rearranged profisetinidins was succeeded by recognition of additional members of this class of condensed tannins from the heartwoods of *Colophospermum mopane*,[46,79-81] *Guibourtia coleosperma*,[19,80-82] *Baikiaea plurijuga*,[19,81,82] and *Julbernardia globiflora*.[83] The usual methods of differentiating regio-isomeric biflavanoids, i.e. those based on the absolute chemical shifts of 'residual' D-ring singlets of methyl ether acetates in CDCl$_3$ at ambient temperatures[44], and by observation of NOE effects between the hydrogen- and methoxy protons of the D-ring,[45] are less reliable for the methyl ether acetates of analogous phlobatannins. When taken in conjunction with the problems associated with assignment of the absolute configuration, these features initiated an investigation into the behaviour of a series of proanthocyanidin oligoflavanoids under mild basic conditions with a view to applying the observed principles to a concise synthesis of their related pyran-rearranged analogues.

Base-catalyzed Pyran Rearrangements of Profisetinidin-type Oligoflavanoids

Treatment of the (-)-fisetinidol-(4α,8)-(+)-catechin (4) with 0.025M Na$_2$CO$_3$-0.025M NaHCO$_3$ buffer (pH 10) for 5h at 50°C in a nitrogen atmosphere, gave significant (ca. 75%) conversion into five products of C-ring isomerization (Scheme 4).[82] These comprise the anticipated 8,9-trans-9,10-cis-3,4,9,10-tetrahydro-2H,8H-pyrano[2,3-h]chromene (92) (J$_{8,9}$ 10.0; J$_{9,10}$ 6.0 Hz) as a product of C-ring isomerization of biflavanoid (4); its C-2(F) epimer (94) representing conversion of the (+)-catechin moiety of (92) into (+)-epicatechin; the corresponding [2,3-f]-isomeric pair (96) and (97), and a single [3,2-g]-regiomer (98).

Substitution of resorcinol A-ring by phloroglucinol D-ring functionality most likely occurs via a B-ring quinone-methide (90).[72] Pyran recyclization involving 7-OH(D) and the re-face at C-2 requires rotation about the C3-C4 bond, which will invariably lead from the 3,4-trans to a 3,4-cis configuration, but with retention of the absolute configuration at C-2 (C-ring) [C-8(C) in phlobatannin (92)] for biflavanoids of the 2R-series with 3,4-trans configuration. An E-ring quinone-methide (91) could undergo rotation about the C3-C4 bond and recyclization via either 5-OH(D) or the D-ring phenoxide ion, hence simultaneously achieving the observed regio- and configurational isomerizations.

In contrast to the lability of the interflavanyl bond in procyanidin biflavanoids,[36,37] the corresponding linkage in the profisetinidin (4) is more stable as was evidenced by the formation of trace quantities of (+)-catechin (3). Since oxygen is a prerequisite for the base-catalyzed epimerization of flavan-3-ols at C-2, the mechanism for the formation of the B- and E-ring quinone-methides (90) and (91) in Scheme 4 is an oversimplification and should involve more correctly the initial formation of a radical anion at the pyrocatechol ring(s) (cf. Scheme 3).[73] As this will have little effect on the stereochemical course of the pyran-rearrangement reactions, the routes to the quinone-methides in Scheme 4 will be retained for simplicity reasons.

To prevent the side reactions associated with an E-ring quinone-methide (91), biflavanoid precursors of type (4) had to be protected selectively at 4-OH(E). This was achieved by selective methylation[84] of (+)-catechin and subsequent acid-catalyzed coupling of 4'-O-methyl-(+)-catechin and (+)-mollisacacidin (1) to give the (-)-fisetinidol-(4,8)- and (4,6)-(+)-catechin mono-O-methyl ethers (99)-(102), following chromatography on Sephadex LH-20 and Fractogel TSK HW-40(S) in ethanol.

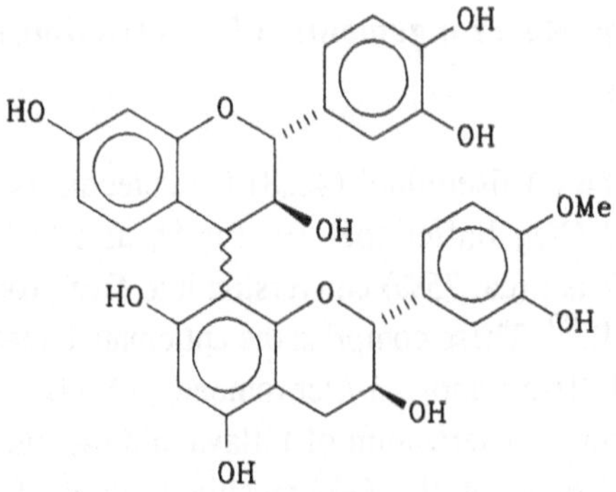

(99)   ≷ = ⋮
(100) ≷ = |

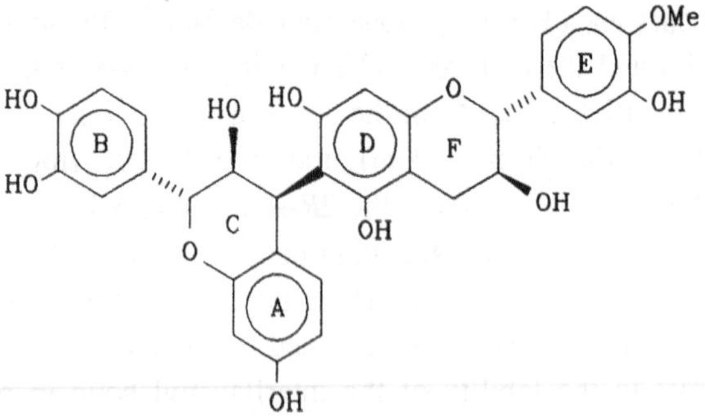

(101) ≷ = ⋮
(102) ≷ = |

Treatment of the (-)-fisetinidol-(4α,8)-(+)-catechin-*O*-methyl ether (99) with base, as mentioned above gave the tetrahydropyrano[2,3-*h*]chromene (104) in 58% yield (Scheme 5).[82] Phlobatannin (104), resulting from highly stereoselective ring isomerization at C-2 in quinone-methide (103), is accompanied by the dehydro-(-)-fisetinidol-(+)-catechin (105), apparently representing the alternative mode of cyclization via C-6(B) in (103) followed by oxidative removal of hydride during work-up. The dehydro-analog may, however, originate alternatively via the oxidative formation of a B-ring *o*-quinone which may then serve as an internal electrophile. All efforts to trap any of the proposed intermediates with powerful nucleophiles such as phenyl sulphide and selenide ions have, thus far, failed. Absence of products resulting

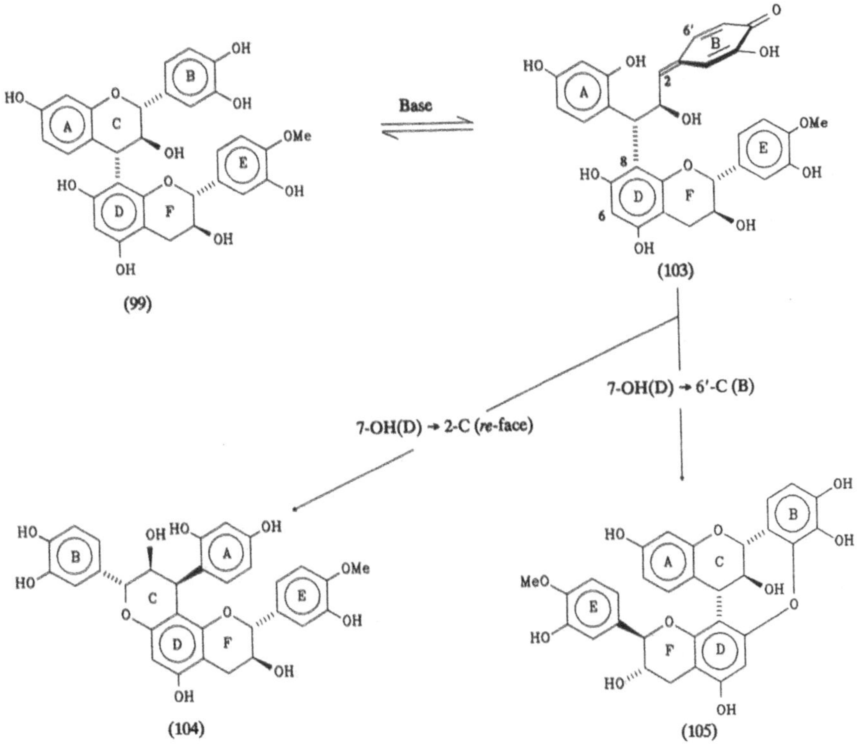

Scheme 5. Base-catalyzed pyran rearrangement of (-)-fisetinidol-(4α,8)-(+)-catechin mono-O-methyl ether (99).

from a 'migrating' flavanyl moiety in the 'protected' biflavanoid (99) confirms the conjecture regarding the mechanism of such a migration in the uncontrolled synthesis (cf. Scheme 4).

Base treatment of the(-)-fisetinidol-(4α,6)-(+)-catechin-O-methyl ether (101) afforded the anticipated products of pyran rearrangement i.e., the 6,7-trans-7,8-cis-tetrahydropyrano[2,3-f]chromene (107) and the [3,2-g]-regiomer (108) as the minor product (Scheme 6).[82] The apparent preference for ring isomerization of biflavanoid (101) involving 5-OH(D) and C-2 (re-face) in an intermediate quinone-methide (106) presumably reflects a solvent-dependent, preferred interflavanyl conformation favouring participation of 5-OH(D) in the cyclization step. Such an assumption is supported by the observation that the two rotational

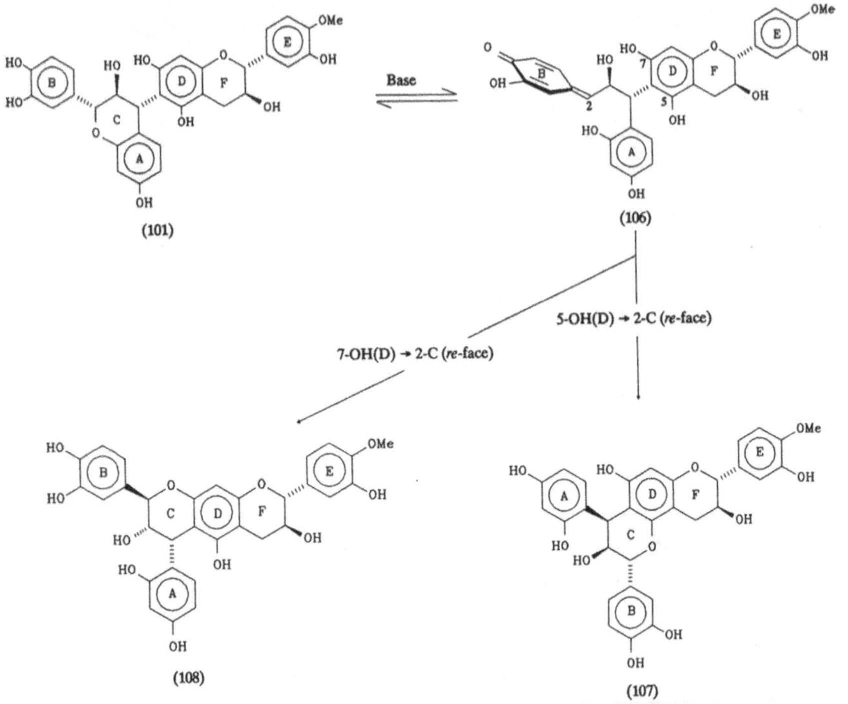

Scheme 6. Base-catalyzed pyran rearrangement of (-)-fisetinidol-(4α,6)-catechin mono-O-methyl ether (101).

isomers at the interflavan bond are unevenly populated in the procyanidins.[85]

Similar treatment of the (-)-fisetinidol-(4β,8)-(+)-catechin-O-methyl ether (100) led to a mixture comprising four ring-isomerized products (Scheme 7).[19] These included the 8,9-cis-9,10-trans- and 8,9-trans-9,10-trans-tetrahydropyrano[2,3-h]chromenes (110) and (111) ($J_{8,9}$ ca. 1.0, 9.5; $J_{9,10}$ ca. 2.0, 8.0 Hz), as well as an additional pair of cis-trans- and all-trans analogues (113) and (114), with ^{1}H NMR characteristics of their heptamethyl ether diacetates closely resembling those of the same derivatives of the former pair of compounds. Prominent NOE associations between 8-H(C) and 6-H(A) in the cis-trans heptamethyl ether diacetate of phlobatannin (110) not only confirmed this configuration and thus differentiated it from an all-cis structure, but also indicated a preferred sofa conformation for the C-ring in which the resorcinol moiety at C-10 occupies a near-axial (α) orientation.

Extensive NOE-, spin decoupling-, and 2D-heteronuclear correlation experiments[19] indicated an interchange of the resorcinol A- and pyrocatechol B-rings in compounds (113) and (114) relative to the positions of these rings in the 'normal' isomers (110) and (111). A notable feature in the 1H NMR

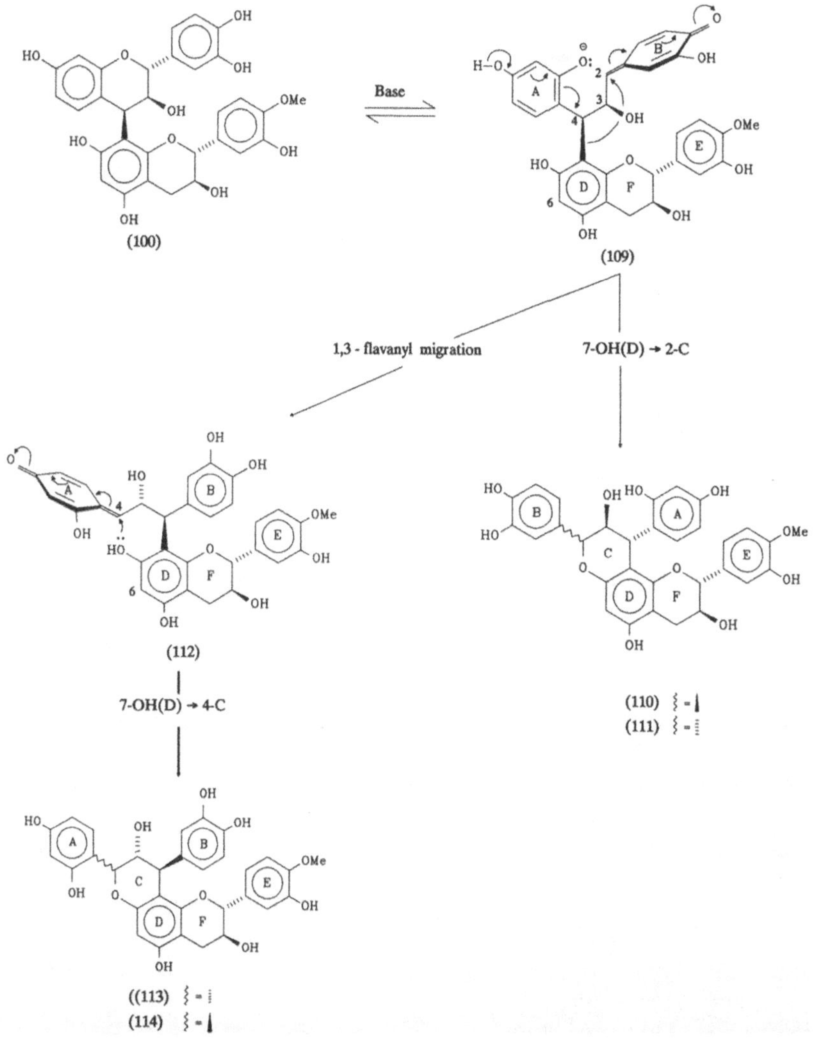

Scheme 7. Proposed route to the formation of tetrahydropyrano[2,3-h]chromenes (110) and (111) and those of the ring interchanged analogues (113) and (114).

spectra of the groups (110), (111), and (113), (114) is the conspicuous deshielding of 6-H(A) [-0.74 and -0.80 for the heptamethyl ether diacetates of (113) and (114) respectively] in the latter pair relative to its chemical shift in the *cis-trans* and all-*trans* isomers (110) and (111). Such a feature is a very useful characteristic of all phlobatannins belonging to the classes (113) and (114), i.e., those with interchanged resorcinol A- and pyrocatechol B-rings.

Monitoring of the base-catalyzed conversion of the biflavanoid (100) indicated that it serves as a direct precursor to both groups of phlobatannins in Scheme 7. Formation of the former pair may hence be rationalized by stereoselective recyclization involving 7-OH(D) and both the *re-* and *si*-faces at C-2 in the intermediate quinone-methide (109).

The unique conversion, (100)→ (113) + (114), is explicable in terms of initial migration of the (+)-catechin DEF moiety to the *re*-face at C-2 in quinone-methide (109). Stereoselective pyran recyclization of the resulting quinone-methide (112) then generates the tetrahydropyrano[2,3-*h*]chromenes (113) and (114), enantiomerically related to analogues (110) and (111) with respect to their C-rings.

The heptamethyl ether diacetates of compounds (110) and (111) exhibit intense negative Cotton effects in the 220-240 nm region of their CD spectra,[19] hence indicating a 10α-aryl substituent and *R* absolute configuration at this stereogenic centre.[49,52-54] The same derivatives of the ring-interchanged analogs (113) and (114) showed similar CD characteristics as those above, thus presumably reflecting similar 9*S*,10*R* absolute configuration for ring C. Such a contradiction may result from significant contributions of A-conformers[23] (F-ring) reversing the sign of the low-wavelength Cotton effect for 10β-aryl groups. The 9*R*,10*S* absolute configurations depicted in formulations (113) and (114), and thus unambiguous proof for the inversion of the absolute configuration at C-9 associated with the mechanism leading to ring interchange, were confirmed by similar transformations of appropriate 4-arylflavan-3-ol model compounds where the structural features adversely affecting the sign of the low-wavelength Cotton effect are absent.[20]

Principles similar to those advanced for the (4β,8)-profisetinidin (100) also govern the base-catalyzed conversion of the (-)-fisetinidol-(4β,6)-(+)-catechin mono-*O*-methyl ether (102) affording four ring-isomerized products with *cis-trans-* (115), (117), (118), and (120), and two with all-trans configuration (116) and (119) of their C-ring (Scheme 8). Differentiation of these and other tetrahydropyrano[2,3-*f*]- and [2,3-*h*] chromenes and the regiomeric [3,2-*g*] analogs is effected by NOE experiments on the heptamethyl ether diacetates which indicate selective association of the D-ring singlet and the methoxy group of this

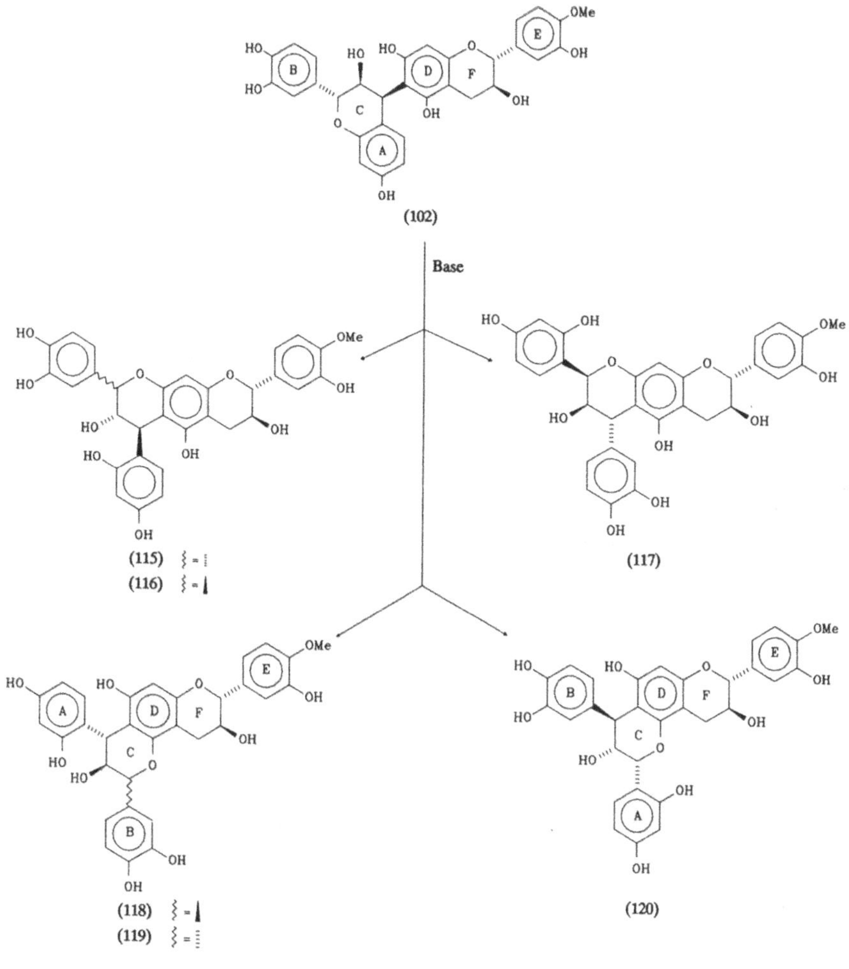

Scheme 8. Base-catalyzed pyran rearrangement of (-)-fisetinidol-4β,6)-(+) catechin mono-0-methyl ether.

ring for the [2,3-*f*]- and [2,3-*h*]-isomers only. The latter groups are accordingly distinguished by the selective NOE association of the D-ring methoxy and the C-ring proton adjacent to the resorcinol A-ring moiety for the [2,3-f]analogues only. Confirmation of the tetrahydropyrano[2,3-*h*]chromene arrangement for those analogues having 2- and 10β-aryl groups or *vice versa* is available via

NOE association of 10-H(C) with 2- and 6-H(E).[19,78,82]

Application of the same protocol to a variety of the different classes of oligoflavanoids, e.g., (+)-fisetinidol-(+)-catechins,[87] (-)-fisetinidol-(-)-epicatechins,[81] bis-fisetinidols,[46] bis-(-)-fisetinidol-(+)-catechin triflavanoids,[80] proguibourtinidin bi- and triflavanoids,[78,80,83] and procyanidins B-2 (**87**) and B-3 (**88**), revealed similar behaviour to those described above for the (-)-fisetinidol-(+)-catechins. Thus, whereas 'upper' 2,3-*trans*-3,4-*trans*-flavan-3-ol units are susceptible to slower but highly stereoselective pyran rearrangements, those moieties with 2,3-*trans*-3,4-*cis* configuration react stereoselectively and are furthermore subject to interchange of resorcinol A- and pyrocatechol B-rings. The rate determining step in these ring isomerizations presumably involves the reversible generation of a quinone-methide of type (**90**) (cf. Scheme 4). In 3,4-*cis* biflavanoids, e.g. (**100**), 7-OH(D) is favourably aligned to anchimerically assist cleavage of the O-C2 bond, thus enhancing both the rate of quinone-methide formation and pyran rearrangement. Once formed, quinone-methides derived from 3,4-*trans*-flavan-3-ol units are favourably orientated for rapid and highly stereoselective recyclization via 7-OH(D). The near-axial (+)-catechin unit in 3,4-*cis* quinone-methide, e.g. (**109**) (Scheme 7) would 'relax' to a more equatorial orientation, thus facilitating stereoselective pyran recyclization with preference for attack of 7-OH(D) at the *si*- and *re*-faces in the 2R and 2S series of proanthocyanidins, respectively. This would presumably result in sufficient life-times to allow for secondary rearrangement to the A/B-ring interchanged products. Owing to the proper alignment of the sp^3 bonding orbital at C-4 and the *p*-orbital at C-2 in quinone-methide (**109**) (Scheme 7), the latter compounds may well originate via a concerted mechanism.

CONCLUSION

The advances that have been made in the utilization of condensed tannins, in e.g. wood adhesive systems, are in fact remarkable because so much has been achieved with little fundamental understanding of the chemistry of these processes. It is therefore encouraging that the emphasis of research on the chemistry of these metabolites has, of late, taken a distinct shift from a predominantly analytical approach to a focus on their *in vitro* behaviour under a variety of conditions. The diversity amongst the C-ring isomerized condensed tannins which emanated from our work presumably indicates ubiquity similar to those of their 'conventional' bi- and triflavanoid precursors. The apparent conformational stability of the pyran-rearranged compounds and the relative planarity of the central 'core' after dual isomerization of triflavanoids (cf. ref. 80) will possibly contribute to reduced solubility in aqueous media and thus enhancement of their affinity for collagen substrates. The 'liberation' of resorcinol- or phloroglucinol-type A-rings via facile ring isomerization in bi- and triflavanoid units, present in commercially-available condensed tannins, may facilitate their activation for use in 'cold-set' adhesive applications.

ACKNOWLEDGEMENTS

We are grateful for the enthusiastic support of our co-workers, E.V. Brandt, J.A. Steenkamp, A. Cronje, P.J. Steynberg, S.L. Bonnet, D.A. Young and J.C.S. Malan. Financial support by the Foundation for Research Development, Pretoria, the Sentrale Navorsingsfonds of this University, and the Marketing Committee, Wattle Bark Industry of South Africa, is gratefully acknowledged.

REFERENCES

1.　PORTER, L.J. 1988. Flavans and proanthocyanidins. In: The Flavanoids-Advances in Research since 1980. (J.B. Harborne, ed.) Chapman and Hall, London, pp. 21-62, and references cited therein.

2.　ROUX, D.G., FERREIRA, D. 1982. Structure and function in the biomimetic synthesis of linear, angular and branched condensed tannins. Pure and Appl. Chem. 54:2465-2478.

3.　ROUX, D.G., FERREIRA, D. 1982. The direct biomimetic synthesis, structure and absolute configuration of angular and linear condensed tannins. Fortschr. Chem. Org. Naturst. 41:47-76.

4.　DREWES, S.E., ROUX, D.G., EGGERS, S.H., FEENEY J. 1967. Three diastereoisomeric 4,6-linked bileucofisetinidins from the heartwood of Acacia mearnsii. J. Chem. Soc. (C). 1217-1227.

5.　DREWES, S.E., ROUX, D.G., SAAYMAN, H.M., EGGERS, S.H., FEENEY, J. 1967. Some stereochemically identical biflavanols from the bark tannins of Acacia mearnsii. J. Chem Soc. (C). 1302-1308.

6.　WEINGES, K., KALTENHAUSER, W., MARX, H-D., NADER, E., NADER, E., PERNER, J., SEILER, D. 1968. Procyanidine aus Früchten. Liebigs Ann. Chem. 711:184-204.

7.　WEINGES, K., GÖRITZ, K., NADER, F. 1968. Konfigurationsbestimmung von $C_{30}H_{26}O_{12}$-Procyanidinen und Structuraufklarung eines neuen Procyanidin. Liebigs Ann. Chem. 715:164-171.

8. GEISSMAN, T.A., YOSHIMURA, N.N. 1966. Synthetic proanthocyanidin. Tetrahedron Lett. 24:2669-2673.

9. WEINGES, K., PERNER, J. 1970. Die Rotationsbehinderung an der $C(sp^3)$-$C(sp^2)$-Bindung der 4-arylsubstituierten Polymethoxyflavane. Chem. Ber. 103:2336-2343.

10. PIZZI, A. 1983. Tannin-based adhesives. In: Wood Adhesives: Chemistry and Technology, Marcel Dekker, New York. pp. 177-246.

11. KREIBICH, R.E., HEMINGWAY, R.W. 1985. The use of tannins in structural laminating adhesives. In: Proceedings of IUFRONTRI Symposium on Wood Adhesives 6:17-5.

12. CREASY, L.L., SWAIN, T. 1965. Structure of condensed tannins. Nature. 208:151-153.

13. BOTHA, J.J., FERREIRA, D., ROUX, D.G. 1978. Condensed tannins: Direct synthesis, structure, and absolute configuration of four biflavanoids from black wattle bark ('Mimosa') extract. J. Chem. Soc. Chem. Commun. 700-702; 1982. Synthesis of Condensed Tannins. Part 4. A Direct Biomimetic Approach to [4,6]- and [4,8]-Biflavanoids. J. Chem. Soc. Perkin Trans. 1:1235-1245.

14. BOTHA, J.J., FERREIRA, D., ROUX, D.G., HULL, W.E. 1979. Condensed tannins: Condensation mode and sequence during formation of synthetic and natural triflavanoids. J. Chem. Soc. Chem. Commun. 510-512.

15. BOTHA, J.J., VIVIERS, P.M., YOUNG, D.A., DU PREEZ, I.C., FERREIRA, D., ROUX, D.G., HULL, W.E. 1982. Synthesis of Condensed Tannins. Part 5. The first angular [4,6:4,8]-triflavanoids and their natural counterparts. J. Chem. Soc. Perkin Trans. 1:527-533.

16. YOUNG, D.A., FERREIRA, D., ROUX, D.G., HULL, W.E. 1985. Synthesis of condensed tannins. Part 15. Structure of natural 'angular' profisetinidin tetraflavanoids: Asymmetric induction during oligomeric synthesis. J. Chem. Soc. Perkin Trans. 1:2537-2544.

17. YOUNG, D.A., KOLODZIEJ, H., FERREIRA, D., ROUX, D.G. 1985. Synthesis of condensed tannins. Part 16. Stereochemical differentiation of the first 'angular' $(2S,3R)$-profisetinidin tetraflavanoids from Rhus lancea (Karree) and the varying dynamic behaviour of their derivatives. J. Chem. Soc. Perkin Trans. 1:2537-2544.

18. BROWN, B.R., SHAW, M.R. 1974. Reactions of flavanoids and condensed tannins with sulphur nucleophiles. J. Chem. Soc. Perkin Trans. 1:2036-2049.

19. STEYNBERG, J.P., BURGER. J.F.W., YOUNG, D.A., BRANDT, E.V., STEENKAMP, J.A., FERREIRA, D. 1988. Novel base-catalyzed rearrangements of (-)-fisetinidol-(+)-catechin profisetinidins with 2,3-*trans*-3,4-*cis* flavan-3-ol constituent units. J. Chem. Soc., Chem. Commun. 1055-1057. 1988. Oligomeric Flavanoids. Part 4. Base-catalyzed conversions of (-)-fisetinidol-(+)-catechin profisetinidins with 2,3-*trans*-3,4-*cis*-flavan-3-ol constituent units. J. Chem. Soc. Perkin Trans. 1:3331-3338.

20. STEYNBERG, J.P., BURGER, J.F.W., YOUNG, D.A., BRANDT, E.V., FERREIRA, D. 1988. Oligomeric flavanoids. Part 6. Evidence supporting the inversion of absolute configuration at 3-C associated with base-catalyzed A-/B-ring interchange of precursors having 2,3-*trans*-3,4-*cis*-flavan-3-ol constituent units. Heterocycles. 28:923-935.

21. STEENKAMP, J.A., MALAN, J.C.S., FERREIRA, D. 1988. Stereospecific C4-bromination of flavan-3-ols. The significance of conformational mobility of the flavan heterocycle in stereoselectivity at the benzylic carbon. J. Chem. Soc. Perkin Trans. 1:2179-2183.

22. BAERT, F., FOURET, M., SLIWA, M., SLIWA, H. 1980. Structural studies of flavan derivatives: Effect of overcrowding on conformation and hydrogen bonding of *cis*- and *trans*-diphenyl flavan-3-yl carbinols. Tetrahedron. 36:2765-2774.

23. PORTER, L.J., WONG, R.Y., BENSON, M., CHAN, B.G., VISHWANADHAN, V.N., GANDOUR, R.D., MATTICE, W.L. 1986. Conformational analysis of flavans: 1H NMR and molecular mechanical (MM2) studies of the benzopyran ring of 3',4',5,7-tetrahydroxyflavan-3-ols: the crystal and molecular structure of the procyanidin:(2*R*,3*S*,4*R*)-3',4',5,7-tetramethoxy-4(2,4,6-trimethoxyphenyl)-flavan-3-ol. J. Chem. Res. 830.

24. VIVIERS, P.M., BOTHA, J.J., FERREIRA, D., ROUX, D.G., SAAYMAN, H.M. 1983. Synthesis of condensed tannins. Part 7. Angular [4,6:4,8]-prorobinetinidin triflavanoids from black wattle ('Mimosa') bark extract. J. Chem. Soc. Perkin Trans. 1:17-28.

25. MALAN, J.C.S., STEYNBERG, P.J., STEYNBERG, J.P., YOUNG, D.A., BEZUIDENHOUDT, B.C.B., FERREIRA, D. 1990. Oligomeric flavanoids. Part 14. Proguibourtinidins based on (-)-fisetinidol and (+)-epifisetinidol units. Tetrahedron. 46:2883-2890.

26. VAN HEERDEN, F.R., BRANDT, E.V., FERREIRA, D., ROUX, D.G.
     1981. Metabolites from the purple heartwoods of the Mimosoideae.
     Part 4. *Acacia fascidulifera* F. Muell ex. Benth: Fasciculiferin,
     fasciculiferol, and the synthesis of 7-aryl- and 7-flavanyl-
     peltogynoids. J. Chem. Soc. Perkin Trans. 1:2483-2490.

27. MALAN, J.C.S., YOUNG, D.A., STEYNBERG, J.P., FERREIRA, D.
     1990. Oligomeric flavanoids. Part 9. The first biflavanoids based on
     mopanol and peltogynol as inceptive electrophiles. J. Chem. Soc.
     Perkin Trans. 1:219-226.

28. THOMPSON, R.S., JACQUES, D., HASLAM, E., TANNER, R.J.N.
     1972. Plant proanthocyanidins. Part 1. Introduction; the isolation,
     structure, and distribution in nature of plant procyanidins. J. Chem.
     Soc. Perkin Trans. 1:1387-1399.

29. FLETCHER, A.C., PORTER, L.J., HASLAM, E., GUPTA, R.K.
     1977. Plant proanthocyanidins. Part 3. Conformational and
     configurational studies of natural proanthocyanidins. J. Chem. Soc.
     Perkin Trans. 1:1628-1637.

30. STEYNBERG, J.P., BURGER, J.F.W., MALAN, J.C.S., CRONJ, A.,
     YOUNG, D.A., FERREIRA, D. 1990. Natural (-)-fisetinidol-(4,8)-
     (-)-epicatechin profisetinidins. Phytochemistry. 29:275-277.

31. SAAYMAN, H.M., ROUX, D.G. 1965. The origins of tannins and
     flavonoids in black-wattle barks and heartwoods and their associated
     'non-tannin' components. Biochem. J. 97:794-801.

32. VIVIERS, P.M., YOUNG, D.A., BOTHA, J.J., FERREIRA, D.,
     ROUX, D.G., HULL, W.E. 1982. Synthesis of condensed tannins.
     Part 6. The sequence of units, coupling positions and absolute
     configuration of the first linear [4,6:4,6]-triflavanoid with terminal
     3,4-diol function. J. Chem. Soc. Perkin Trans. 1:535-540.

33. MALAN, J.C.S., YOUNG, D.A., STEENKAMP, J.A., FERREIRA, D.
     1988. Oligomeric flavanoids. Part 2. The first profisetinidins with
     dihydroflavonol constituent units. J. Chem. Soc. Perkin Trans.
     1:2567-2572.

34. STEENKAMP, J.A., FERREIRA, D., ROUX, D.G., HULL, W.E.
     1983. Synthesis of condensed tannins. Part 8. The first 'branched'
     [4,6:4,8:4,6]-tetraflavanoid. coupling sequence and absolute
     configuration. J. Chem. Soc. Perkin Trans. 1:23-28.

35. JACQUES, D., HASLAM, E., BEDFORD, G.R., GREATBANKS, D.
     1974. Plant proanthocyanidins. Part II. Proanthocyanidin-A2 and its
     derivatives. J. Chem. Soc. Perkin Trans. 1:2663-2671.

36. HEMINGWAY, R.W., FOO, L.Y. 1983. Condensed tannins: quinone methide intermediates in procyanidin synthesis. J. Chem. Soc. Chem. Commun. 1035-1036.

37. FOO, L.Y., HEMINGWAY, R.W. 1984. Condensed tannins: synthesis of the first 'branched' procyanidin trimer. J. Chem. Soc. Chem. Commun. 85-86.

38. DELCOUR, J.A., SERNEELS, E.J., FERREIRA, D., ROUX, D.G. 1985. Synthesis of condensed tannins. Part 13. The first 2,3-*trans*-3,4-*cis*-procyanidins: sequence of units in a 'trimer' of mixed stereochemistry. J. Chem. Soc. Perkin Trans. 1:669-676.

39. KOLODZIEJ, H. 1985. The first 2,3-*trans*-3,4-*cis* procyanidin. Phytochemistry. 24:2460-2462.

40. KOLODZIEJ, H. 1986. Synthesis and characterization of procyanidin dimers as their peracetates and octamethylether diacetate. Phytochemistry. 25:1209-1215.

41. SCHLEEP, S., FRIEDRICH, H., KOLODZIEJ, H. 1986. The first natural procyanidin with a 2,3-*cis* configuration. J. Chem. Soc. Chem. Commun. 392-393.

42. HASLAM, E. 1974. Biogenetically patterned synthesis of procyanidins. J. Chem. Soc. Chem. Commun. 594-595.

43. ATTWOOD, M.R., BROWN, B.R., LISSETER, S.G., TORRERO, C.L., WEAVER, P.M. 1984, Spectral evidence for the formation of quinone methide intermediates from 5- and 7-hydroxyflavonoids. J. Chem. Soc. Chem. Commun. 177-179.

44. HUNDT, H.K.L., ROUX, D.G. 1981. Synthesis of condensed tannins. Part 3. Chemical shifts for determining the 6- and 8-bonding positions of 'terminal' (+)-catechin units. J. Chem. Soc. Perkin Trans. 1:1227-1234.

45. YOUNG, E., BRANDT, E.V., YOUNG, D.A., FERREIRA, D., ROUX D.G. 1986. Synthesis of condensed tannins. Part 17. Oligomeric (2R,3S)-3,3',4',7,8,-pentahydroxyflavans: atropisomerism and conformation of biphenyl and *m*-terphenyl analogues from *Prosopis glandulosa* ('Mesquite'). J. Chem. Soc. Perkin Trans. 1:1737-1749.

46. MALAN, J.C.S., STEENKAMP, J.A., STEYNBERG, J.P., YOUNG, D.A., BRANDT, E.V., FERREIRA, D. 1990. Oligomeric flavanoids. Part 8. The first profisetinidins and proguibourtinidins based on C-8 substituted (-)-fisetinidol units and related C-ring isomerized analogues. J. Chem. Soc. Perkin Trans. 1:209-218.

47. STEENKAMP, J.A., MALAN, J.C.S., ROUX, D.G., FERREIRA, D.

1988. Oligomeric flavanoids. Part 1. Novel dimeric profisefinidins from *Colophospermum mopane*. J. Chem. Soc. Perkin Trans. 1:1325-1330.

48.   STEYNBERG, J.P., FERREIRA, D., ROUX, D.G. 1987. Synthesis of condensed tannins. Part 18. Stilbenes as potent nucleophiles in regio- and stereospecific condensations: novel guibourtinidol-stilbenes from *Guibourtia coleosperma*. J. Chem. Soc. Perkin Trans. 1:1705-1712.

49.   BOTHA, J.J., YOUNG, D.A., FERREIRA, D., ROUX, D.G. 1981. Synthesis of condensed tannins. Part 1. Stereoselective and stereospecific syntheses of optically pure 4-arylflavan-3-ols, and assessment of their absolute stereochemistry at C-4 by means of circular dichroism. J. Chem. Soc. Perkin Trans. 1:1213-1219.

50.   PORTER, L.J., FOO, L.Y. 1982. Leucocyanidin: Synthesis and Properties of (2R,3S,4R)-(+)-3,4,5,7,3',4'-hexahydroxyflavan. Phytochemistry. 21:2947-2952.

51.   DELCOUR, J.A., FERREIRA, D., ROUX, D.G. 1983. Synthesis of Condensed tannins. Part 9. The condensation sequence of leucocyanidin with (+)-catechin and with the resultant procyanidins. J. Chem. Soc. Perkin Trans. 1:1711-1717.

52.   BARRETT, M.W., KLYNE, W., SCOPES, P.M., FLETCHER, A.C., PORTER, L.J., HASLAM, E. 1979. Plant proanthocyanidins. Part 6. Chiroptical studies. Part 95. Circular dichroism of procyanidins. J. Chem. Soc. Perkin Trans. 1:2375-2377.

53.   DE ANGELIS, G.G., WILDMAN, W.C. 1969. Circular dichroism studies - I. A quadrant rule for the optically active aromatic chromophore in rigid polycyclic systems. Tetrahedron. 25:5099-5112.

54.   VAN DER WESTHUIZEN, J.H., FERREIRA, D., ROUX, D.G. 1981. Synthesis of condensed tannins. Part 2. Synthesis by photolytic rearrangement, stereochemistry, and circular dichroism of the first 2,3-*cis*-3,4-*cis*-4-arylflavan-3-ols. J. Chem. Soc. Perkin Trans. 1:1220-1226.

55.   CLARK-LEWIS, J.W., JACKMAN, L.M., SPOTSWOOD, T.M. 1964. Nuclear magnetic resonance spectra, stereochemistry, and conformation of flavan derivatives. Austral. J. Chem. 17:632-648.

56.   FRONCZEK, F.R., GRANNUCH, G., TOBIASON, F.L., BROEKER, J.L., HEMINGWAY, R.W., MATTICE, W.L. 1984. Dipole moment, solution, and solid state structure of (-)-epicatechin, a monomer unit of procyanidin polymers. J. Chem. Soc. Perkin Trans. 2:1611-1616.

57. ENGEL, D.W., HATTING, M., HUNDT, H.K.L., ROUX, D.G. 1978. X-ray structure, conformation, and absolute configuration of 8-bromotetra-O-methyl-(+)-catechin. J. Chem. Soc. Chem. Commun. 695-696.

58. HASLAM, E. 1975. Natural proanthocyanidins. In: The Flavanoids. (J.B. Harborne, T.J. Mabry, H. Mabry, eds.). Chapman and Hall, London, pp. 505-559.

59. PORTER, L.J., WONG, R.Y., CHAN, B.G. 1985. The molecular and crystal structure of (+)-2,3-trans-3,4-trans-leucocyanidin[(2R,3S,4R)-(+)-3,3',4,4',5,7-hexahydroxyflavan] dihydrate, and comparison of its heterocyclic ring conformation in solution and the solid state. J. Chem. Soc. Perkin Trans. 1:1413-1418.

60. FLETCHER, A.C., PORTER, L.J., HASLAM, E. 1976. Hindered rotation and helical structures in natural procyanidins. J. Chem. Soc. Chem. Commun. 627-629.

61. FOO, L.Y., PORTER, L.J. 1983. Synthesis and conformation of procyanidin diastereoisomers. J. Chem. Soc. Perkin Trans. 1:1535-1543.

62. BERGMANN, W.R., BARKLEY, M.D., HEMINGWAY, R.W., MATTICE, W.L. 1987. Heterogeneous fluorescence decay of (4-6)- and (4-8)-linked dimers of (+)-catechin and (-)-epicatechin as a result of rotational isomerism. J. Am. Chem. Soc. 109:6614-6619.

63. GAFFIELD, W., FOO, L.Y., PORTER, L.J. 1989. Exciton split Cotton effect of dimeric procyanidins. J. Chem. Research (S), 144-145.

64. HEMINGWAY, R.W. 1989. Biflavonoids and proanthocyanidins. In: Natural Products of Woody Plants I. (J.W. Rowe, ed.). Springer-Verlag, Berlin, 571-650.

65. HASLAM, E. 1974. Polyphenol-protein interactions. Biochem. J. 139:285-288.

66. TILSTRA, L.F., MAEADA, H., MATTICE, W.L. 1988. Interaction of (+)-catechin with the edge of the a sheet formed by poly-(S-carboxymethyl-L-cysteine). J. Chem. Soc. Perkin Trans. 2:1613-1616, and refs. therein.

67. CHO, D., TIAN, R., PORTER, L.J., HEMINGWAY, R.W., MATTICE, W.L. 1990. Variations in the heterogeneity of the decay of the fluorescence of six procyanidin dimers. J. Am. Chem. Soc. 112:4273-4277.

68. SEARS, K.D., CASEBIER, R.L., HERGERT, H.L., STOUT, G.H., McCANDLISH, L.E. 1974. The structure of catechinic acid. A base rearrangement product of catechin. J. Org. Chem. 39:3244-3247.

69. KIATGRAJAI, P., WELLONS, J.D., GOLLOB, L., WHITE, J.D. 1982. Kinetics of epimerization of (+)-catechin and its rearrangement to catechinic acid. J. Org. Chem. 47:2910-2912.

70. FREUDENBERG, K., BOHME, O., PURRMANN, L. 1911. Raumisomere catechine, II. Chem. Ber. 55:1734-1738.

71. FREUDENBERG, K., PURRMANN, L. 1923. Raumisomere Catechine, III. Chem. Ber. 56:1185-1194. 1924. Raumisomere catechine, IV. Liebigs Ann. Chem. 437:274-285.

72. MEHTA, P.P., WHALLEY, W.B. 1963. The stereochemistry of some catechin derivatives. J. Chem. Soc. 5327-5332.

73. KENNEDY, J.A., MUNRO, M.H.G., POWELL, H.K.J., PORTER, L.J. 1984. The protonation reactions of catechin, epicatechin and related compounds. Aust. J. Chem. 37:885-892.

74. KUHNLE, J.A., WINDLE, J.J., WAISS, A.C. 1969. Electron paramagnetic resonance spectra of flavonoid anion-radicals. J. Chem. Soc. (B) 613-616.

75. JENSEN, O.H., PEDERSEN, J.A. 1983. The oxidative transformations of (+)-catechin and (-)-epicatechin as studied by ESR. Tetrahedron. 39:1609-1615.

76. LAKS, P.E., HEMINGWAY, R.W. 1987. Condensed tannins. Base-catalyzed reactions of polymeric procyanidins with toluene-α-thiol. Lability of the interflavanoid bond and pyran ring. J. Chem. Soc. Perkin Trans. 1:465-470.

77. LAKS, P.E., HEMINGWAY, R.W., CONNER, A.H. 1987. Condensed tannins. Base-catalyzed reactions of polymeric procyanidins with phloroglucinol: intramolecular rearrangements. J. Chem. Soc. Perkin Trans. 1:1875-1881.

78. STEENKAMP, J.A., STEYNBERG, J.P., BRANDT, E.V., FERREIRA, D., ROUX, D.G. 1985. Phlobatannins, a novel class of ring-isomerized condensed tannins. J. Chem. Soc. Chem. Commun. 1678-1679.

79. MALAN, J.C.S., YOUNG, D.A., STEYNBERG, J.P., FERREIRA, D. 1990. Oligomeric flavanoids. Part 10. Structure and synthesis of the first tetrahydropyrano[3,2-g]chromenes related to (4,6)-bis-(-)-fisetinidol profisetinidins. J. Chem. Soc. Perkin Trans. 1:227-234.

80. STEYNBERG, J.P., STEENKAMP, J.A., BURGER, J.F.W., YOUNG, D.A., FERREIRA, D. 1990. Oligomeric flavanoids. Part 11. Structure and synthesis of the first phlobatannins related to (4α,6:4α,8)-bis-(-)-fisetinidol-(+)-catechin profisetinidin tri-flava-

noids. J. Chem. Soc. Perkin Trans. 1:235-240.

81. STEYNBERG, J.P., BURGER, J.F.W., CRONJ, A., BONNET, S.L., MALAN, J.C.S., YOUNG, D.A., FERREIRA, D. 1990. Structure and synthesis of phlobatannins related to (-)-fisetinidol-(-)-epicatechin profisetinidins. Phytochemistry. 29:2979-2989.

82. STEYNBERG, J.P., BURGER, J.F.W., YOUNG, D.A., BRANDT, E.V., STEENKAMP, J.A., FERREIRA, D. 1988. Oligomeric Flavanoids. Part 3. Structure and synthesis of phlobatannins related to (-)-fisetinidol-(4α,6)- and (4α,8)-(+)-catechin profisetinidins. J. Chem. Soc. Perkin Trans. 1:3323-3329.

83. STEYNBERG, P.J., BURGER, J.F.W., BEZUIDENHOUDT, B.C.B., STEYNBERG, J.P., VAN DYK, M.S., FERREIRA, D. 1990. The first natural condensed tannins with (-)-catechin 'terminal' units. Tetrahedron Lett. 31:2059-2062.

84. SWEENY, G.J., IACOBUCCI, G.A. 1979. Regiospecificity of (+)-catechin methylation. J. Org. Chem. 44:2298-2299.

85. BERGMANN, W.R., VISWANADHAN, V.N., MATTICE, W.L. 1988. Conformations of polymeric proanthocyanidins composed of (+)-catechin or (-)-epicatechin joined by 4→ 6 interflavan bonds. J. Chem. Soc. Perkin Trans. 2:45-47.

86. BURGER, J.F.W., STEYNBERG, J.P., YOUNG, D.A., BRANDT, E.V., FERREIRA, D. 1989. Oligomeric flavanoids. Part 5. Base-catalyzed C-ring isomerization of (+)-fisetinidol-(+)-catechin profisetinidins. J. Chem. Soc. Perkin Trans. 1:671-681.

87. BURGER, J.F.W., KOLODZIEJ, H., HEMINGWAY, R.W., STEYNBERG, J.P., YOUNG, D.A., FERREIRA, D. 1990. Oligomeric flavanoids. Part 15. Base-catalyzed pyran rearrangements of procyanidin B-2, and evidence for the oxidative transformation of B- to A-type procyanidins. Tetrahedron. 46:5733-5740.

88. STEYNBERG, J.P., BEZUIDENHOUDT, B.C.B., BURGER, J.F.W., YOUNG, D.A., FERREIRA, D. 1990. Oligomeric flavanoids. Part 7. Novel base-catalyzed rearrangments of procyanidins. J. Chem. Soc. Perkin Trans. 1:203-208.

Chapter Ten

ENZYMATIC SYNTHESIS OF GALLOTANNINS AND RELATED COMPOUNDS

Georg G. Gross

Universität Ulm, Abteilung Allgemeine Botanik
D-7900 Ulm, Germany

INTRODUCTION

The onset of the 20th century marks the period when plant tannins, as many other natural products, became the subject of intense investigations aimed at the elucidation of their chemical structures, properties, and distribution in the Plant Kingdom. Among many researchers of that time, the name of E. Fischer deserves special attention as his eminent work provided fundamental, lasting insights into the constitution of Chinese and Turkish hydrolyzable tannins and their eventual biogenetic precursors. Although his work was continued by

Phenolic Metabolism in Plants, Edited by H.A. Stafford
and R.K. Ibrahim, Plenum Press, New York, 1992

excellent scientists, such as K. Freudenberg and P. Karrer, the whole challenging field was increasingly ignored due to the growing insight that the then available analytical armament was insufficient to tackle the evidently tremendous complexity of these plant constituents. Fortunately, a renaissance in the 1950s can be recorded, starting with the outstanding investigations of O.Th. Schmidt and W. Mayer on ellagitannins, and continued by other research groups, mainly in England and Japan, such as the laboratories of E.C. Bate-Smith, T. Swain, E. Haslam, T. Okuda and I. Nishioka.

The results of the combined efforts of these and many other scientists have now provided us with deep insights into the occurrence and structures of innumerable gallotannins and related compounds.[1-4] It is obvious that emphasis began to be placed on related challenges, being directed either to more practical questions such as the role of tannins in traditional medicine, and in ecological systems, or to considerations of the biosynthetic routes responsible for the production of these complex molecules.

Concerning this latter question, traditionally biogenetic pathways have been examined by tracer experiments, i.e. by applying suitably labeled potential precursors to living plants or plant parts, followed by analysis of the reaction products. A major limitation of such comparatively simple experiments, however, is the difficulty of evaluating the significance of isolated metabolites with respect to individual pathways. This problem has often been solved by the use of mutants with defined genetic blocks in the metabolic sequences to be studied—a strategy that works nicely with procaryotes but is impracticable with higher plants. In view of this dilemma, studies at the enzyme level are the method of choice in many instances. This technique not only allows the unequivocal identification of obligate metabolic intermediates, but also provides otherwise inaccessible information about trace amounts of labile 'activated' precursors as a prerequisite for the elucidation of biochemical reaction mechanisms.

It should not be forgotten that specific and serious problems are often encountered in the extraction of enzymes from plant tissues. Most of these sources not only require exceedlingy strong mechanical forces to disrupt their rigid, often heavily lignified cell walls, but are also characterized by the widespread occurrence of high concentrations of organic acids and phenolics that effectively denature released enzymes. It is obvious that these negative factors apply particularly to studies with those plants that actively produce tannins, constituents that are notorious for their pronounced tendency to combine with proteins thus causing their precipitation and denaturation, including the inactivation of enzymes.[2,5] As these effects are not restricted to complex gallotannins but occur also with most of their biogenetic precursors (only mono-

and digalloylglucoses exhibit no tannin properties[3,5]), it was considered somewhat daring to attempt investigations on the biosynthesis of gallotannins just by means of enzymatic studies. Fortunately, as documented in this chapter, these fears did not hold true—the enzymes of the pathways examined were remarkably resistant to their unfavorable substrates, whereas major problems were encountered only in the preparation and characterization of enzyme substrates and products.

CLASSIFICATION AND STRUCTURAL PRINCIPLES

Before discussing the biochemical events involved in the formation of hydrolyzable tannins, it is appropriate to briefly describe their structural principles as an indispensable prerequisite for the understanding of the questions, and their solutions, related to the biogenesis of these natural products.

According to a classical definition,[6] plant tannins are commonly divided into *condensed* tannins (nowadays often referred to as proanthocyanidins owing to their behavior in the presence of acids) and *hydrolyzable* tannins. The latter are characterized by a central polyol moiety (usually β-D-glucopyranose, but also hamamelose, shikimic acid, quinic acid or cyclitols have been identified[4]) whose hydroxyl groups are typically esterified with gallic acid (Fig. 1) (**1**).

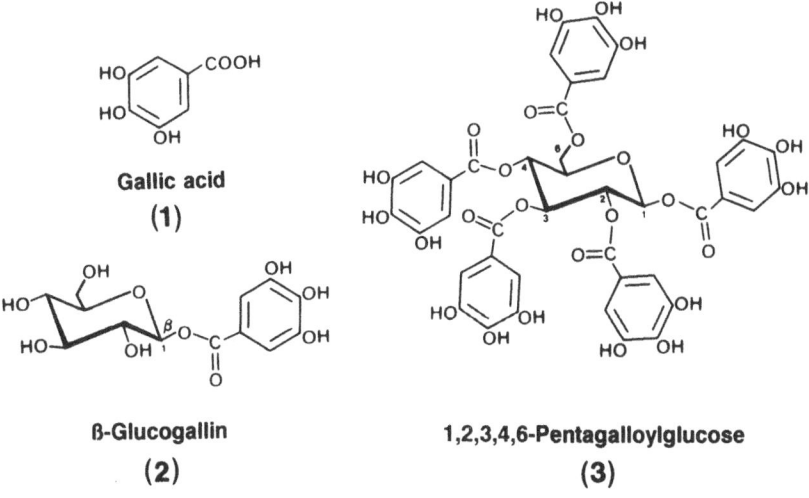

Gallic acid
(**1**)

β-Glucogallin
(**2**)

1,2,3,4,6-Pentagalloylglucose
(**3**)

Fig. 1. Structures of important gallotannin components.

However, other acyl residues, e.g. cinnamic acids, have also been observed sporadically, as well as the occurrence of phenolic glucosides.[2] Stepwise substitution, beginning with monogalloylglucoses, such as glucogallin (2), leads to 1,2,3,4,6-pentagalloylglucose (3) which is regarded as the common precursor of the two subclasses of hydrolyzable tannins, i.e. *gallotannins* and *ellagitannins* (Fig. 2). The former compounds are characterized by the introduction of additional galloyl residues that are attached to the pentgalloylglucose core (3) via so-called *meta*-depside bonds (cf. 4); they can range up to deca- or dodecagalloyl-glucoses, as has been shown for the tannins from *Rhus semialata* (Chinese gallotannin),[8] *Quercus infectoria* (Turkish gallotannin)[8] or *Paeonia albiflora* (syn. *P. lactiflora*).[9,10] (It should be noted that NMR spectroscopy studies[7] indicated the existence of both *meta* and *para*-depsides in gallotannins; use of the traditional term '*meta*-depside' in this article may thus be an over-simplification.)

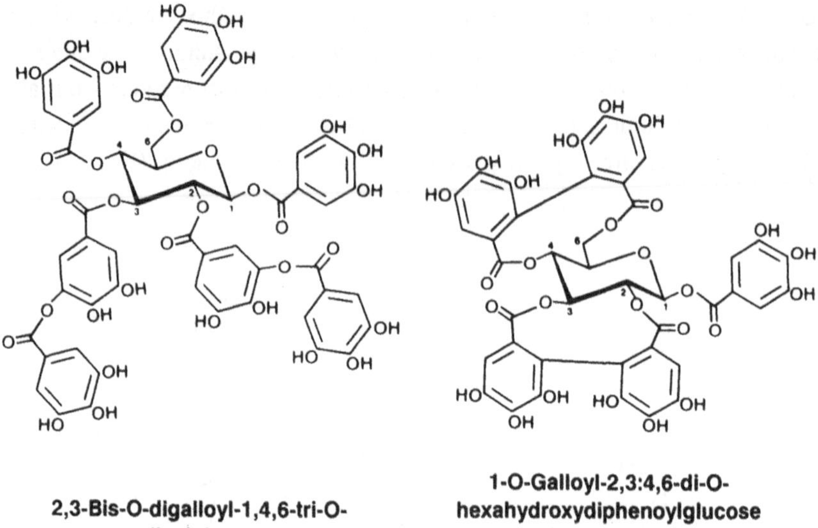

2,3-Bis-O-digalloyl-1,4,6-tri-O-galloylglucose

(4)

1-O-Galloyl-2,3:4,6-di-O-hexahydroxydiphenoylglucose (casuarictin)

(5)

Fig. 2. Characteristic structures of gallotannins and ellagitannins. The heptagalloylglucose (4), a gallotannin, has been isolated from Chinese gallotannin[7] and is characterized by galloyl residues attached as *meta*-depsides to the pentagalloylglucose (3) core. The ellagitannin, casuarictin (5), is characterized by two hexahydroxy-diphenoyl (HHDP, 6) groups.

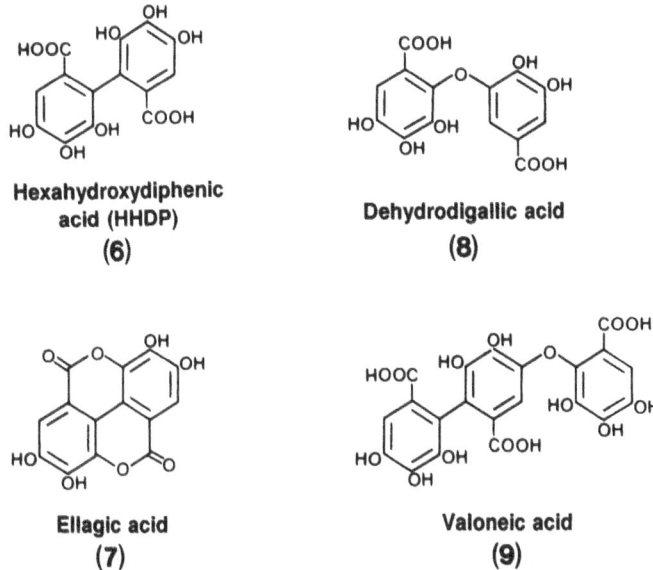

Fig. 3. Phenolic esters formed by oxidative coupling of galloyl units in ellagitannins.

Ellagitannins, in contrast, are the result of oxidative processes that lead to the introduction of one or two C–C linkages between adjacent galloyl groups of pentagalloylglucose (3). Depending on the configuration of the central glucose moiety (usually $^4C_1$ or $^1C_4$), a wide variety of different derivatives can be constructed that have indeed been found in nature; one typical example, casuarictin (5), is depicted in Figure 2. In all these reactions, characteristic hexahydroxydiphenoyl (HHDP) residues are formed (Fig. 3) (6). After their eventual release from the tannin core, the resulting free acids rearrange spontaneously to the stable name-giving dilactone, ellagic acid (7). It should be added that these *intra*-molecular coupling reactions often continue *inter*-molecularly, yielding ellagitannin dimers (and subsequently oligomers) that are interconnected via nonahydroxytriphenoyl linkages. More common, however, is the participation of a galloyl-OH in such oxidation reactions that leads, depending on the nature of the reactants, to dehydrodigallic acid (8) and valoneic acid (9) or its isomer, sanguisorbic acid, which all are characterized by aryl C-O-C bridges.[1-4]

Innumerable variations of these fundamental structural principles have been discovered in higher plants,[1-4] but their discussion is beyond the scope of this article. In any event, this short essay should have elucidated the focal points of eventual considerations on the mechanisms involved in the biogenetic pathways from gallic acid to highly complex hydrolyzable tannins: (1) the origin of gallic acid, (2) the biosynthesis of β–glucogallin as the first specific intermediate on the pathway to hydrolyzable tannins, (3) the conversion of this monoester to pentagalloylglucose, and finally the secondary transformations of this pivotal intermediate to yield (4) gallotannins and (5) ellagitannins. The results obtained from enzyme studies on these questions are discussed in the subsequent sections.

ORIGIN OF GALLIC ACID

It is now generally accepted that benzoic acids (phenylcarboxylic acids) of higher plants are the result of degradative processes, either by true catabolism of complex natural products (e.g., flavonoids) or, in anabolic sequences, by side-chain degradation of cinnamic acids (phenylacrylic acids). In the latter case an acetyl group is split off from the precursor. The exact details of this conversion, however, are still a matter of dispute, involving a coenzyme A-dependent β-oxidation sequence versus a coenzyme A-independent 'nonoxidative' pathway.[11] In contrast, particular problems are encountered with gallic acid (3,4,5-trihydroxybenzoic acid, 1). In spite of numerous investigations during the past decades, the biosynthesis of this wide-spread plant constituent is still one of the major enigmas of plant phenolic metabolism. Some essentials of the conflicting proposals are depicted in Figure 4 (routes a, b, or c). A conventional pathway (a) based on the above considerations was formulated by Zenk[12] in which 3,4,5-trihydroxy-cinnamic acid (12) undergoes β-oxidation to gallic acid (1). The major objection to this proposal was the fact that the precursor, thought to be formed by hydroxylation of caffeic acid (11), has never been identified as natural product. A related proposition (route b) avoided this problem, favoring the sequence caffeic (11) –> protocatechuic (13) –> gallic acid (1).[13] Supporting evidence for this view arose from a preliminary report[14] that cell-free extracts from *Pelargonium* were able to catalyze the second step of these reactions.

A quite different approach resulted from tracer studies with fungi[15] and higher plants[16,17] that were indicative of a direct aromatization of shikimic or dehydroshikimic acid (10) to gallic acid (route c in Fig. 4). Also, a conversion

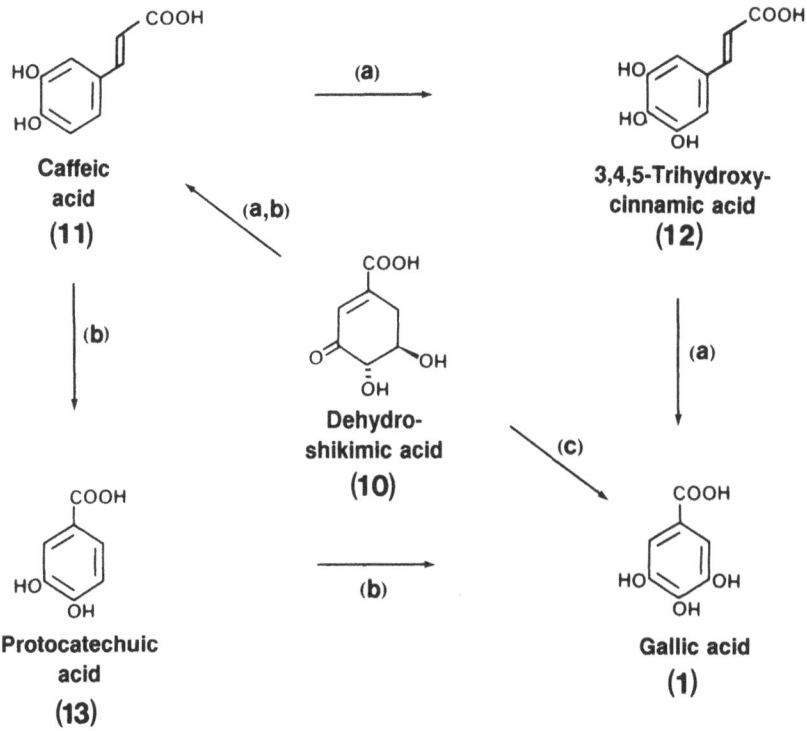

Fig. 4. Proposed biosynthetic pathways to gallic acid.

of (10) to protocatechuic acid (13) was observed with crude enzyme preparations from mung bean seedlings.[18] Additional evidence for the operation of this short-cut to gallic acid arose from studies with the herbicide glyphosate [*N*-(phosphonomethyl)glycine], a phosphoenolpyruvate analog. This compound competitively inhibits 5-enolpyruvylshikimate-3-phosphate synthase, a key enzyme of the shikimate pathway, which blocks the formation of phenolics via the subsequent phenylalanine-cinnamate pathway. After application of this agent to various plants, it has been observed that not only the levels of shikimic acid but also those of gallic acid (and of the structurally related *p*-hydroxybenzoic and protocatechuic acid) increased considerably, indicating that the direct aromatization of (dehydro)shikimic acid represents a significant, if not the major, route.[19-21]

At least in theory, the latter route (c) could be distinguished easily from the other routes by experiments with carboxyl-labeled (dehydro)shikimic acid, since this functional group would be lost along the phenylalanine-cinnamate pathway but retained in the case of a direct conversion to gallic acid. Such studies have recently shown that p-hydroxybenzoic acid was exclusively formed via cinnamate precursors in *Lithospermum* (route a or b).[22] On the other hand, young tea shoots converted carboxyl-labeled shikimic, dehydroshikimic and dehydroquinic acids directly to gallic acid (route c), but utilized also C_6C_3-precursors (route a or b).[23] It must be concluded that at least two pathways for gallic acid biosynthesis exist, with the preferential route depending on the plant species and its developmental stage.[24,25]

BIOSYNTHESIS OF β-GLUCOGALLIN AND RELATED ESTERS

β-Glucogallin (1-O-galloyl-β-D-glucose, **2**) was first isolated from Chinese rhubarb (*Rheum officinale*) in 1903,[2,3] and is considered the primary metabolite in the biosynthesis of hydrolyzable tannins.[26] For thermodynamic reasons, the participation of an 'activated' intermediate has to be postulated in the formation of such an ester. This requirement can be met by two ways, either by reaction of an energy-rich galloyl derivative with free glucose, or via an activated glucose derivative (most likely the common UDP-glucose) that combines with the free acid. When the studies discussed in this chapter were begun it was already well documented that esterification of many aromatic acids with different polyols proceeded via acyl-CoA intermediates, and this list has since been expanded considerably. Since the first report on the caffeoyl-CoA dependent synthesis of the ubiquitous plant depside chlorogenic acid (3-O-caffeoylquinic acid) in a cell-free system by Stöckigt and Zenk in 1974,[27] a wide variety of hydroxycinnamoyl esters (mainly p-coumaroyl and caffeoyl derivatives) have been recognized to be analogously formed via the corresponding acyl-CoA's. These esters included numerous hydroxylated compounds that were utilized as alcoholic moieties, e.g. quinate,[28,29] shikimate,[30] 3,4-dihydroxyphenyl-lactate,[31] tartronate,[32] isocitrate,[33] sugar acids (gluconate, glucarate, galactarate, glucarolactone),[34] myo-inositol[35] or flavonoids.[36,37]

By analogy, it appeared conceivable that galloyl-CoA might be involved in the biosynthesis of gallotannins. As a prerequisite for the intended investigations, this then unknown thioester was synthesized chemically via the N-succinimidyl derivative of 4-O-β-D-glucosidogallic acid.[38] Enzyme studies with cell-free extracts from oak leaves soon indicated that this thioester was inactive in the formation of either β-glucogallin or its higher galloylated deriva-

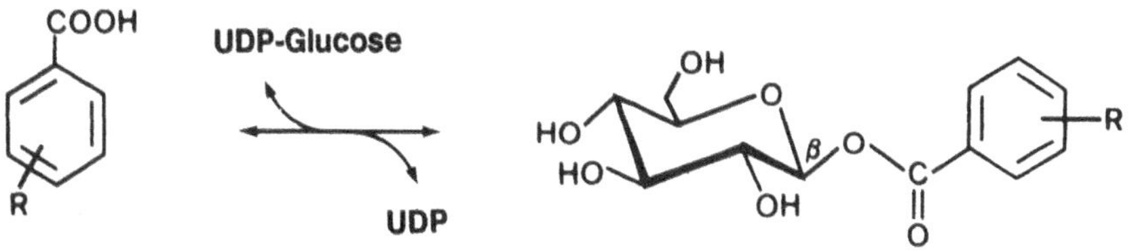

Fig. 5. Biosynthesis of β-glucogallin and related 1-O-phenyl-carboxylglucoses, catalyzed by UDP-glucose: vanillate 1-O-glucosyltransferase (EC 2.4.1.-).

tives. Instead, the second alternative, i.e., the reaction of UDP-glucose with free gallic acid, could be verified with these enzyme preparations.[40] The partially purified glucosyltransferase[41] catalyzed the formation of β-glucogallin and related esters as depicted in Figure 5. UDP-glucose was found to act as the exclusive sugar donor while numerous benzoic and, at significantly lower rates, cinnamic acids could serve as acceptors. In addition, this enzyme was successfully employed for the preparation of related 1-O-phenylcarboxyl-glucoses.[42]

The detection of this enzyme activity was not too surprising because several laboratories obtained in vitro evidence that 1-O-glucose esters of phenolic acids were formed throughout in plants by this mechanism. Confirming a preliminary communication,[43] it was reported that the conjugation of glucose with numerous benzoic and cinnamic acids,[44-49] or with the plant hormone indolylacetic acid,[50] occurred by this means. (In this context it should be emphasized that glucose esters must not be confused with glucosides; different enzymes have been shown to be involved in the biosynthesis of these two types of phenolic glucose derivatives[51]). Summarizing these data, it appears that UDP-glucose is the general activated sugar donor required for the synthesis of phenolic acid glucose esters, whereas the reaction with other hydroxylated moieties proceeds via acyl-CoA thioesters as the energy-rich component.[52]

THE PATHWAY TO PENTAGALLOYLGLUCOSE

β-Glucogallin-Dependent Acyltransferases

According to recent suggestions,[2,26] the primary metabolite β-glucogallin (2) should undergo a series of galloylation steps to yield 1,2,3,4,6-

pentagalloylglucose (3). Elucidation of the details of this hypothetical biogenetic sequence was established with enzyme preparations from young leaves of oak (*Quercus robur, Q. rubra*) and sumach (*Rhus typhina*). In initial experiments it was demonstrated that, by analogy to the preceding synthesis of β–glucogallin, galloyl-CoA was not required.[39] Instead, both di- and trigalloylglucose were formed in the presence of β-glucogallin as the sole substrate, a surprising finding that led to the conclusion that the comparatively low group-transfer potential of β-glucogallin was still sufficient to permit subsequent transacylation reactions.[39] To the author's knowledge, thermodynamic data for β-glucogallin do not exist. However, a $\Delta G_0'$ of about 5 kcal/mol has been reported for the comparable hemiacetal phosphate of glucose-1-phosphate, whereas the rather inert ester linkage of glucose-6-phosphate has only 2.5-3 kcal/mol. In contrast, the most common 'activated' acids, acyl-CoA thioesters, have $\Delta G_0'$ values of 7-8 kcal/mol.[53]

The above mentioned observations on the role of β–glucogallin in the biosynthesis of gallotannins coincided with similar results from other laboratories on the intermediacy of 1-*O*-acylglucoses in the enzymatic esterification of numerous phenolic acids. In many *Brassicaceae*, sinapoylglucose was found to serve as donor of the acyl moiety of sinapoylcholine (sinapine)[54,55] and sinapoylmalate.[56,57] *p*-Coumaroylglucose acylated *meso*-tartaric acid in spinach,[58] other hydroxycinnamoylglucoses acted as donors in the acylation of glucaric acid[59] and betalains,[60] and indolylacetylglucose was converted in *Zea mays* to indolylacetyl-*myo*-inositol.[61,62] An enzyme from *Ipomoea*[63] was found to produce chlorogenic acid and related depsides from hydroxycinnamoylglucoses and free quinic acid, thus presenting evidence of a novel pathway as an alternative to the long established acyl-CoA dependent biosynthesis of these compounds. Interestingly enough, the reaction product, chlorogenic acid, was in turn utilized as caffeoyl donor in the acylation of glucaric and galactaric acid in tomato.[64,65]

It is evident from these data that phenolic glucose esters occupy a prominent position in the secondary metabolism of higher plants, and this applies particularly to the biogenesis of hydrolyzable tannins which depends exclusively on the participation of such acyl derivatives. Some characteristics of the enzymes and reactions related to this pathway are reported below.

β–Glucogallin-glucose exchange. In initial studies on the origin of galloylglucoses, an enzyme catalyzing a very unusual galloyl exchange reaction was discovered in cell-free extracts from oak:[39]

$$\beta\text{-Glucogallin} + {}^*\text{D-glucose} = {}^{(*)}\text{D-glucose} + {}^*\beta\text{-glucogallin}$$

(The asterisk symbolizes an appropriate label, e.g. [14]C, to allow measurement of the reaction.) The partially purified enzyme, whose physiological role is still unclear, was active with various 1-O-acylglucoses, while D-glucose was the exclusive acceptor; the reaction was of practical value for the economic preparation of labeled β-glucogallin required in subsequent investigations.[66,67]

Formation of 1,6-digalloylglucose (**14**). Preliminary experiments[39] suggested the existence of an enzyme that catalyzed the formation of digalloyl-glucose from two molecules of β–glucogallin. This proposal was confirmed with a partially purified enzyme (EC 2.3.1.90) from young oak leaves that was found to specifically produce 1,6-digalloylglucose (**14**).[68] After immobiliztion on phenyl-Sepharose, this acyltransferase was employed for the preparation of this rare natural product as a convenient alternative to its isolation from plant drugs.[69] In the unusual reaction catalyzed by this enzyme, two identical substrate molecules react together, one acting as donor and the other two as acceptor, while glucose is concomitantly released as the deacylated by-product (cf. Fig. 6). Analogous 'disproportionations'[70] have later been identified in the biosynthesis of 1,2-disinapoylglucose in *Raphanus*[70,71] and of 3,5-dicaffeoyl-quinic acid (isochlorogenic acid) in *Ipomoea*.[72,73]

1,2,6-Trigalloylglucose (**15**). The long proposed[39] continuation of the above sequence, i.e., the conversion of digalloylglucose to trigalloylglucose with β-glucogallin as donor, has been substantiated with enzyme extracts from staghorn sumach (*Rhus typhina*) leaves that catalyzed the highly position-specific galloylation of the 2-OH of the substrate to yield 1,2,6-trigalloylglucose (**15**) (cf. Fig. 7).[74] With partially purified enzyme, it was found that, besides several unphysiological substrates, β-glucogallin and its dihydroxy analog, 1-protocatechuoylglucose, were potent acyl donors; similarly, the corresponding 1,6-diacylated derivatives proved to be efficient acceptor substrates.[75] (For the β-glucogallin-*independent* synthesis of trigalloylglucose, see below.)

Formation of tetra- and pentagalloylglucose. On the basis of the above described reactions of the pathway to pentagalloylglucose, it is almost trivial to report that β-glucogallin was also found to function as galloyl donor in the acylation of 1,2,6-trigalloylglucose (**15**) to 1,2,3,6-tetragalloylglucose (**16**; Hagenah and Gross, unpublished), followed by the analogous conversion of this intermediate to the final metabolite of the entire sequence, 1,2,3,4,6-pentagalloylglucose (**3**) (cf. Fig. 8).[76] Nevertheless, some details of these reactions deserve to be mentioned. First, it was found that a partially purified enzyme from *R. typhina* was about 10-times more active with 1,3,6-trigalloyl-glucose than with its 1,2,6-isomer (**15**); however, only the latter compound was exclusively produced by the preceding galloylation of 1,6-digalloylglucose.[74,75]

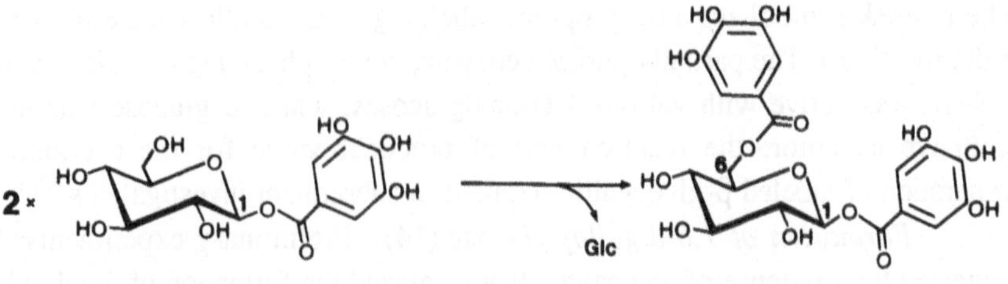

Fig. 6. Enzymatic synthesis of 1,6-digalloylglucose (14) by 'disproportionation' of two molecules of β–glucogallin (2), catalyzed by β–glucogallin: β-glucogallin 6-*O*-galloyltransferase (EC 2.3.1.90). Glc, glucose.

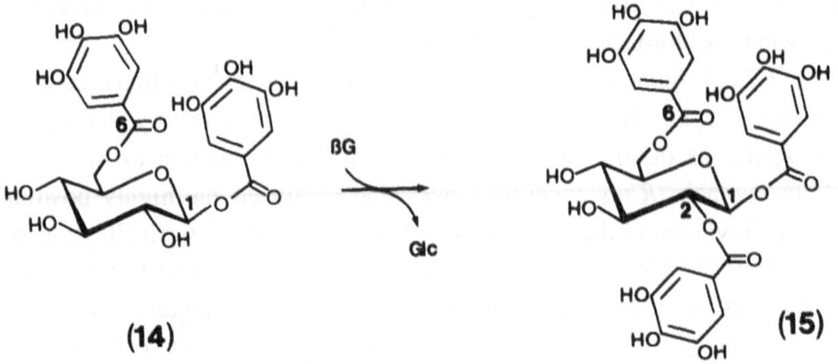

Fig. 7. β-Glucogallin-dependent conversion of 1,6-digalloylglucose (14) to 1,2,6-trigalloylglucose (15) by 2-*O*-galloyltransferase from *Rhus typhina*. βG, β-glucogallin; Glc, glucose.

This finding supports the view that metabolic sequences in plants are governed not only by the reactivities of the often less specific enzymes catalyzing these routes, but by the availability of certain precursors. Second, 1,2,6-trigalloylglucose (15) was specifically acylated to 1,2,3,6-tetragalloylglucose

(16) and the latter, in turn, served as the exclusive substrate for the formation of pentagalloylglucose (3) in oak leaves.[76] The previously discussed ambiguity concerning the intermediacy of 1,2,3,6- and 1,2,4,6-tetragalloylglucose[26] has thus been clarified by recent enzyme studies, at least for *Rhus* and *Quercus* where the 1,2,4,6-isomer plays no role. It is interesting to note that this result is consistent with earlier analyses of the galloylglucoses formed in tissue cultures of *Q. robur* where the 1,2,4,6-isomer was found to occur in only trace amounts, if at all, in contrast to substantial quantities of 1,2,3,6-tetragalloylglucose.[77]

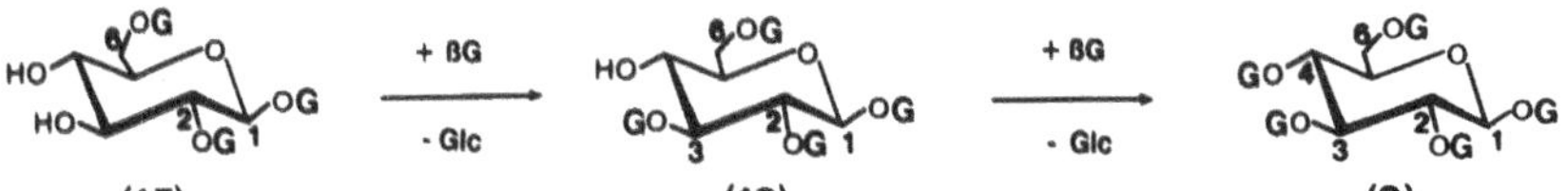

Fig. 8. Enzymatic conversion of 1,2,6-trigalloylglucose (15) to 1,2,3,6-tetragalloylglucose (16) and 1,2,3,4,6-pentagalloylglucose (3) by βglucogallin-dependent galloyltransferases from *Rhus* and *Quercus*. βG, β-glucogallin; Glc, glucose.

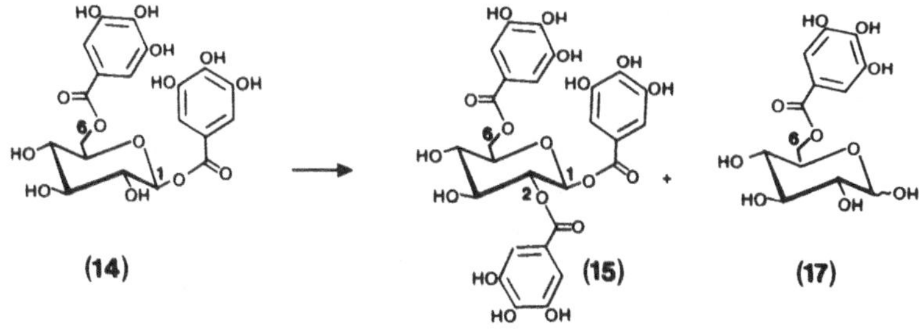

Fig. 9. Enzymatic 'disproportionation' of 1,6-digalloylglucose (14) to 1,2,6-trigalloylglucose (15) and anomeric 6-O-galloylglucose (17).

β-Glucogallin-Independent Pathways

The results of the enzyme studies described above were indicative of a β-glucogallin-dependent pathway from gallic acid to pentagalloylglucose. This clear and logical picture was blurred by the discovery of a parallel and apparently β-glucogallin-*independent* side-reactions. Results from enzyme extracts of sumach leaves have led to the identification of a galloyl-transferase that produced 1,2,6-trigalloylglucose (15) from 1,6-digalloylglucose (14) in the absence of the established galloyl donor, β-glucogallin (2). Detailed investigations of this problem[78] revealed the existence of a novel 'disproportionation' reaction (see above) of two molecules of digalloylglucose (14), yielding trigalloylglucose (15) and anomeric 6-galloylglucose (17) (Fig. 9). Similarly, β-glucogallin could be replaced by 1,6-digalloylglucose (14) as acyl donor in the conversion of 1,2,6-trigalloylglucose (15) to 1,2,3,6-tetragalloylglucose (16). Moreover, tri- and tetragalloylglucoses were found to exhibit some activity as galloyl donors, provided that they had the essential 1-*O*-acyl group; only 1,2,3,4,6-pentagalloylglucose (3) was definitively inactive.

In contrast to previous assumptions,[52] it is thus no longer tenable to assign an exclusive role to the monoester β-glucogallin as the acyl donor. In the light of these new results, it would be expected from thermodynamic considerations that analogous compounds with higher galloylation levels can serve as suitable acyl donors. However, owing to the observed decreasing reactivity in response to increasing substitution,[78] the 1-mono- and 1,6-diesters still represent the predominant acyl donors, whereas the reactions involving higher analogs remain more or less negligible. These results are consistent with recent theoretical considerations of the origin of the relatively wide-spread galloylglucoses in which the anomeric position of the sugar is unacylated;[3] however, interference of the above processes with simple hydrolysis of the comparatively labile 1-*O*-acyl bond of glucosyl esters cannot so far be excluded. It should be mentioned finally that galloylation of acylglucoses bearing a free anomeric OH-group has never been observed in vitro, a fact indicating that the existence of an acyl residue at this specific position is indispensable, not only for thermodynamic reasons but also to ensure the correct identification of these esters as galloyltransferase substrates.

Some general remarks on the enzymes involved in the pathway from gallic acid to pentagalloylglucose appear appropriate at this stage. In the preceding sections, enzymes catalyzing the individual steps of this sequence have been isolated from leaves of oak and sumach. Although not explicitly stated, it should be emphasized that the entire pathway was found to operate identically in both plant species. In addition, the general properties of the involved enzymes were similar; their pH-optima lay around pH 6-6.5, they were stable in slightly acidic media (pH 4-6), temperature optima were at about 40-50°C, and a common cold tolerance with residual activities of 10-25% at 0°C was observed. Also, a general trend for unusually high molecular weights is apparent, with M_r values between ca. 300,000 and 750,000. However, significant differences were determined only for the UDPG-dependent glucosyltransferase from oak[40,41] and the β-glucogallin-independent 1,2,6-trigalloylglucose-forming enzyme from sumach, which had molecular weights of 68,000 and 56,000, respectively.

Certainly one of the most striking features of the pathway to pentagalloylglucose is the pronounced position-specificity of the individual galloylation steps, being 1-OH > 6-OH > 2-OH > 3-OH > 4-OH for the enzymatic substitution of β-D-glucose. Interestingly, an identical sequence of reactivities has been determined for the *chemical* esterification of the hydroxyl groups of β-glucose.[79,80] Possible reasons for this behavior are that, after the 1-OH group, the primary 6-OH is more reactive than the secondary hydroxyls; among these, the 2-OH group is the most reactive one due to an activating effect of the neighboring anomeric center. Finally, access to the 4-OH group adjacent to the (already substituted) bulky 6-position is thought to be sterically hindered, resulting in a higher relative reactivity of the 3-OH group.

PENTAGALLOYLGLUCOSE-DERIVED HYDROLYZABLE TANNINS

As already mentioned, 1,2,3,4,6-pentagalloylglucose (3) is regarded as the principal common precursor in the biogenesis of both gallotannins and ellagitannins. Some recent observations on these pathways are discussed below.

Gallotannins

Fortunately, the disappointing statement of a recent review article, published only two years ago,[52] that "virtually nothing is known about the formation of the characteristic *meta*-depside bond" of gallotannins was superseded by the discovery that enzyme extracts from *R. typhina* leaves catalyzed the efficient galloylation of 1,2,3,4,6-pentagalloylglucose (3), yielding a mixture of numerous hexa-, hepta-, octa- and nonagalloylglucoses, together with traces of decagalloylglucose.[81] By analogy to the preceding reactions leading to pentagalloylglucose, β-glucogallin (2) was again found to serve as the specific galloyl donor in these conversions, indicating the versatility of β-glucogallin to act not only as acylating agent in the substitution of aliphatic hydroxyls but also in the acylation of the chemically different phenolic OH-groups. The supposed gallotannin nature of the reaction products was unequivocally proven by high-performance liquid chromatography (HPLC) on silica gel,[9] hydrolysis with the fungal esterase tannase (yielding only glucose and gallic acid as cleavage products) and methanolysis of the isolated hexa-, hepta-and octagalloylglucose fractions, affording 1,2,3,4,6-penta-galloylglucose and methyl gallate in the expected stoichiometric ratios of 1:1, 1:2, and 1:3, respectively.[81]

Conclusive proof of the proposed reaction was obtained with the three enzymatically formed hexagalloylglucoses that were purified from the reaction mixtures by chromatography on Sephadex LH-20 and reversed-phase HPLC.[81] As a result of detailed analyses by ^{1}H and ^{13}C NMR spectroscopy, structures **18-20** (Fig. 10) were assigned to these compounds (Gross, Hofmann and Schilling, unpublished), and these findings were fully consistent with those obtained earlier for the *in vivo* produced hexagalloylglucoses from the related species *R. semialata*.[7] As shown in Table 1, the prominent signals of the aryl-hydrogen atoms of the galloyl residues (H-2,6) allow the facile and unequivocal identification of these tannins, in particular by comparison of the ppm values of

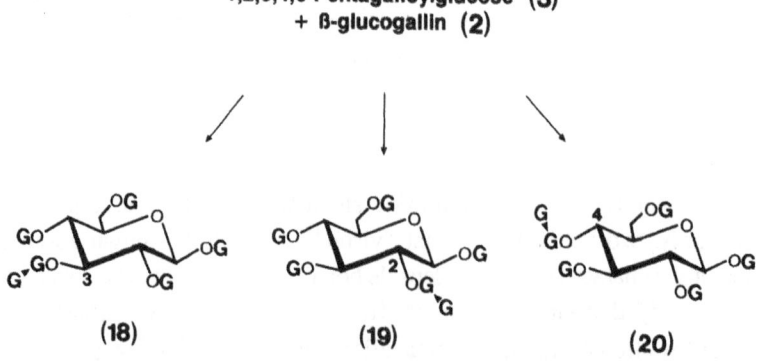

1,2,3,4,6-Pentagalloylglucose (3)
+ ß-glucogallin (2)

(18)   **(19)**   **(20)**

Fig. 10. Structures of the three hexagalloylglucoses (**18-20**) formed from pentagalloylglucose (**3**) and the acyl donor β-glucogallin (**2**) by enzyme preparations from leaves of *Rhus typhina*.

Table 1. ^{1}H NMR chemical shifts of galloyl residues of pentagalloylglucose (3) and the enzymatically derived hexagalloylglucoses **18-20** (δ_H values in ppm TMS).[a]

| Compound[b] | Aryl H-2,6 of galloyl group located at glucose carbon | | | | | |
	1	2	3	4	6	X[c]
3	7.10	7.00	6.96	7.04	7.17	--
19	7.16	7.34, 7.28	6.96	7.03	7.22	7.10
18	7.15	7.00	7.30, 7.23	7.05	7.22	7.10
20	7.16	6.99	6.96	7.35, 7.33	7.25	7.09

[a] Measured in d$_6$-acetone.

[b] For structures, see Figures 1 and 10.

[c] X denotes the sixth galloyl residue, depsidically attached to the pentagalloylglucose (3) core.

the two doublets that originate from the *meta*-depsidic attachment of the newly introduced sixth galloyl residue to the pentagalloylglucose core.

Ellagitannins

In spite of many attempts to elucidate the mechanism of the oxidative conversion of pentagalloylglucose (3) to ellagitannins, this old challenge remains highly enigmatic. In previous experiments, gallic acid and numerous derivatives (e.g. methyl gallate, β-glucogallin, 3,6-digalloylglucose, pentagalloylglucose) were treated *in vitro* with the fungal enzyme laccase or peroxidase from higher plants.[82-86] It is common to all these investigations that ellagic acid (7) was formed as a typical product, indicating the preceding formation of hexahydroxydiphenic acid (6). However, this latter compound has never been detected in the form of a glucose ester and it thus remains very questionable whether true ellagitannins have ever been produced in these studies. Doubts whether the rather unspecific enzyme systems used in these experiments (i.e. laccase or other polyphenol oxidases + O_2, or peroxidase + H_2O_2, which

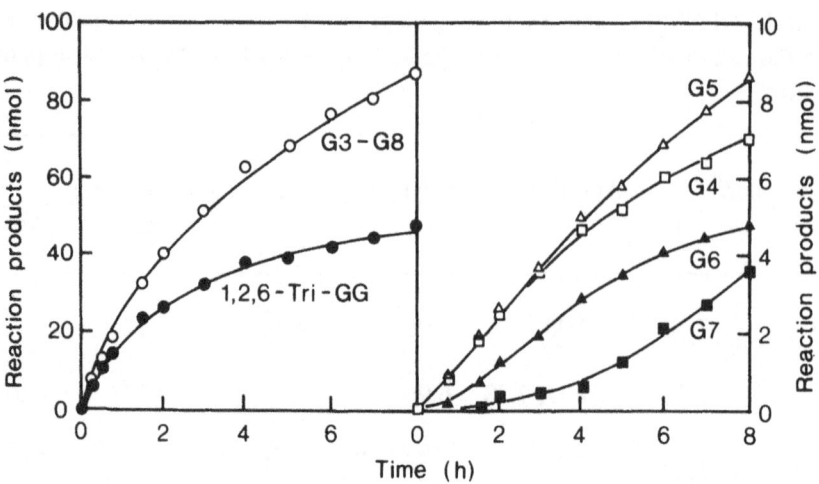

Fig. 11. Conversion of β-glucogallin (**2**) and 1,6-digalloylglucose (**14**) to 1,2,6-trigalloylglucose (1,2,6-Tri-GG) and higher galloylated derivatives by an enzyme preparation from leaves of *Rhus typhina*. G3, G4,... G8, tri-, tetra-,... octagalloylglucose.

all depend on intermediary oxygen radicals) really reflect natural conditions were supported by recent investigations on the radical scavenging effects of tannins and related polyphenols having *ortho*-trihydroxyphenyl structures.[87] However, upon incubation of enzyme extracts from different oak tissues (leaves, cotyledons, cupulae) together with various oxidoreductase coenzymes (e.g. NAD, NADP, FAD, FMN, ubiquitin) no conclusive evidence of the formation of glucose-bound HHDP residues, i.e. true ellagitannins, was obtained (Hofmann and Gross, unpublished). It is evident that sophisticated techniques and unconventional new strategies will be required for the eventual clarification of this challenging question.

REGULATORY ASPECTS

The foregoing discussions were all based on the tacit assumption that these individual biogenetic steps were catalyzed by specific individual enzymes. This applies certainly to the catalytic activities effecting the mere chemical reactions but not necessarily to their physical state. Indications of some sort of strong cooperativity arose from the time-course experiments depicted in Figure 11. The β–glucogallin-dependent conversion of 1,6-di- to 1,2,6-trigalloyl-glucose, catalyzed by a partially purified enzyme from sumach,[74] was found to proceed subsequently to higher substituted derivatives, with an apparent trend for the interim accumulation of pentagalloylglucose before this ester was further metabolized to depsidic gallotannins after a pronounced lag-phase of 1-2 h (Denzel and Gross, unpublished). These data pointed to the possible existence of enzyme complexes, an interpretation that was corroborated by similar studies with an enzyme preparation from green acorns of $Q. robur$ (Denzel and Gross, unpublished). Again, an accumulation of pentagalloylglucose was observed upon prolonged incubation periods (Fig. 12). More important, however, was the finding that the primary reaction product, 1,2,6-trigalloylglucose, was formed at significant rates only after a lag-phase of about 1-2 h; after this time, previously accumulated 1,6-digalloylglucose (that obviously had been formed from the substrate β-glucogallin) began to disappear from the reaction mixture. One possible explanation for this unusual behavior is that this 'endogenously' synthesized digalloylglucose was the preferred acceptor substrate, instead of the 'exogenously' supplied ester, and this interpretation would lead again to the assumption of at least loose enzyme complexes that cooperated in the formation of hydrolyzable tannins and their precursors.

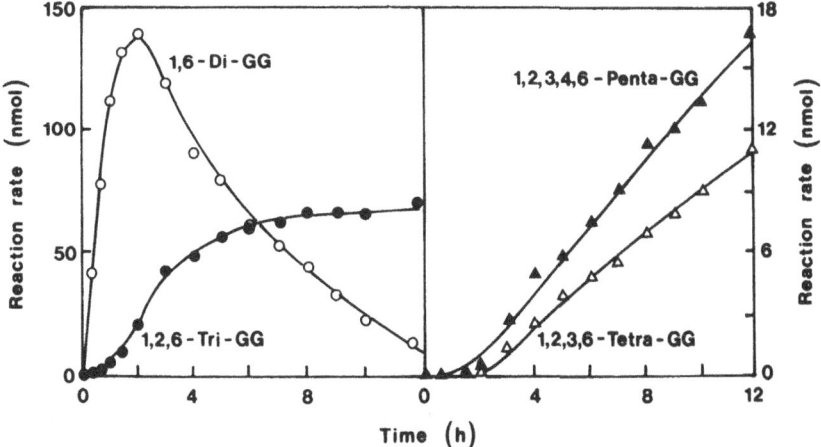

Fig. 12. Formation of 1,6-di-, 1,2,6-tri-, 1,2,3,6-tetra- and 1,2,3,4,6-pentagalloylglucose in the presence of the substrates β-glucogallin (2) and 1,6-digalloylglucose (14) by an enzyme preparation from green acorns of $Quercus robur$. GG, galloylglucose.

Conceivably, these ideas are highly speculative at present, but are notimprobable if one considers that many other pathways of plants secondary metabolism, including the important routes to phenolic constituents, have been recognized to operate as more or less organized multienzyme complexes.[88-90]

CONCLUSION

In this chapter, the current status of the investigation of the pathways to gallotannins and ellagitannins has been described. Considering the deleterious effects exerted by these substances on the reactivities of enzymes, it is surprising that considerable progress in the elucidation of the underlying mechanisms related to these sequences has been achieved just by the intense application of enzymatic studies. It is hoped that this article will reduce unfounded reservations against this technique, thus stimulating further and fruitful research in this scientific area with its many challenges that still require clarification.

ACKNOWLEDGMENTS

I am indebted to the colleagues and coworkers who contributed to the research reported from my laboratory, and to the Deutsche Forschungs-gemeinschaft and the Fonds der Chemischen Industrie for financial support.

REFERENCES

1.   HASLAM, E. 1982. The metabolism of gallic acid and hexahydroxydi-phenic acid in higher plants. Fortschr. Chem. Org. Naturst. 41:1-46.
2.   HASLAM, E. 1989. Gallic acid derivatives and hydrolyzable tannins. In: Natural Products of Woody Plants I. Chemicals extraneous to the lignocellulosic cell wall. (J.W. Rowe, ed.) Springer, Berlin, pp. 399-438.
3.   HASLAM, E. 1989. Plant polyphenols. Vegetable Tannins Revisited. Cambridge University Press, Cambridge, 230 pp..
4.   PORTER, L.J. 1989. Tannins. In: Methods in Plant Biochemistry. Vol. 1. Plant Phenolics. (J.B. Harborne, ed.) Academic Press, London, pp. 389-419.
5.   HASLAM, E. 1988. Twenty-second Procter Memorial lecture. Vegetable

Tannins—renaissance and reappraisal. J. Soc. Leather Technol. Chem. 72:45-64.

6. FREUDENBERG, K. 1920. Die Chemie der Natürlichen Gerbstoffe. Springer, Berlin, 161 pp.

7. NISHIZAWA, M., YAMAGISHI, T., NONAKA, G., NISHIOKA, I. 1982. Tannins and related compounds. 5. Isolation and characterization of polygalloylglucoses from chinese gallotannin. J. Chem. Soc. Perkin Trans. 1:2963-2968.

8. NISHIZAWA, M., YAMAGISHI, T., NONAKA, G., NISHIOKA, I. 1983. Tannins and related compounds. 9. Isolation and characterization of polygalloylglucoses from Turkish galls (*Quercus infectoria*). J. Chem. Soc. Perkin Trans. 1:961-965.

9. NISHIZAWA, M., YAMAGISHI, T., NONAKA, T., NISHIOKA, I. 1980. Structure of gallotannins in *Paeoniae* radix. Chem. Pharm. Bull. 28:2850-2852.

10. NISHIZAWA, M., YAMAGISHI, T., NONAKA, G., NISHIOKA, I., NAGASAWA, T., OURA, H. 1983. Tannins and related compounds. XII. Isolation and characterization of galloylglucoses from *Paeoniae* radix and their effect on urea-nitrogen concentration in rat serum. Chem. Pharm. Bull. 31:2593-2600.

11. GROSS, G.G. 1985. Biosynthesis and metabolism of phenolic acids and monolignols. In: Biosynthesis and Biodegradation of Wood Components. (T. Higuchi, ed.) Academic Press, Orlando, pp. 229-271.

12. ZENK, M.H. 1964. Zur Frage der Biosynthese der Gallussäure. Z. Naturforsch. 19b:83-84.

13. NEISH, A.C., TOWERS, G.H.N., CHEN, D., EL-BASYOUNI, S.Z., IBRAHIM, R.K. 1964. The biosynthesis of hydroxybenzoic acids in higher plants. Phytochemistry 3:485-492.

14. KATO, N. SHIROYA, M., YOSHIDA, S., HASEGAWA, M. 1968. Biosynthesis of gallic acid by a homogenate from the leaves of *Pelargonium inquinans*. Bot. Mag. Tokyo 81:506-507.

15. HASLAM, E., HAWORTH, R.D., KNOWLES, P.F. 1961. Gallotannins. IV. The biosynthesis of gallic acid. J. Chem. Soc. 1854-1859.

16. CONN, E.E., SWAIN, T. 1961. Biosynthesis of gallic acid in higher plants. Chem. Ind. 592-593.

17. CORNTHWAITE, D.C., HASLAM, E. 1965. Gallotannins. IX. The biosynthesis of gallic acid in *Rhus typhina*. J. Chem. Soc. 3008-3011.

18. TATEOKA, T.N. 1968. Formation of protocatechuic acid from 5-dehydroshikimic acid in the extract of mung bean seedling. Bot. M~g. Tokyo 81:103-104.

19. AMRHEIN, N., TOPP, H., JOOP, O. 1984. The pathway of gallic acid biosynthesis in higher plants. Plant Physiol. 75s:18.

20. LYDON, J., DUKE, S.O. 1988. Glyphosate induction of elevated levels in higher plants. J. Agric. Food Chem. 813-818.

21. BECERRIL, J.M., DUKE, S.O., LYDON, J. Glyphosate effects on shikimate pathway products in leaves and flowers of velvetleaf. Phytochemistry 28:695-699.

22. HEIDE, L., FLOSS, H.G., TABATA, M. 1989. Incorporation of shikimic acid into p-hydroxybenzoic acid in Lithospermum erythrorhizon cell cultures. Phytochemistry 28:2643-2645.

23. SAIJO, R. 1983. Pathway of gallic acid biosynthesis and its esterfication with catechins in young tea shoots. Agric. Biol. Chem. 47:455-460.

24. ISHIKURA, N. 1975. Incorporation rate of shikimic acid-^{14}C and phenylalanine-^{14}C into gallic acid in Rhus and Acer leaves. Experientia 31:1407-1408.

25. ISHIKURA, N., HAYASHIDA, S., TAZAKI, K. 1984. Biosynthesis of gallic and ellagic acids with ^{14}C-labeled compounds in Acer and Rhus leaves. Bot. Mag. Tokyo 97:355-367.

26. HADDOCK, E.A., GUPTA, R.K., AL-SHAFI, S.R.K., LAYDEN, K., HASLAM, E., MAGNOLATO, D. 1982. The metabolism of gallic acid and hexahydroxydiphenic acid in plants: biogenetic and molecular taxonomic considerations. Phytochemistry 21:1049-1062.

27. STÖCKIGT, J., ZENK, M.H. 1974. Enzymatic synthesis of chlorogenic acid from caffeoyl coenzyme A and quinic acid. FEBS Lett. 42:131-134.

28. ULBRICH, B., ZENK, M.H. 1979. Partial purification and properties of hydroxycinnamoyl-CoA: quinate hydroxycinnamoyl transferase from higher plants. Phytochemistry 18:929-933.

29. RHODES, M.J.C., WOOLTORTON, L.S.C., LOURENÇO, E.J. 1979. Purification and properties of hydroxycinnamoyl CoA quinate hydroxycinnamoyl transferase from potatoes. Phytochemistry 18:1125-1129.

30. ULBRICH, B., ZENK, M.H. 1980. Partial purification and properties of p-hydroxycinnamoyl-CoA: shikimate-p-hydroxycinnamoyl transferase from higher plants. Phytochemistry 19:1625-1629.

31. PETERSEN, M., ALFERMANN, A.W. 1988. Two new enzymes of

rosmarinic acid biosynthesis from cell cultures of *Coleus blumei*: hydroxyphenylpyruvate reductase and rosmarinic acid synthase. Z. Naturforsch. 43c:501-504.

32. STRACK, D., RUHOFF, R., GRWE, W. 1986. Hydroxycinnamoyl-coenzyme A: tartronate hydroxycinnamoyltransferase in protein preparations from mung bean. Phytochemistry 25:833-837.

33. STRACK, D., LEICHT, P., BOKERN, M, WRAY, V., GROTJAHN, L. 1987. Hydroxycinnamic acid esters of isocitric acid: accumulation and enzymatic synthesis in *Amaranthus cruentus*. Phytochemistry 26:2919-2922.

34. STRACK, D., KELLER, H., WEISSENBÖCK, G. 1987. Enzymatic synthesis of hydroxycinnamic acid esters of sugar acids and hydroaromatic acids by protein preparations from rye (*Secale cereale*) primary leaves. J. Plant Physiol. 131:61-73.

35. HEILEMANN, J., WRAY, V., STRACK, D. 1990. Synthesis of 2-*O*-(4-coumaroyl)-*myo*-inositol by a protein preparation from needles of *Taxus baccata*. Phytochemistry 29:3487-3489.

36. SAYLOR, M.H., MANSELL, R.L. 1977. Hydroxycinnamoyl-coenzyme A transferase involved in the biosynthesis of kaempferol 3-(*p*-coumaroyl triglucoside) in *Pisum sativum*. Z. Naturforsch. 32c:765-768.

37. TEUSCH, M., FORKKMANN, G., SEYFFERT, W. 1987. Genetic control of hydroxycinnamoyl-coenzyme A: anthocyanidin 3-glycoside hydroxycinnamoyltransferase from petals of *Matthiola incana*. Phytochemistry 26:991-994.

38. GROSS, G.G. 1982. Synthesis of galloyl-coenzyme A thioester. Z. Naturforsch. 37c:778-783.

39. GROSS, G.G. 1983. Synthesis of mono-, di- and trigalloyl-β-D-glucose by β-glucogallin-dependent galloyltransferase from oak leaves. Z. Naturforsch. 38c:519-523.

40. GROSS, G.G. 1982. Synthesis of β-glucogallin from UDP-glucose and gallic acid by an enzyme preparation from oak leaves. FEBS Lett. 148:67-70.

41. GROSS, G.G. 1983. Partial purification and properties of UDP-glucose:vanillate 1-*O*-glucosyl transferase from oak leaves. Phytochemistry 22:2179-2182.

42. WEISEMANN, S., DENZEL, K., SCHILLING, G., GROSS, G.G. 1988. Enzymatic synthesis of 1-*O*-phenylcarboxyl-β-D-glucose esters. Bioorg. Chem. 16:29-37.

43. CORNER, J.J., SWAIN, T. 1965. Enzymatic synthesis of the sugar esters of hydroxy-aromatic acids. Nature 207:634-635.

44. MACHEIX, J.J. 1977. Biosynthèse des esters glucosés des acides hydroxy-cinnamique à partir d'uridine diphospho-glucose et des acides libres chez la pomme. C. R. Acad. Sci. Ser. D. 284:33-36.

45. FLEURIET, A. MACHEIX, J.J., SUEN, R., IBRAHIM, R.K. 1980. Partial purification and some properties of a hydroxycinnamoyl glucosyltransferase from tomato fruits. Z. Naturforsch. 35c:967-972.

46. STRACK, D. 1980. Enzymatic synthesis of 1-sinapoylglucose from free sinapic acid and UDP-glucose by a cell-free system from *Raphanus sativus* seedling. Z. Naturforsch. 35c:204-208.

47. NURMANN, G., STRACK, D. 1981. Formation of 1-sinapoylglucose by UDP:glucose: sinapic acid glucosyltransferase from cotyledons of *Raphanus sativus*. Z. Pflanzenphysiol. 102:11-17.

48. NAGELS, L., MOLDEREZ, M. PARMENTIER, F. 1981. UDPG-*p*-coumarate glucosyltransferase activity in enzyme extracts from higher plant. Phytochemistry 20:965-967.

49. SHIMIZU, T., KOJIMA, M. 1984. Partial purification and characterization of UDPG: *t*-cinnamate glucosyltransferase in the root of sweet potato, *Ipomoea batatas* Lam. J. Biochem. 95:205-212.

50. MICHALZUK, L., BANDURSKI, R.S. 1980. UDP-glucose: indoleacetic acid glucosyl transferase and indoleacetyl-glucose: *myo*-inositol indoleacetyl transferase. Biochem. Biophys. Res. Commun. 93:588-592.

51. BÄUMKER, P.A., JÜTTE, M., WIERMANN, R. 1987. The separation of two different enzymes catalyzing the formation of hydroxycinnamic acid glucosides and esters. Z. Naturforsch. 42c:1223-1230.

52. GROSS, G.G. Enzymology of gallotannin biosynthesis. 1989. In: Plant cell wall polymers: biogenesis and biodegradation. (N.G. Lewis, M.G. Paice, eds.) ACS Symp. Ser. 399:108-121.

53. ATKINSON, M.L., MORTON, R.K. 1960. Free energy and the biosynthesis of phosphates. In: Comparative Biochemistry. Vol. II. Free energy and biological function. (M. Florkin, H.S. Mason, eds.) Academic Press, New York- London, pp. 1-95.

54. STRACK, D., KNOGGE, W., DAHLBENDER, B. 1983. Enzymatic synthesis of sinapine from 1-*O*-sinapoyl-β-D-glucose and choline by a cell-free system from developing seeds of red radish (*Raphanus sativus* L. var. *sativus*). Z. Naturforsch. 38c:21-27.

55. REGENBRECHT, J., STRACK, D. 1985. Distribution of 1-

sinapoylglucose: choline sinapoyltransferase activity in the *Brassicaceae*. Phytochemistry 24:407-410.

56.   TKOTZ, N., STRACK, D. 1980. Enzymatic synthesis of sinapoyl-*L*-malate from 1-sinapoylglucose and L-malate by a protein preparation from *Raphanus sativus* cotyledons. Z. Naturforsch. 35c:835-837.

57.   STRACK, D. 1982. Development of 1-*O*-sinapoyl-β-D-glucose L-malate sinapoyltransferase activity in cotyledons of red radish (*Raphanus sativus* L. var. sativus). Planta 155:31-36.

58.   STRACK, D., HEILERMANN, J., BOEHNERT, B., GROTJAHN, L., WRAY, V. 1987. Accumulation and enzymatic synthesis of 2-*O*-acetyl-3-*O*-(*p*-coumaroyl)-*meso*-tartaric acid in spinach cotyledons. Phytochemistry 26:107-111.

59.   STRACK, D., GROSS, W., HEILERMANN, J., KELLER, H., OHM, S. 1988. Enzymic synthesis of hydroxycinnamic acid esters of glucaric acid and hydroaromatic acids from the respective 1-*O*-hydroxycinnamoylglucoside and hydroxycinnamoyl-coenzyme A thioester as acyl donors with a protein preparation from *Cestrum elegans* leaves. Z. Naturforsch. 43c:32-36.

60.   BOKERN, M., STRACK, D. 1988. Synthesis of hydroxycinnamic acid esters of betacyanins via 1-*O*-acylglucosides of hydroxycinnamic acids by protein preparations from cell suspension cultures of *Chenopodium rubrum* and petals of *Lampranthus sociorum*. Planta 174:101-105.

61.   MICHALZUK, L., BANDURSKI, R.S. 1982. Enzymic synthesis of 1-*O*-indol-3-ylacetyl-β-D-glucose and indol-3-ylacetyl-*myo*-inositol. Biochem. J. 207:273-281.

62.   KESY, J.M., BANDURSKI, R.S. 1990. Partial purification and characterization of indol-3-ylacetylglucose:*myo*-inositol indol-3-ylacetyltransferase (indoleacetic acid-inositol synthase). Plant Physiol. 94:1598-1604.

63.   VILLEGAS, R.J.A., KOJIMA, M. 1986. Purification and characterization of hydroxycinnamoyl D-glucose quinate hydroxy-cinnamoyl transferase in the root of sweet potato, *Ipomoea batatas*. Lam. J. Biol. Chem. 261:8729-8733.

64.   STRACK, D., GROSS, W., WRAY, V., GROTJAHN, L. 1987. Enzymic synthesis of caffeoylglucaric acid from chlorogenic acid and glucaric acid by a protein preparation from tomato cotyledons. Plant Physiol. 83:475-478.

65.   STRACK, D., GROSS, W. 1990. Properties and activity changes of

chlorogenic acid: glucaric acid caffeoyltransferase from tomato (*Lycopersicon esculentum*). Plant Physiol. 92:41-47.

66.   GROSS, G.G., SCHMIDT, S.W., DENZEL, K. 1986. β–Glucogallin-dependent acyltransferase from oak leaves. I. Partial purification and characterization. J. Plant Physiol. 126:173-179.

67.   DENZEL, K., WEISEMANN, S., GROSS, G.G. 1988. β–Glucogallin-dependent acyltransferase from oak leaves. II. Application for the synthesis of 1-O-phenylcarboxyl-β-D-[^{14}C] glucose esters. J. Plant Physiol. 133:113-115.

68.   SCHMIDT, S.W., DENZEL, K., SCHILLING, G., GROSS, G.G. 1987. Enzymatic synthesis of 1,6-digalloylglucose from β-glucogallin by β–glucogallin 6-O-galloyltransferase from oak leaves. Z. Naturforsch. 42c:87-92.

69.   GROSS, G.G., DENZEL, K., SCHILLING, G.G. 1990. Enzymatic synthesis of di-O-phenylcarboxyl-β-D-glucose esters by an acyltransferase from oak leaves. Z. Naturforsch. 45c:37-41.

70.   DAHLBENDER, B., STRACK, D. 1984. Enzymatic synthesis of 1,2-disinapoylglucose from 1-sinapoylglucose by a protein preparation from cotyledons of *Raphanus sativus* grown in the dark. J. Plant Physiol. 116:375-379.

71.   DAHLBENDER, B., STRACK, D. 1986. Purification and properties of 1-(hydroxycinnamoyl)-glucose: hydroxy-cinnamoyltransferase from radish seedlings. Phytochemistry 25:1043-1046.

72.   KOJIMA, M., KONDO, T. 1985. An enzyme in sweet potato root which catalyzes the conversion of chlorogenic acid, 3-caffeoylquinic acid, to isochlorogenic acid, 3,5-di-caffeoylquinic acid. Agric. Biol. Chem. 49:2467-2469.

73.   VILLEGAS, R.J.A., SHIMOKAWA, T., OKUYAMA, H., KOJIMA, M. 1987. Purification and characterization of chlorogenic acid: chlorogenate caffeoyl transferase in sweet potato roots. Phytochemistry 26:1577-1581.

74.   DENZEL, K., SCHILLING, G., GROSS, G.G. 1988. Biosynthesis of gallotannins. Enzymatic conversion of 1,6-digalloylglucose to 1,2,6-trigalloylglucose. Planta 176:135-137.

75.   GROSS, G.G., DENZEL, K. 1991. Biosynthesis of gallotannins. β-Glucogallin-dependent galloylation of 1,6-digalloyl-glucose to 1,2,6-trigalloylglucose. Z. Naturforsch. 46c:389-394.

76.   CAMMANN, J., DENZEL, K., SCHILLING, G., GROSS, G.G. 1989. Biosynthesis of gallotannins: β-glucogallin-dependent formation of

1,2,3,4,6-pentagalloylglucose by enzymatic galloylation of 1,2,3,6-tetragalloylglucose. Arch. Biochem. Biophys. 273:58-63.

77. KRAJCI, I., GROSS, G.G. 1987. Formation of gallotannins in callus cultures from oak (*Quercus robur*). Phytochemistry 26:141-143.

78. DENZEL, K., GROSS, G.G. 1991. Biosynthesis of gallotannins. Enzymatic 'disproportionation' of 1,6-digalloylglucose to 1,2,6-trigalloylglucose and 6-galloylglucose by an acyltransferase from leaves of *Rhus typhina* L. Planta 184:285-289.

79. WILLIAMS, J.M., RICHARDSON, A.C. 1967. Selective acylation of pyranosides. I. Benzoylation of methyl β-D-glycopyranosides of mannose, glucose and galactose. Tetrahedron 23:1369-1378.

80. REINEFELD, E., AHRENS, D. 1971. Einfluss der Konfiguration auf die partielle Veresterung von D-Glucopyranosiden. Liebigs Ann. Chem. 747:39-44.

81. HOFMANN, A.S., GROSS, G.G. 1990. Biosynthesis of gallotannins: Formation of polygalloylglucoses by enzymatic acylation of 1,2,3,4,6-penta-O-galloylglucose. Arch. Biochem. Biophys. 283:530-532.

82. MAYER, S., HOFFMANN, E.H., LÖSCH, N., WOLF, H., WOLTER, B., SCHILLING, G. 1984. Dehydrierungsreaktionen mit Gallus-säureestern. Liebigs. Ann. Chem. 929-938.

83. HATHAWAY, D.E. 1957. The transformation of gallates into ellagate. Bichem. J. 67:445-450.

84. KAMEL, M.Y., SALEH, N.A., GHAZY, A.M. 1977. Gallic acid oxidation by turnip peroxidase. Phytochemistry 16: 521- 524.

85. POSPÍŠIL, F., CVIKROVÁ, M. HRUBCOVÁ, M. 1983. Oxidation of gallic acid by an enzyme preparation isolated from the culture medium of *Nicotiana tabacum* cell suspension. Biol. Plantarum 25:373-377.

86. FLAIG, W., HAIDER, K. 1961. Reaktionen mit oxydierenden Enzymen aus Mikroorganismen. Planta Med. 9:123-139.

87. HATANO, T., EDAMATSU, R., HIRAMATSU, M., MORI, A., FUJITA, Y., YASUHARA, T., YOSHIDA, T., OKUDA, T. 1989. Effects of the interaction of tannins with co-existing substances. VI. Effects of tannins and related polyphenols on superoxide anion radical, and on 1,1-diphenyl-2-picrylhydrazyl radical. Chem. Pharm. Bull. 37:2016-2021.

88. STAFFORD, H.A. 1981. Compartmentation in natural product biosynthesis by multienzyme complexes. In: The Biochemistry of

Plants. Vol. 7. Secondary plant products. (E.E. Conn, ed.) Academic Press, New York, pp. 117-137.

89.   HRAZDINA, G., WAGNER, G.J. 1985. Compartmentation of plant phenolic compounds; sites of synthesis and accumulation. In: The Biochemistry of Plant Phenolics. (C.F. van Sumere, P.J. Lea, eds.) Ann. Proc. Phytochem. Soc. Europe 25:119-133.

90.   BOUDET, A.M., GRAZIANA, A., RANJEVA; R. 1985. Recent advances in the regulation of prearomatic pathways. In: The Biochemistry of Plant Phenolics. (C.F. van Sumere, P.J. Lea, eds.) Ann. Proc. Phytochem. Soc. Europe 25:135-159.

Chapter Eleven

PHENYLPROPANOID METABOLISM: BIOSYNTHESIS OF MONOLIGNOLS, LIGNANS AND NEOLIGNANS, LIGNINS AND SUBERINS

Laurence B. Davin and Norman G. Lewis
Institute of Biological Chemistry

Washington State University
Pullman, WA 99164-6340

Phenolic Metabolism in Plants, Edited by H.A. Stafford
and R.K. Ibrahim, Plenum Press, New York, 1992

INTRODUCTION

During the growth and development of vascular plants, specific phenyl-propanoid metabolites differentially accumulate in particular tissues and/or cells of specialized function(s). Examples include the deposition of lignins in xylem tissue and vascular bundles and suberins in suberized cells, the accumulation of flavonoids in vacuolar and sometimes in wall compartments, and the ester attachment of hydroxycinnamic acids to arabinoxylans in the cell wall. Many of these metabolites confer unique properties to particular tissues or cells without which the competitive survival of vascular plants would be severely, if not fatally, compromised.

Such properties include enhancing the structural integrity of cell walls in order to withstand compressive forces and the formation of "water-impermeable" tissues suitable for translocation of nutrients and fluid retention. Phenylpropanoid metabolites also serve in many defense roles: for example, the *de novo* synthesis of bioresistant phenylpropanoid polymers (i.e., lignins, suberins) at or near the sites of infection and which act as a barrier to encroach-ment by pathogens, and the formation of miscellaneous compounds having phytotoxin, phytoalexin, and UV protectant properties.

Figure 1 shows the pathway (enzymatic conversions and intermediates) from phenylalanine (Phe) to the hydroxycinnamic acids (*E-p*-coumaric, ferulic and sinapic acids) and the monolignols (*E-p*-coumaryl, coniferyl and sinapyl alcohols).[1] Offshoots of the pathway give rise to the formation of lignins, suberins, lignans, neolignans, flavonoids, etc., in specific tissues, but how such processes are regulated or controlled is unknown. Consequently, this last decade has witnessed a growing application of "molecular" approaches in an attempt to address and resolve such questions. As evident later in the text, although progress has been made, it has been punctuated with a number of false starts.

This chapter, therefore, attempts to summarize and critically evaluate our current knowledge regarding the synthesis, regulation, structure and proposed functions of several important classes of phenylpropanoids, i.e., monolignols, lignans and neolignans, lignins and suberins. Flavonoid biosynthesis, requiring malonyl-CoA in addition to phenylpropanoids, is not addressed as various aspects of this process are described in Chapters 1, 4 and 5.

ENZYMOLOGY AND REGULATION OF HYDROXYCINNAMIC ACID AND MONOLIGNOL FORMATION

Phenylalanine Ammonia-Lyase

The first step involving deamination of phenylalanine (Phe) by phenylalanine ammonia-lyase, PAL (EC 4.3.1.5) (Fig. 1, 1) has been extensively studied, and it is often *speculated* to be a key regulatory step in phenylpropanoid metabolism. This is an intriguing contention, since it is yet to be demonstrated as to how this conversion controls subsequent metabolic flux into the different classes of phenylpropanoids.

PAL, discovered by Koukol and Conn in 1961,[2] acts by stereospecifically abstracting the pro-*S* proton from C_3 of L-Phe, with concomitant deamination to afford *E*-cinnamic acid.[3] The enzyme exists as a tetramer with a MW range of 240,000 to 330,000 depending upon the species, but can be dissociated into subunits of MW ~ 55,000 to 85,000.[4] PAL can be induced in response to infection,[5,6] wounding[5-8] and UV-illumination.[5,6,9] Interestingly, in dicots, PAL only uses Phe as a substrate,[2,7,10,11] whereas in grasses, both Phe and tyrosine are often employed.[5,12,13] The reasons for such substrate specificity differences have never been satisfactorily explained.

PAL has been purified from a number of plant sources, and is reported to exist as a single isoenzyme in some species and as multiple forms in others. It apparently exists as a single form (based on single isoelectric point(s) and linear Michaelis-Menten kinetics) in bamboo shoots (*Bambusa oldhami* Munro),[14] sunflower hypocotyls (*Helianthus annuus* L.),[15] strawberry fruit (*Fragaria ananassa* Duch.)[16] and loblolly pine developing xylem (*Pinus taeda*).[17] In the latter case, the protein was cloned and it was considered to be encoded by a single gene. Clearly, formation of the different classes of phenylpropanoid metabolites (flavonoids, lignins, etc.) cannot be differentially regulated simply on the basis of expression of a single gene encoding a single isoenzyme.

Alternatively, distinct isoenzymes of PAL have been observed, e.g., in cell suspension cultures of fungal-elicitor treated alfalfa (*Medicago sativa* L.)[18] and faba bean (*Vicia faba*) leaves.[19] Although each purified isoenzyme displayed linear Michaelis-Menten kinetics, the mixture of the three isoenzymes in each case did not and curvilinear kinetics were observed instead. Interestingly, no specific function for any of these isoenzymes has yet been identified and, thus, the reasons for their formation are unclear. PAL isoenzymes also apparently exist as products of a small (3-4) gene family in bean (*Phaseolus vulgaris* L.),[20,21] parsley (*Petroselinum crispum* L.),[22,23] rice (*Oryza sativa* L.)[24] and

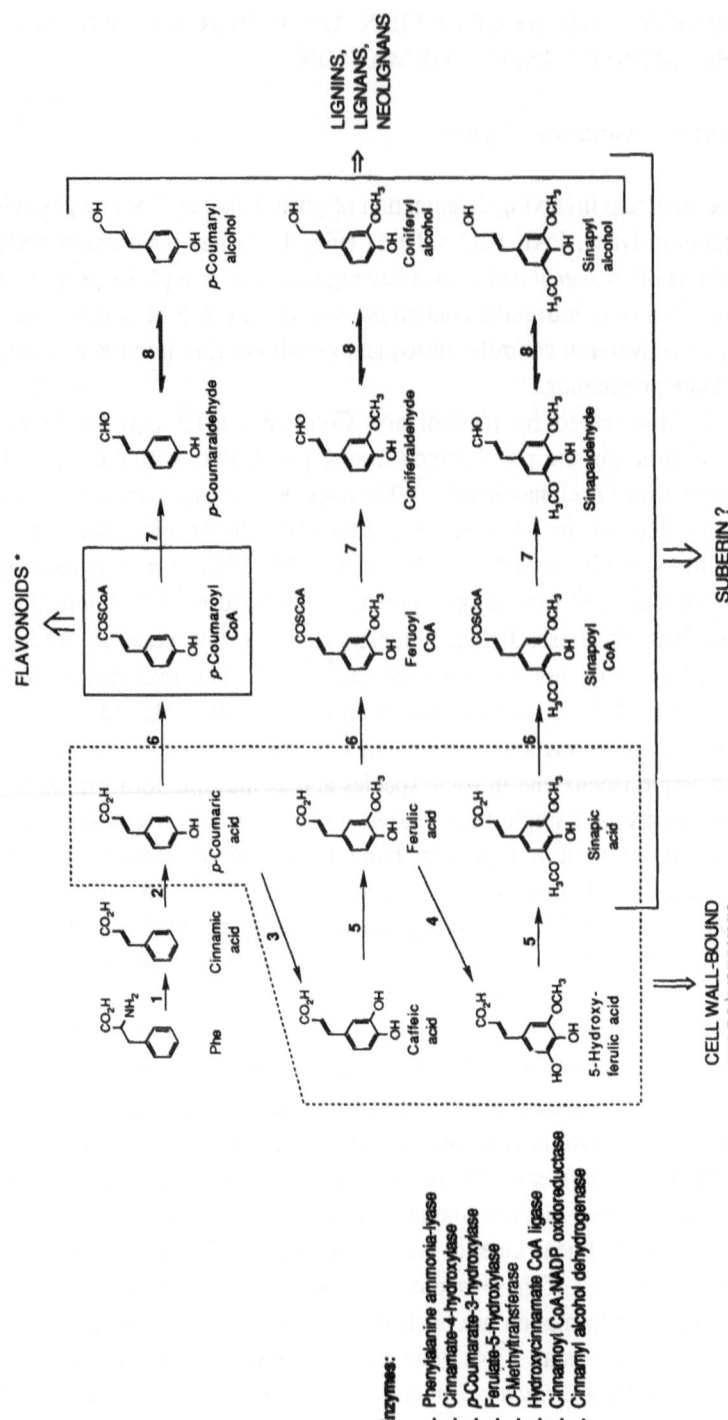

Fig. 1. Biosynthetic pathway to the main classes of phenylpropanoid metabolites. (*Note: plus acetate pathway for both flavonoids and suberin).

Enzymes:

1. Phenylalanine ammonia-lyase
2. Cinnamate-4-hydroxylase
3. p-Coumarate-3-hydroxylase
4. Ferulate-5-hydroxylase
5. O-Methyltransferase
6. Hydroxycinnamate CoA ligase
7. Cinnamoyl CoA:NADP oxidoreductase
8. Cinnamyl alcohol dehydrogenase

Arabidopsis thaliana.[25] These isoenzymes are synthesized *de novo* in response to UV irradiation, fungal challenges and wounding.[21,23] In bean, it has been reported that the transcript-elicitor of individual PAL genes apparently exhibit different patterns of accumulation in response to these stresses,[21,26] thus suggesting specific roles/functions for each.

However, attempts to precisely identify specific metabolic or developmental roles for individual isoenzymes have met with little success. For example, in the case of *Arabidopsis thaliana*, which contains several PAL isoenzymes, the gene encoding PAL-1 was cloned and its 1.8 kb promoter was introduced into a construct containing a promoterless β-glucuronidase (GUS) reporter gene.[25] In the corresponding transformed *Arabidopsis thaliana* plants, GUS tissue-specific staining was used to indicate likely sites where the PAL-1 promoter was being (specifically) induced, thereby implying, but not proving, PAL-1 expression at these sites.

Initial staining was observed following emergence from the seed coat, but rapidly dissipated. Staining was next noted at the base of the radicle, the base and tip of the cotyledons and in areas thought to be associated with vascular tissue; the authors rationalized that cotyledon staining was due to the induction of PAL by light, and this was presumed to result in anthocyanin formation. "GUS" staining was also observed some 4-5 days beyond germination but was now restricted to "vascular" tissue. Staining was not observed in the root tip and apical meristem, in contrast to tobacco where these tissues stained heavily with a bean PAL-2 GUS reporter gene.[27]

It should be evident from these results that induction of the presumed PAL-1 promotor seems to have no specific effect, i.e., no particular metabolic or developmental role was identified. Indeed, for such studies to be more meaningful, these GUS "staining" experiments need to be correlated with respect to (a) onset of actual PAL activity and location thereof, and (b) the precise identity of the metabolite(s) being formed.

In the case of parsley, it should be recognized that a cDNA clone, designated LF 53[28,29] and presumed PAL-specific, has no significant sequence similarity with other parsley PAL cDNAs more recently obtained.[23] Nor did it cross-hybridize with PAL cDNA from either bean or potato. This emphasizes the need for more rigorous biochemical controls to be carried out in order to prove that a putative cDNA clone is indeed encoding the particular enzyme of interest.

More recent studies of PAL have been directed towards determining its subcellular location in potato (*Solanum tuberosum*) tubers[30] and loblolly pine (*Pinus taeda*) (unpublished results, this laboratory) using immunocytochemical techniques. Such investigations are important since the subcellular organization and location of the phenylpropanoid enzymes is essentially unknown. In both cases, indirect immunogold labeling of PAL antibodies revealed that this enzyme was mainly in the cytoplasm, although it was also associated with various membranous organelles, including the plasma membrane in *Pinus taeda*.

Beyond cinnamic acid, the remaining steps to the monolignols (i.e., *p*-coumaryl, coniferyl and sinapyl alcohols) involve hydroxylation of the aromatic ring, methylation of diphenolics (e.g., caffeic acid), ligation of hydroxycinnamic acids to give the corresponding CoA esters and a two-step reduction to the monolignols via the corresponding aldehydes. Each of these steps and their substrate specificities are described below.

Hydroxylases

These catalyze three different types of aromatic ring hydroxylations resulting in the formation of various hydroxycinnamic acids: the first, cinnamate-4-hydroxylase (Fig. 1, 2) introduces a *p*-hydroxyl group at C_4 of the aromatic ring of cinnamic acid to generate *p*-coumaric acid, whereas the other two, *p*-coumarate-3-hydroxylase and ferulate-5-hydroxylase catalyze the formation of the substituted catechols, caffeic and 5-hydroxyferulic acids, respectively (Fig. 1, 3,4). The latter two enzymes function as branching points proper in monolignol/hydroxycinnamic acid synthesis.

All of the monooxygenases are found in microsomal fractions, and are cytochrome P-450 dependent with requirements for NAD(P)H and molecular oxygen.[31-34] None have been cloned. The first enzyme, cinnamate-4-hydroxylase, has been detected in many different plant species (see review by Gross, 1985)[34] and acts only on *E*-, rather than *Z*-cinnamic acid.

The next conversion, affording caffeic acid from *p*-coumaric acid, still awaits further clarification in terms of the precise enzyme(s) involved. Initially, it was demonstrated that phenolase ($\equiv$ polyphenol oxidase, tyrosinase and catechol oxidase), as well as mixed function oxidases from *Silene dioica*[35] and potato (*Solanum tuberosum*)[36] catalyzed this conversion. Since the phenolase-catalyzed transformations were of low substrate specificity, this implied that a highly specific *p*-coumarate-3-hydroxylase did not exist. However, when mung bean (*Vigna mungo*) seedlings were treated with tentoxin, a fungal toxin inhibiting phenolase, it was found that the accumulation of caffeic acid derived substances, such as rutin, was unaffected.[37]

More detailed investigations with mung bean seedlings have since revealed the presence of a specific *p*-coumarate-3-hydroxylase, which is not inhibited by tentoxin. In this case, only *p*-coumaric acid served as a substrate, and not other closely related compounds.[33] Caffeic acid formation was established by co-chromatography (five different solvent systems) and comparison of its UV spectrum to that of caffeic acid itself. It now needs to be established whether this apparently specific p-coumarate-3-hydroxylase is involved in general phenylpropanoid metabolism (i.e. to lignins, suberins, etc.) and is thus widespread throughout the plant kingdom, or whether it has only a limited biochemical role in a highly specific branch (e.g., to flavonoids, such as rutin). Thus, the involvement of polyphenol oxidase in catalyzing this conversion in most plant species cannot summarily be dismissed. In this context, a recent survey[38] revealed polyphenol oxidase activities in most angiosperms and ferns examined. Curiously, very few gymnosperms assayed showed polyphenol oxidase activity; it needs to be proven that this apparent lack of activity is due to the absence of the enzyme rather than its inactivation, e.g., by phenolics during the assay procedures utilized. Consequently, a resolution to the question of the enzymology involved in caffeic acid formation is required.

It would also be instructive to locate subcellularly both the phenolase(s) and p-coumarate-3-hydroxylase using immunocytochemical techniques. Preliminary work has shown that the p-coumarate-3-hydroxylase is present in a membrane fraction sedimenting in a sucrose gradient between the mitochondria and endoplasmic reticulum,[33] whereas phenolases are considered to be mainly associated with chloroplast or plastids based on both organelle fractionation[39] and cytochemical localization.[38]

As far as the last hydroxylation step is concerned, both ferulate-5-hydroxylase[31] (Fig.1, 4) and its product, 5-hydroxyferulic acid,[40] escaped detection until recently. The enzyme was discovered in poplar (*Populus X euramericana*), where it not only catalyzed the formation of 5-hydroxyferulic acid (apparent K_m=6.3 μM) but it could also convert cinnamic acid into p-coumaric acid (apparent K_m=1.9 μM).[31] Significantly, assays of both cinnamate-4-hydroxylase and ferulate-5-hydroxylase activities in xylem and sclerenchyma-enriched tissue from poplar stems revealed notable differences. The ratio of cinnamate 4-hydroxylase to ferulate-5-hydroxylase activities was 3.4:1 in xylem and 1.6:1 in sclerenchyma, respectively.[31] These differences in relative hydroxylase activities may *partially* account for the higher sinapyl alcohol content of the lignin in sclerenchyma cells, when compared to xylem tissue.

O-Methyltransferases

O-Methyltransferases (OMTs; E.C. 2.1.1.6), from 50 plant species in 43 different families, were examined in order to evaluate their relative abilities to methylate caffeic and 5-hydroxyferulic acids respectively (Fig. 1, **5**).[41] While substrate specificities varied widely between species, the following generalizations could be made: gymnosperms tended to be monofunctional, but were nevertheless capable of methylating 5-hydroxyferulic acid (albeit poorly) with ratios of sinapic (SA) to ferulic (FA) acid being formed ranging from 0.03 to 0.83. Such ratios may be quite misleading since, at least in the case of *Pinus thunbergii* (SA/FA ratio = 0.037), the presence of caffeic acid completely inhibits the methylation of 5-hydroxyferulic acid.[42] This fact, together with the *presumed* absence of ferulate-5-hydroxylase in most gymnosperms, provides a convincing explanation as to why sinapic acid and sinapyl alcohol are not normally formed. On the other hand, angiosperm OMTs tend to be bifunctional; monocot OMTs gave a range of SA/FA ratios of 0.9 to 2.2, whereas with dicot OMTs the ratios were larger (1.2 to 6.4).[41] In no case was it possible to separate OMT activities into caffeic or 5-hydroxyferulic acid-specific forms. Thus, in different species, OMTs seem to exist as single enzymes having widely varying substrate specificities.

Recent OMT studies have focused on immunochemistry and cloning: antibodies were raised against three distinct OMTs putatively required for lignification in tobacco.[43] Next, using parsley (*Petroselinum crispum* L.) cell cultures, caffeoyl-CoA 3-*O*- methyltransferase was purified, and subsequently cloned, thereby providing the first *bona fide* example of a cloned protein specific to phenylpropanoid metabolism.[44] In an analogous manner, and in an extension of the original work by Kuroda *et al.*,[45] a bifunctional aspen (*Populus tremuloides*) OMT was purified to apparent homogeneity; antibodies were raised and the enzyme was putatively cloned.[46] The aspen OMT antibodies were used in the isolation of a functionally active cDNA clone, encoding caffeic acid-3-*O*-methyltransferase in alfalfa (*Medicago sativa* L.).[47] The clone was expressed in *E. coli*, and gave a 2.1:1 substrate methylation ratio with 5-hydroxyferulic and caffeic acids, respectively.

Hydroxycinnamate:CoA Ligase

The activation of cinnamic acids as their CoA esters is catalyzed by hydroxycinnamate:CoA ligase, sometimes referred to as 4-coumarate:CoA ligase (E.C. 6.2.1.12) (Fig. 1, **6**). This transformation has received much attention, since it has long been speculated that specific isoenzymes catalyze entry into particular branches of the phenylpropanoid pathway, i.e., to lignins, flavonoids chlorogenic acid, etc. Consequently, several studies using soybean,[48] pea,[49] petunia,[50] poplar[51] and parsley[52] have been directed towards separating (presumed) hydroxycinnamate:CoA ligase isoenzymes, and attempting to ascribe particular functions to each on the basis of relative substrate specificities. This has met with limited success.

Thus, with soybean (*Glycine max*) cell suspension cultures, two isoenzymes were separated and designated I and II.[48] While both catalyzed ligation of *p*-coumaric, caffeic and ferulic acids, only isoenzyme I had the ability to engender sinapoyl-CoA formation. A similar situation was observed for pea (*Pisum sativum*) seedling isoenzymes.[49] This suggests that isoenzyme II in both species may have a preferred role in flavonoid/chlorogenic acid formation. Alternatively, both isoenzymes may be required for lignin formation, with isoenzymes I and II modulating sinapyl and coniferyl alcohol syntheses, respectively.

Three CoA ligase isoenzymes were found in petunia leaves[50] (*Petunia hybrida*) and poplar stems[51] (*Populus X euroamericana*), respectively. In petunia,[50] isoform Ia only ligated *p*-coumaric acid and caffeic acid and was consequently ascribed a role in quinic ester formation (e.g., chlorogenic acid). Form Ib, by contrast, catalyzed the formation of sinapoyl-CoA and *p*-coumaroyl-CoA and was speculated to be required for sinapyl alcohol synthesis. The last isoenzyme, so-called form II, engendered the synthesis of feruloyl-CoA and *p*-coumaroyl-CoA; it has been proposed roles in both flavonoid and coniferyl alcohol synthesis.

The three isoenzymes in poplar showed not only distinctive substrate specificities, but also different levels of activity in the various tissue types.[51] Isoenzyme I catalyzed the ligation of *p*-coumaric, ferulic, 5-hydroxyferulic, and sinapic acids. Isoenzyme II catalyzed the synthesis of feruloyl-CoA and *p*-coumaroyl-CoA, whereas isoenzyme III engendered the synthesis of *p*-coumaroyl-, caffeoyl-, and feruloyl-CoA. Neither isoenzyme II or III had any measurable effect on 5-hydroxyferulic or sinapic acids. In terms of tissue specificity, only isoenzyme I was detectable in the sclerenchyma cells. On the other hand, all three were detected in xylem and parenchyma cells, but isoenzyme I predominated in the xylem whereas isoenzyme III was the major form found in the parenchyma tissue. These results provide further explanation as to the presence of sinapyl alcohol-rich lignins in sclerenchyma cells.

Finally, two CoA ligase isoenzymes have been observed in parsley and cDNAs from each were obtained.[52,53] No specific role to either has yet been given, as they show similar catalytic activities.

Cinnamoyl-CoA:NADP Oxidoreductase

The conversion of hydroxycinnamate-CoA esters to the monolignols occurs via a two-step reduction process. The first step is catalyzed by cinnamoyl-CoA:NADP oxidoreductase (E.C. 1.2.1.44) which converts hydroxycinnamate CoA esters into the corresponding aldehydes (Fig. 1, 7). This oxidoreductase has been purified from three sources, namely soybean cell suspension cultures (*Glycine max*),[54] spruce cambial sap (*Picea abies* L.)[55] and poplar stems (*Populus X euramericana*);[56] the molecular weights of all preparations ranged from 36,000 to 38,000 with no subunits being detected. The oxidoreductases exclusively use NADPH as co-factor and is a type B reductase (see ref. 34).

Regarding substrate specificity, the spruce cinnamoyl-CoA:NADP oxidoreductase used feruloyl CoA as the preferred substrate (V/Km, 78 nkat 1 mg^{-1} μmol^{-1}); both sinapoyl- and *p*-coumaroyl-CoA (V/Km, 1.5 and 0.6 nkat 1 mg^{-1} μmol^{-1}, respectively) were less efficiently used.[55] By contrast, soybean cinnamoyl-CoA:NADP oxidoreductase catalyzed the reduction of both feruloyl- and sinapoyl-CoA (V/Km, 512 and 221 nkat 1 mg^{-1} μmol^{-1}), respectively; as before, *p*-coumaroyl-CoA was the least preferred (V/Km, 3.5 nkat 1 mg^{-1} μmol^{-1}).[55]

A similar trend was observed with poplar stem tissue. Interestingly, the specificity of the cinnamoyl-CoA:NADP oxidoreductase remained unchanged, irrespective of whether it was obtained from sclerenchyma or xylem tissues;[56] this strongly suggests that this enzyme does not play any crucial role in defining monolignol composition of the lignin in these tissues.

Cinnamyl Alcohol Dehydrogenase

The oxidoreductase, cinnamyl alcohol dehydrogenase (CAD; E.C. 1.1.1.195), catalyzes the final reductive step of monolignol synthesis, and has been detected in various gymnosperm and angiosperm species (Fig. 1, 8). It was discovered by Mansell *et al*[57] in 1974 in extracts of the angiosperm *Forsythia suspensa*, and it was concluded to be a type A reductase of MW ~80,000 (using NADPH as cofactor and coniferyl aldehyde as a substrate). By taking advantage of the reversible nature of this reaction, it was found capable of oxidizing a number of cinnamyl alcohol substrates, including *p*-coumaryl, coniferyl and sinapyl alcohols.

Then followed isolation of two CAD isoenzymes from cell suspension cultures of soybean (*Glycine max* L.) of MW ~43,000 (isoenzyme 1) and MW

~69,000 (isoenzyme 2), respectively.[58] Isoenzyme 1 only oxidized coniferyl alcohol, whereas isoenzyme 2 oxidized all three monolignols; with isoenzyme 1, the K_m for coniferyl alcohol oxidation was 75 times higher than for isoenzyme 2. No explanation for these differences has been given.

CAD was also purified to apparent homogeneity from *Populus X euramericana*[56] and reported to be of MW 40,000; V/K_m ratios for coniferyldehyde, *p*-coumaraldehyde and sinapaldehyde were 67.8, 13.8 and 4.1 nkat 1 mg^{-1} μmol^{-1}, respectively. As for cinnamoyl CoA:NADP oxidoreductase, no differences in specificity were observed between the CADs extracted from sclerenchyma or xylem tissue. This suggests that lignin heterogeneity observed in these tissues is not influenced by differences in substrate-specific CAD isoenzymes. Although the angiosperm species were capable of oxidizing all three monolignols, the CAD isolated from two gymnosperms, spruce (*Picea abies*, MW ~72,000)[55] and Japanese black pine (*Pinus thunbergii*,[59] MW ~67,000) were not; i.e., only coniferaldehyde and *p*-coumaraldehyde were readily reduced. Sinapaldehyde did not serve as a substrate.

Next followed a period of great confusion regarding cloning of a putative bean CAD, which again highlights the dangers involved in cloning without sound biochemical controls. Antibodies[60] were raised against the poplar CAD (MW ~40,000) previously reported to be pure on the basis of Coomassie blue staining.[56] These antibodies were then used to immunoprecipitate a putative CAD (MW ~ 65,000) from bean (*Phaseolus vulgaris* L.) cell suspension culture induced by a fungal elicitor from *Colletotrichum lindemuthianum*.[60] Ultimately, a cDNA clone supposedly encoding CAD was obtained and findings were described regarding its mRNA induction by fungal elicitor.[61] However, no proof that the clone had any cinnamyl alcohol dehydrogenase activity was provided; more detailed studies revealed the clone to be a malic enzyme[62,63] and not CAD, as claimed!

Since then, cinnamyl alcohol dehydrogenase has been purified to apparent homogeneity from *Pinus taeda*, where only one isoenzyme was observed.[64] Monoclonal antibodies were raised against rigorously purified protein and used to immunochemically determine the subcellular location of CAD in ultrathin sections of *Pinus taeda* cell suspension cultures and young seedlings (unpublished results, this laboratory). In the case of the suspension cultures, CAD was mainly present in the plastid, whereas in the seedlings it was found in the chloroplast. Immunoresponses, albeit small, were also observed for the endoplasmic reticulum and mitochondria/cytoplasm; the latter presumably reflect sites of protein synthesis/transport. This finding, together with that obtained for polyphenolase activity,[38,39] suggest that the biochemical machinery to afford the monolignols may reside within the chloroplast and/or plastids.

The next sections address progress regarding the fate of monolignol metabolites: biochemical mechanisms involved in Z-monolignol formation, proposed functions and biosynthetic pathways in lignan and neolignan biosynthesis, and processes of lignification and suberization.

Z-MONOLIGNOL FORMATION

In most plant species, monolignols are present only in minute amounts. An apparent exception to this trend is found in the monolignol-rich bark tissue of the *Fagaceae*,[65-67] where the expected *E*-isomers are not detected but only the corresponding Z-analogues are found, i.e., Z-coniferyl[66] and Z-sinapyl[67] alcohols in *Fagus grandifolia* Ehrh (Fig. 2). This finding raises questions about the mechanism of Z-monolignol formation, and the long-assumed exclusivity of *E*-monolignols in lignification (see ref. 1).

As far as the first question is concerned, two isomerization mechanisms accounting for Z-monolignol formation can be envisaged, i.e., by either photochemically or enzymatically mediated isomerizations of hydroxycinnamic acid, aldehyde or monolignol intermediates. However, based on experimental evidence gathered to date, we have tentatively concluded that *E*-monolignols are first formed, which then undergo an enzyme-mediated (rather than photochemical) isomerization to yield the Z-isomers.[68-70] This conclusion was based on the following lines of evidence: (1) when either $[2-^{14}C]$-*E*-coniferyl alcohol or *E*-ferulic acid were incubated with *F. grandifolia* bark, both were converted into Z-coniferyl alcohol.[68] These transformations occurred under conditions where photochemical isomerism was not observed in vitro; (2) CAD from *F. grandifolia* bark tissue displayed a marked substrate specificity for *E*-coniferyl alcohol over its corresponding Z-analogue (ratio of V_{max} for *E*:*Z* isomers = 15:1),[68] strongly implying that *E*-monolignol formation first occurs in *F. grandifolia* ; and (3) photochemical reactions of both *E*- and Z-monolignols (and their glucosides) in vitro afforded mixtures of *E*- and Z-isomers,[69] whereas in *F. grandifolia* bark tissue only the Z-monolignols are found.[66,67] Taken together, these results provide strong although indirect evidence for the involvement of a novel $E \rightarrow Z$ monolignol isomerase.[70] Clearly, isolation and characterization of the putative isomerase is now required.

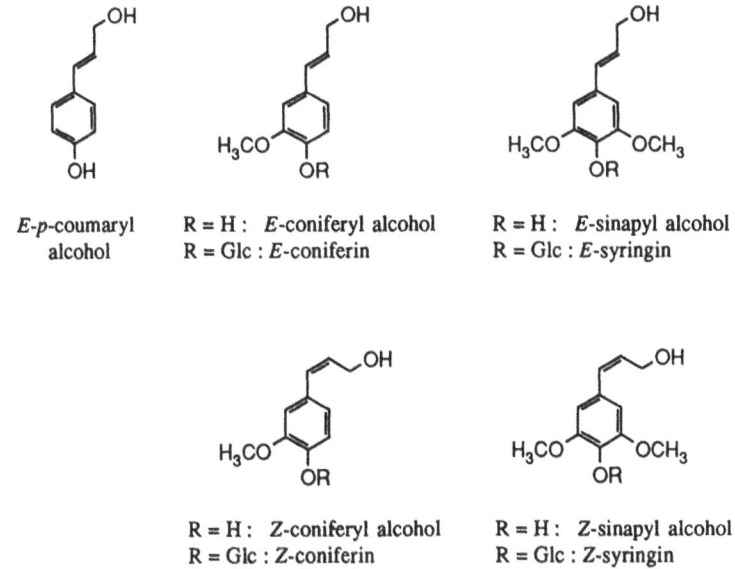

Fig. 2. *E*- and Z-Monolignols and their corresponding glucosides.

LIGNANS AND NEOLIGNANS

Occurrence and Proposed Functions

Lignans and neolignans represent the next level of structural complexity beyond the monolignols. They are a diverse array of compounds widely distributed throughout the plant kingdom (ranging from woody plants to constituents of vegetable fiber and grain).

Typically, they are found in "dimeric" form, but higher oligomers continue to be discovered.[71] Lignans are often linked through 8,8' bonds, and normally occur as substituted dibenzylbutanes, dibenzylbutyrolactones, furans, furanofurans, aryltetrahydronaphthalenes and, less often, as O,O bridged biphenyls.[71] Alternatively, neolignans are found linked via 3,3', 8,3', and 8-O-4' bonds, although other structural types (e.g., benzofurans and dihydrobenzofurans) have also been reported. Some representative examples are shown in Figure 3.

Over the last decade, there has been a growing realization of the importance of lignans and neolignans in plant defense functions. For example,

Lignans :

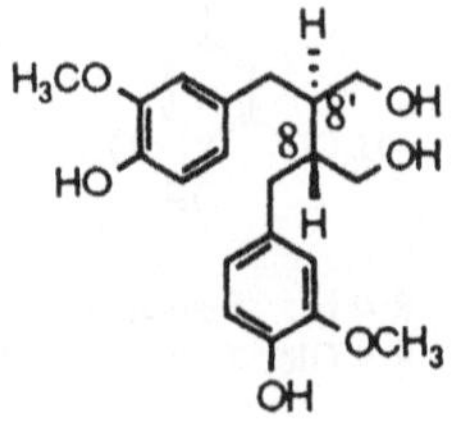

| Secoisolariciresinol | Matairesinol |
| (dibenzylbutane) | (dibenzylbutyrolactone) |

| Pinoresinol | Podophyllotoxin |
| (furanofuran) | (aryltetrahydronaphthalene) |

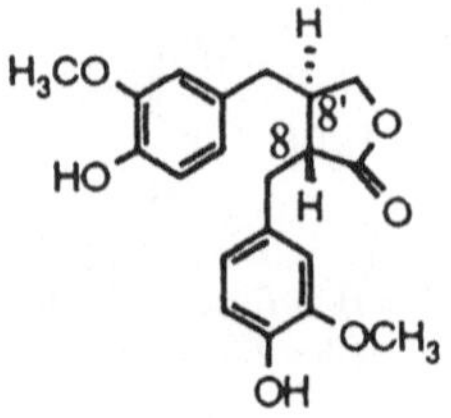

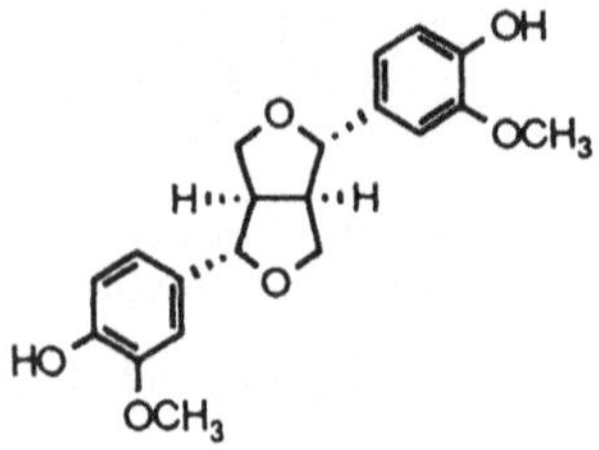

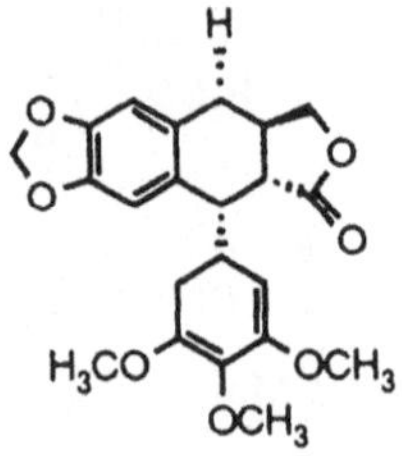

Neolignans :

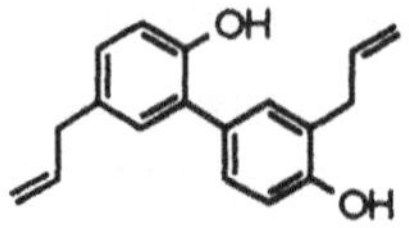

| Honokiol | Carinatol |
| (3-3') | (8-3') |

| Virolin | Dehydrodiconiferyl |
| (8-O-4') | glucoside |

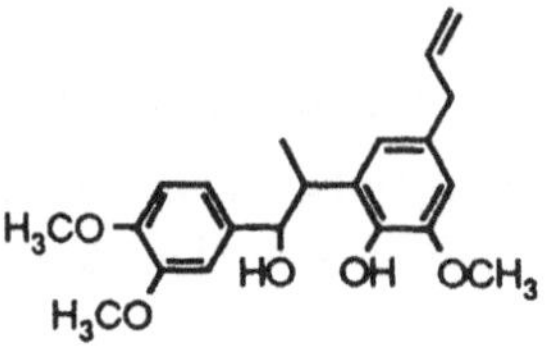

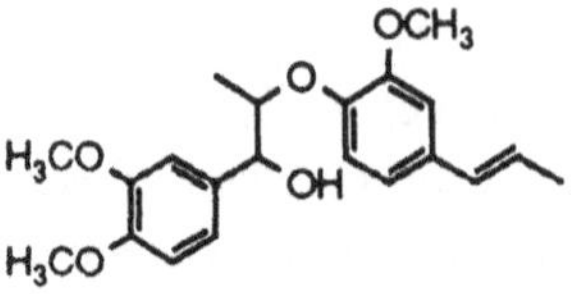

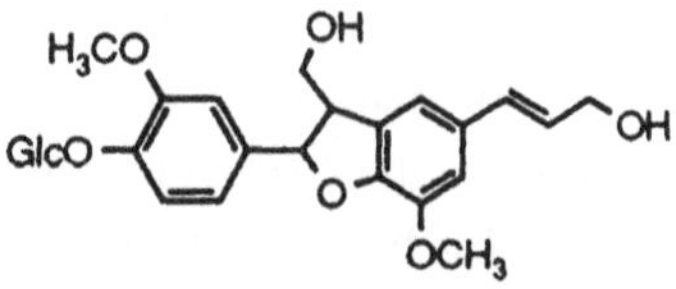

Fig. 3. Representative examples of lignans and neolignans.

antimicrobial, antifungal, and antifeedant properties have been documented, and lignans/neolignans are of undisputed significance in containing bacterial infection, fungal attack, and predation (see refs. 71-74). Examples will suffice to illustrate their importance (see Fig. 4): magnolol and isomagnolol inhibit the growth of the bacteria *Staphylococcus aureus*, *Mycobacterium smegmatis*, *Saccharomyces cerevisiae* and *Trichophyton mentagrophytes*,[75] and *meso*-dihydroguaiaretic acid from Papuan mace (*Myristica argentea* Warb.) is active against *Streptococcus mutans*.[76] A number of benzofurans and dihydrobenzo-furans from *Ratanhiae radix* Ph. Eur. (*Krameria triandria* Ruiz et Pavon) also show excellent antibacterial activities.[73,77] Phytoalexin effects have been reported as well, e.g., with the neolignan, magnolol, present in the twig cortical tissue of *Cercidiphyllum japonicum*.[78]

As regards fungistatic properties, a number of 8-*O*-4' linked neolignans show activity against the fungus, *Schistosoma mansoni*.[79]

Antifeedant properties are well-documented; futoquinol, a neolignan from *Piper futokadzura* Sieb. et Zucc.[80] and yatein from *Libocedrus yateensis*,[81] serve as insect antifeedants (Fig. 4), and others show synergistic properties with insecticides.[72] The piscicidal constituents of *Stellera chamaejasme* L.[82] and *Justicia hayatai*[83] have also been identified, these being the lignans, lirioresinol-B (syringaresinol), pinoresinol (from *S. chamaejasme*) and aryltetrahydronaphthalene lignans, justicidin A and B (from *J. hayatai*). The piscicidal activity of the justicidins is comparable in effect to that of the potent piscicide, rotenone.[83]

Cytokinin-like properties have been proposed[84] for the isomers of the neolignan, dehydrodiconiferyl glucoside (Fig. 3) in *Vinca rosea* crown gall tumors, but this needs to be rigorously established. Further, they were reported as being released from cell wall components, but this suggestion is undoubtedly incorrect.

Furthermore, biological activities of various lignans and neolignans in human (and mammalian) medicine have been extensively documented, and exciting findings continue to be reported.[71-74,85-99] These include potent antitumor, antimitotic, antiviral, antihepatotoxic, antihormonal and antistress activities, as well as enzyme inhibitory (against cAMP phosphodiesterase and monooxygenase, cathartic, and cardiovascular effects. A particularly striking finding is the observed high levels of "mammalian" lignans (i.e., enterodiol, enterolactone; Fig. 4) found in human urine, following dietary intake of vegetable fiber and grain.[100-103] There is an apparent correlation with their presence and reductions in breast and prostate cancer incidences. Anti-HIV properties have also been observed for (-)-arctigenin and (-)-trachelogenin,[104] suggesting that these (or related) lignans may find important roles there also.

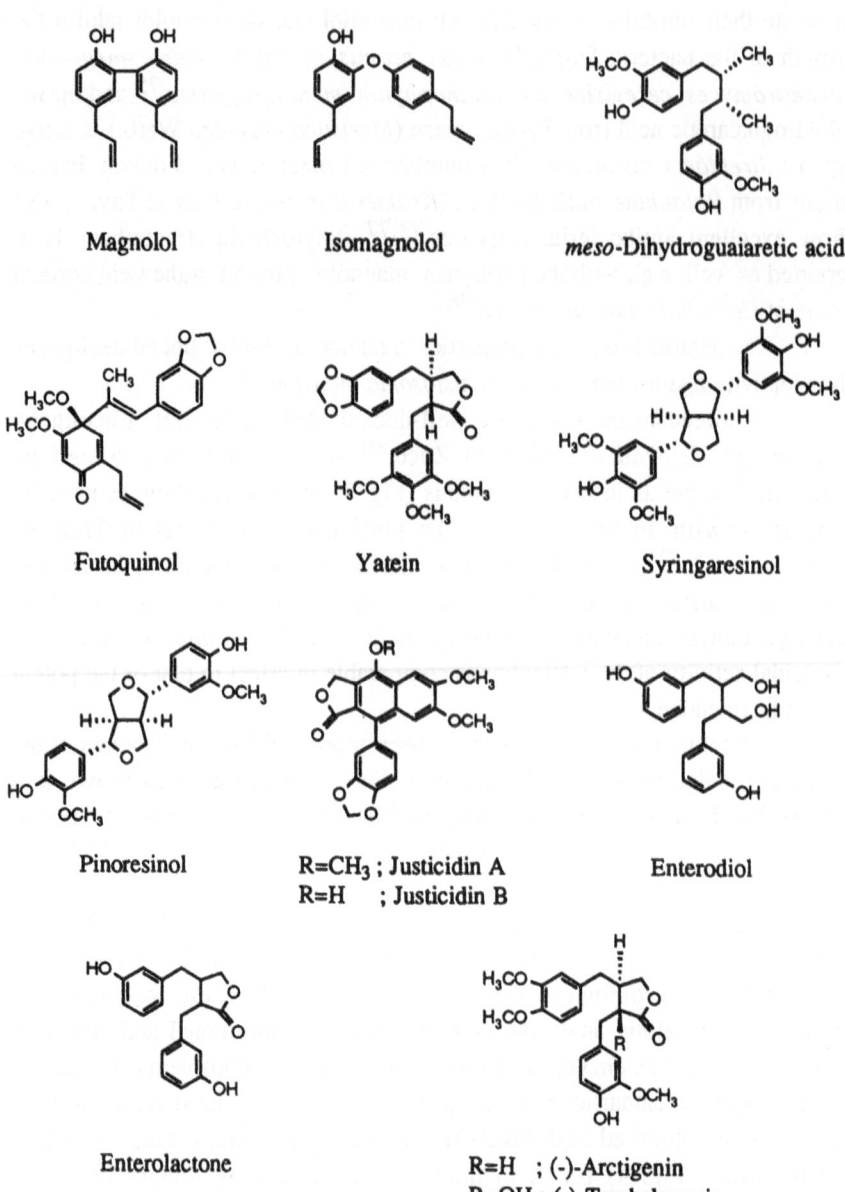

Fig. 4. Lignans with various plant defense and mammalian pharmaceutical properties.

Biosynthesis

Although lignans and neolignans are an abundant class of phytochemicals, little is known about their biochemical pathways (i.e., enzymes or intermediates), or their sites of synthesis and accumulation. Some researchers have proposed that lignans and neolignans serve as precursors of lignins, with the rationale being that lignin substructures embody certain lignan and neolignan structures (see later);[105,106] however, as yet there is no definitive evidence in support of this notion. It is just as likely that they constitute a distinct metabolic class of compounds, sharing with lignin, the monolignols as precursors. However, this question awaits clarification.

Lignans and neolignans differ from their polymeric lignin counterparts in terms of measurable optical activity,[107] i.e., lignans and neolignans are normally found in optically active forms, whereas lignins are viewed to be devoid of any measurable optical rotation. For example, (+)-pinoresinol is found in *Forsythia suspensa*,[108,109] and (-)-secoisolariciresinol and (-)-matairesinol occur in *F. intermedia* (Fig. 5).[110] These lignans were established to be optically pure, using chiral column chromatography coupled with UV detection.[109,110] The corresponding antipodes were not detected. Clearly, the presence of optically pure lignans suggests that the coupling process is stereoselective. That is, it cannot be a typical H_2O_2 requiring peroxidase-catalyzed reaction which would only afford racemic products, e.g., as demonstrated in the coupling of coniferyl alcohol which affords (±)-pinoresinols.[109,111]

Conversely, some lignans e.g., syringaresinol (Fig. 5), are not always found optically pure, and have been isolated from different plant species with widely varying $[\alpha]_D$ values.[112] Table 1 shows the range of $[\alpha]_D$ values reported for syringaresinol; subsequent analyses by chiral column chromatography revealed that these samples differed only in relative amounts of (+)- and (-)-antipodes.[113]

Two possible explanations can be forwarded to account for the presence of more than one enantiomeric form: (a) first, that two stereoselective coupling enzymes are present in certain species. Consequently, the ratios of lignans, such as (+)- and (-)-syringaresinol, are dependent upon the level of the activity of each specific coupling enzyme, or (b) that the phenylpropanoid coupling reaction is not stereoselective, and both antipodes are formed in equal amounts. Then, depending upon the plant species in question, a particular enantiomer may be totally or partially metabolized into some other product, thus leaving the other antipode optically pure or in enantiomeric excess.

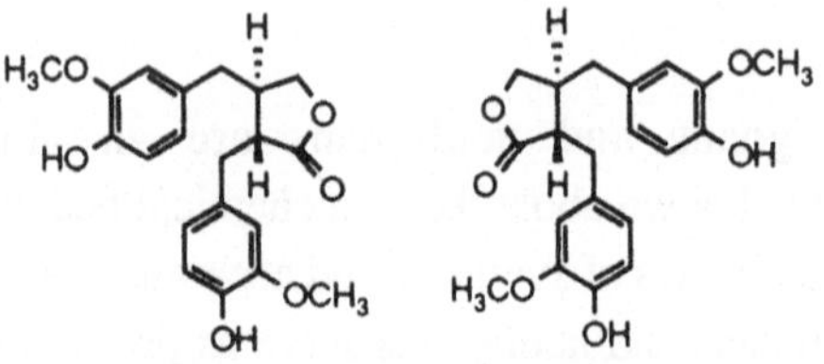

(-)   (+)

Matairesinol

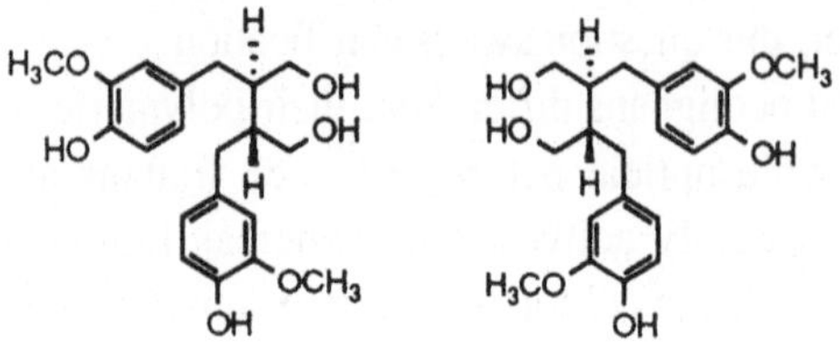

(-)   (+)

Secoisolariciresinol

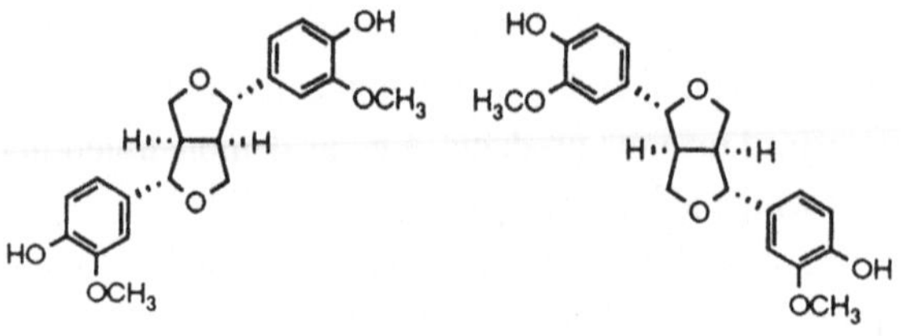

(+)   (-)

Pinoresinol

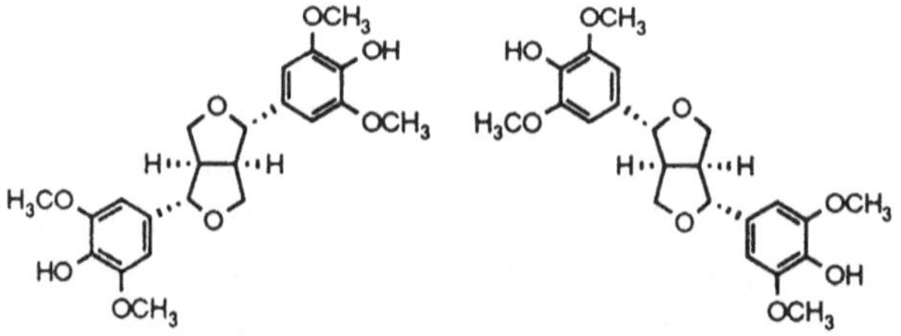

(+)   (-)

Syringaresinol

Fig. 5. Enantiomeric forms of selected lignans.

Table 1. Summary of melting points and $[\alpha]_D$ values for syringaresinol isolated from different species.[112]

Plant Species	$[\alpha]_D$	Melting Point (°C)
Liriodendron tulipifera	+48.9	185-186
Eucommia ulmoides	+44.0	183.5
Cistanche salsa	+21.7	182-183
Passerina vulgaris	+13.1	nd
Sinomenium acutum	+1.93	175
Xanthoxylum inerme	0	179-185.5
Daphne tangutica	-2.1	174-176
Xanthoxylum ailanthoides	-9.6	175-180
Acanthopanax sessiliflorum	-21.5	170-172
Aspidosperma marcgravianum	-34.8	177-183

n.d. = not determined

In order to clarify the precise mechanism of lignan formation in plants, we chose to first examine the biogenetic pathways leading to the optically pure lignans in the *Forsythia* species. Hence, [U-^{14}C]Phe was administered to intact *F. suspensa* stems and allowed to metabolize over a 3 h period.[109] Isolation of the radiolabeled lignans revealed that only (+)-pinoresinol was formed. The corresponding antipode was not detected, even when unlabeled (±)-pinoresinols were added as radiochemical carrier. In an analogous manner, uptake and metabolism of [8-^{14}C]coniferyl alcohol in *F. intermedia* stems ultimately afforded (-)-secoisolariciresinol and (-)-matairesinol, and not the corresponding (+)-forms.[110] On the other hand, when [8-^{14}C]coniferyl alcohol was administered to *F. suspensa*, both (+)- and (-)-pinoresinols were obtained, with the former (slightly) predominating.[109] This result is in stark contrast to that described using [U-^{14}C]Phe as precursor, where only (+)-pinoresinol was formed *de novo*.

Subsequent experiments[109,110,114] with cell-free extracts from both *F. intermedia* and *F. suspensa* confirmed and extended these results. Individual incubations of [8-^{14}C]- and [9,9 [2]H$_2$, OC[2]H$_3$]-coniferyl alcohols with a cell-free preparation from *F. intermedia*, in the presence of H_2O_2 and NAD(P)H, gave the corresponding radio- and deuterolabeled (-)-secoisolariciresinols,

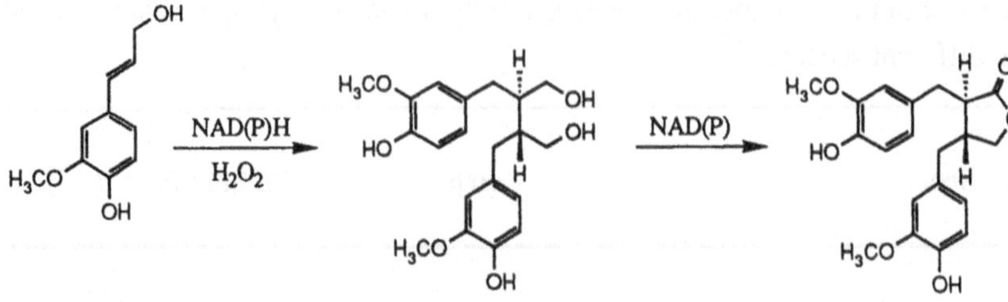

Fig. 6. Biosynthetic pathway to the lignans, (-)-secoisolariciresinol and (-)-matairesinol in *F. intermedia* stems (and cell-free extracts thereof).

respectively; this conversion was stereoselective since the corresponding (+)-antipode was not formed in either case. This cell-free extract also catalyzed the conversion of [Ar-^{3}H]- and [Ar-^{2}H]-secoisolariciresinols into tritiated and deuterated (-)-matairesinols when NADP was added as a co-factor. This reaction was again stereoselective, since (+)-secoisolariciresinol did not serve as a substrate for the formation of either (+)- or (-)-matairesinols. Thus, the biochemical pathway to (-)-secoisolariciresinol and (-)-matairesinol is shown in Figure 6. Lastly, when *F. suspensa* was incubated with [8-^{14}C]coniferyl alcohol, both (+)- and (-)-pinoresinols were again obtained, with the (+)-form slightly predominating (~2%).

These findings can be tentatively explained as follows: (1) following uptake and appropriate subcellular compartmentalization of [U-^{14}C]Phe in *F. suspensa* stems, the [U-^{14}C]coniferyl alcohol formed *de novo* undergoes stereoselective coupling to give only the putative (-)-(8R,8R')-quinone methide, QM (Fig. 7); and not the corresponding (+)(8S,8S')-QM antipode. Subsequent intramolecular cyclization of the (-)-(8R,8R')-QM affords (+)-pinoresinol; (2) by contrast, the [8-^{14}C] coniferyl alcohol, when taken up by *F. suspensa* stems, is not properly compartmentalized in the plant and is in contact with nonspecific peroxidases as well as the stereoselective coupling enzyme. Consequently, both (±)-pinoresinols are formed with the (+)-form predominating. This interference by nonspecific peroxidases also occurs in the cell-free extracts of *F. suspensa* and both (+)- and (-)-forms of pinoresinol are synthesized, with the (+)-form slightly predominating; (3) when *F. intermedia* cell-free extracts are incubated with

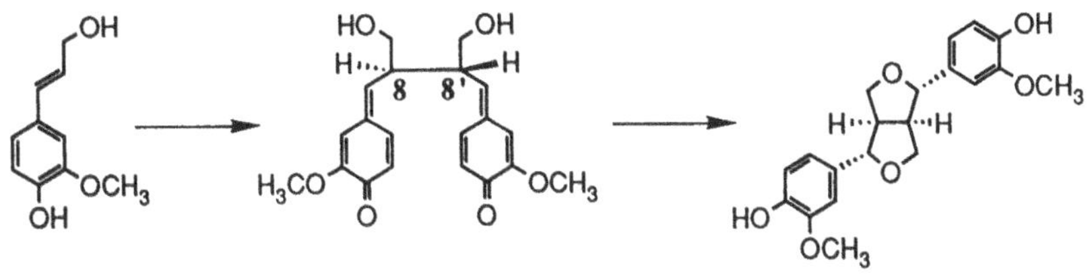

Coniferyl   (-)(8R,8R')-Quinone methide   (+)-Pinoresinol
alcohol

Fig. 7. Proposed biosynthetic pathway to (+)-pinoresinol in *Forsythia suspensa* intact plants.

[8-^{14}C]- and [9,9 ^{2}H$_2$, OC^{2}H$_3$]-coniferyl alcohols, both (+)(8S,8S')- and (-)(8R,8R')-quinone methides are formed when H$_2$O$_2$ is present as a co-factor. The (8R,8R')-antipode is then rapidly reduced by a stereoselective, NAD(P)H-requiring reductase to give (-)-secoisolariciresinol; the (+)(8S,8S')-form is not reduced; (4) lastly, in a stereoselective enzymatic NAD(P)-requiring dehydrogenation, only (-)-secoisolariciresinol, and not its (+)-antipode, is converted to (-)-matairesinol.

A summary of the putative enzymatic transformations affording *Forsythia* lignans in cell-free extracts is shown in Scheme 1. These represent the first experimentally verified enzymatic transformations in lignan biosynthesis. These enzymes now await purification and characterization in order to determine their modes of action and, ultimately, their subcellular location and tissue specificity.

LIGNIFICATION

Lignins are complex phenylpropanoid polymers[1] located mainly in the secondary walls and (compound) middle lamellae of cell walls of tracheids, vessels, fibers, etc., in vascular plants. They are formed by polymerization of monolignols in a reaction catalyzed by peroxidase(s) in the presence of H$_2$O$_2$. Lignification is a highly orchestrated component part of cell wall synthesis and accompanies polysaccharide (cellulose, hemicellulose) deposition.

Since lignification has been comprehensively reviewed recently,[1] only three areas of current interest are addressed: monolignol transport and storage, cell wall β-glucosidase(s) and peroxidase(s) and lignin heterogeneity/lignification in cell suspension culture.

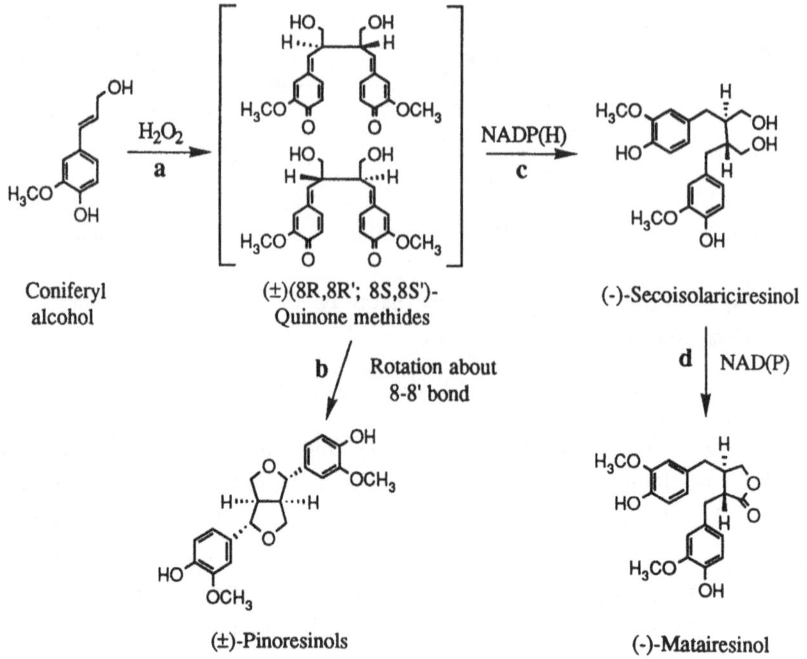

Coniferyl alcohol

(±)(8R,8R'; 8S,8S')-Quinone methides

(-)-Secoisolariciresinol

(±)-Pinoresinols

(-)-Matairesinol

Scheme 1. Proposed biochemical pathway to lignans in *Forsythia* cell-free extracts. Step: (a) H_2O_2/peroxidase coupling to give (±)(8R,8R'; 8S, 8S')-quinone methides due to interference by nonspecific peroxidases, (b) intramolecular cyclization of racemic quinone methides to give (±)-pinoresinols; (c) stereoselective reduction to give (-)-secoisolariciresinol, and (d) stereoselective dehydrogenation to give (-)-matairesinol.

Monolignol Transport and Storage

Almost nothing is known about the mechanism of monolignol transport through the plasma membrane and into the cell wall where lignification occurs.[113,115] Interestingly, monolignols accumulate infrequently in plant tissues, whereas the corresponding glucosides (e.g., coniferin, syringin) are often found in small amounts.

It has been proposed that monolignol glucosides, rather than monolignols themselves, serve as the chemical entities being transported into the cell wall (see ref. 115). There is some literature support to this notion, since β-glucosidases are found in the cell walls and could, therefore, regenerate the monolignols from the corresponding glucosides.[116,117] However, no explanation has been forwarded to account for the fate of the equimolar amounts of glucose liberated during this envisaged process.

Alternatively, monolignol glucosides may simply function as storage products, and are conscripted into the cell wall only as needed. In this case, the monolignols themselves would serve as the major molecular species undergoing transport into the cell wall. Whatever situation exists, a role for monolignol glucosides in lignification can be contemplated.

As one approach to clarify the uncertainties surrounding monolignol glucoside formation and function, we have focused on two glucosyltransferases, one from beech (*Fagus grandifolia*) bark tissue and the other from loblolly pine (*Pinus taeda*) cambial sap, with respect to determining their substrate specificities and subcellular location(s). Thus, in the case of *Fagus grandifolia* bark tissue, which accumulates the Z-monolignol glucosides, Z-coniferin and Z-syringin (Fig. 2),[67] it was established that the preferred substrate for the glucosylation reaction was Z-coniferyl alcohol rather than its E-analogue (5.74 vs. 0.24% conversion).[70] Conversely, in the case of *P. taeda*, E-coniferyl alcohol was only slightly preferred over its Z-counterpart (57.52 vs. 27.54% conversion) (unpublished results, this laboratory).

These interesting substrate specificity differences suggest yet another control mechanism operative in lignification, and its significance needs to be determined. Moreover, these results raise significant concern with respect to the exclusivity of E-monolignols in lignification, and a modified hypothesis utilizing both E- and Z-monomers is proposed (Fig. 8).

The UDPG:coniferyl alcohol glucosyltransferases from both sources are now being purified to homogeneity in order that the subcellular location of each can be determined; this information, together with the localization of the monolignol glucosides themselves, should shed important light on whether they are mainly storage products or are actively transported into the cell wall.

Cell Wall β-Glucosidase(s) and Peroxidase(s)

In support of the hypothesis for a role for monolignol glucosides in lignification,[118] coniferin and syringin specific β-glucosidases associated with tracheids and endo-, epi- and exodermal tissues were found in spruce (*Picea abies*)[116,119,120] and chick pea (*Cicer arietinum* L.) seedlings.[117,121] Interesting substrate specificity differences were also noted: in chick pea, E-coniferin

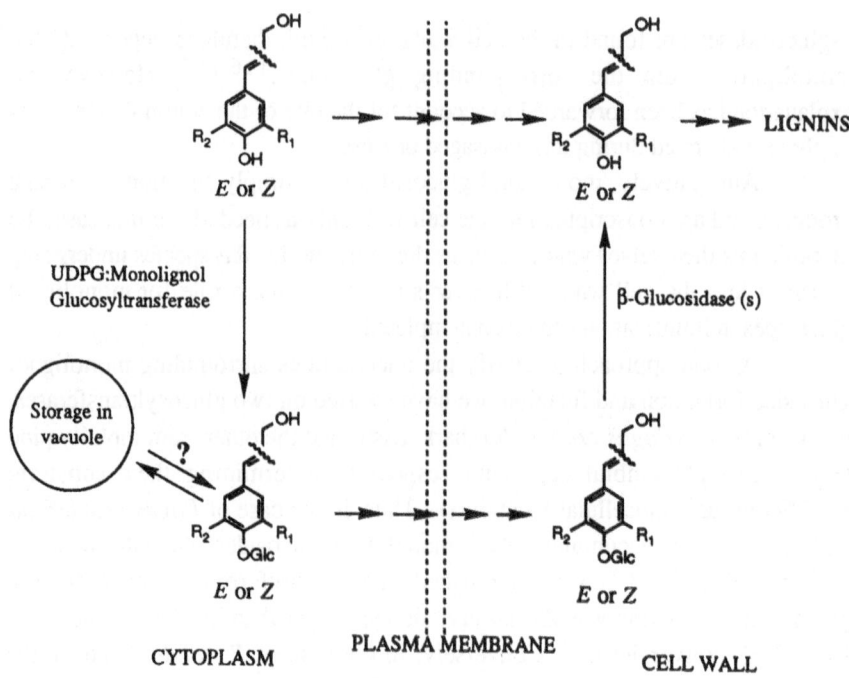

Fig. 8. Hypothetical involvement of both *E*- and *Z*-monolignol (glucosides) in lignification. R_1, R_2 = H or OCH$_3$

was favored as a substrate over *E*-syringin (V/K$_m$ 125 and 6 mkat l kg^{-1} mmol^{-1}, respectively),[120,121] whereas in spruce hypocotyls both coniferin and syringin were efficiently utilized (V/K$_m$ 111 and 67 mkat l kg^{-1} mmol^{-1}, respectively).[119,120] It now remains to be established whether β-glucosidases from different plant sources also display different specificities toward *E*- and *Z*-monolignol glucosides (for example, as previously observed in the glucosylation reactions).

There are other reports, however, suggesting that monolignol glucosides play only a secondary role in monolignol transport: with spruce hypocotyls,[122] there was no direct correlation with coniferin turnover and lignification, suggesting that, in this species, monolignol glucosides were not the preferred chemical moieties for monomer transport. Moreover, attempts to correlate β-glucosidase activity with presumed lignification in a variety of cell cultures were inconclusive.[123,124] Taken together, these results seem to suggest that both monolignols and their glucosides are used in monomer transport into the cell wall, but with the former predominating.

The final step in lignification involving the peroxidase catalyzed, H_2O_2-dependent polymerization of monolignols was comprehensively reviewed recently.[1] Briefly, numerous investigations have suggested that specific "acidic" peroxidases are involved in lignification but, remarkably, such studies invariably did not (a) use monolignols as substrates, (b) determine the subcellular location of the putative isoenzyme(s), (c) determine the primary structure of the peroxidase or (d) correlate enzyme activity/gene expression with lignification. Some of these shortcomings have been addressed in more recent investigations: Lagrimini et al. cloned two strongly anionic (i.e., acidic) peroxidase isoenzymes (pI 3.5 and 3.75, respectively) from Nicotiana tabacum,[125] considered to be identical to cell wall acidic peroxidases previously reported.[126] These had MW 36,000 and 37,000, and were 52% and 46% homologous with peroxidase(s) from horseradish and turnip; both isoenzymes were considered to be a result of translational modification of a single gene product.[125] In apparent agreement with their assumed localization in the cell walls, it was found that the mRNA for this protein was prevalent in lignified stem tissue.[127] Surprisingly, neither wounding nor tobacco mosaic virus infection of N. tabacum induced activity of either strongly anionic isoenzyme (as evidenced by isoelectrofocusing), whereas other isoenzymes (moderately anionic or cationic) were activated, depending upon the type of challenge experienced.[127]

Since then, transgenic N. tabacum plants overexpressing anionic (pI 3.5 and 3.75, respectively) peroxidases have been obtained,[128] and had increased peroxidase levels of 2-10 times the controls. However, the transgenic plants differed from their controls in the sense that they were stunted and underwent extensive wilting at the onset of flowering. On the other hand, partial "silencing" of anionic peroxidase(s) gene expression was carried out using an antisense mRNA approach, which gave transgenic plants with ~20% of their original peroxidase activity.[129] Although it has since been claimed that overexpression results in increased lignification, and underexpression decreased lignin deposition, this conclusion is based only on phloroglucinol staining[130] and, hence, is open to subjective interpretation. Indeed, there is an urgent need to conduct more rigorous and informative analyses to establish if lignification has been affected; this requires the quantitative determination of lignin contents and monomeric composition. Until such information is obtained, the reported modulation of peroxidase activities cannot satisfactorily be correlated with lignin deposition processes. It is highly likely that other metabolic processes are being altered, rather than just lignification, as indicated by lack of growth and wilting.

Lignin Heterogeneity

The subject of lignin heterogeneity was thoroughly reviewed recently,[1] and only the pertinent features need to be discussed. Monomer compositions (i.e., p-coumaryl, coniferyl, and sinapyl alcohol contents) of lignin polymers are highly variable, not only between and within species but also in specific tissues and individual cell wall layers. As noted earlier in the section on monolignol formation, these differences have a biochemical rationale in terms of (a) substrate specificities of certain enzymes, and (b) tissue- or plant type-specific responses. Typically, gymnosperm lignins are derived mainly from coniferyl alcohol, together with relatively small amounts of p-coumaryl alcohol. There are, however, exceptions such as *Tetraclinis articulata*, which also synthesizes sinapyl alcohol. In the case of angiosperms, all three monolignols are generally used. As a further complication, grasses and cereals contain hydroxycinnamic acids that are linked to the lignin macromolecule via ester[131] and ether[132] linkages. We have postulated that cell wall-bound hydroxycinnamic acids might serve as a locus or initiation point for lignification.[115]

From the standpoint of cell wall formation, there appears to be little or no lignin deposited at the primary wall stage. Lignin deposition occurs at the onset of cell wall secondary thickening, this being initiated at the cell corners and middle lamella, and then extends into the secondary wall.[1] Remarkably, the monolignol composition of lignins appears to be both spatially and temporally predetermined. In a general sense, it appears that p-coumaryl alcohol is predominantly deposited during the early stages of lignin formation; this is then followed by coniferyl alcohol and, lastly (with angiosperms), sinapyl alcohol deposition occurs.[133]. Consequently, the cell corners and middle lamella of gymnosperms contain a higher p-coumaryl alcohol content than in secondary wall lignin, which is mainly composed of coniferyl alcohol. In the case of a typical angiosperm, sinapyl alcohol deposition occurs during the latter stages of lignification.[133]

Although these variations in monolignol composition, which ultimately result in lignin heterogeneity, are well established, there is currently no explanation as to how temporal and spatial deposition of monolignols is regulated or controlled.

Lignification in Cell Suspension Culture

The major difficulty in studying lignification in intact plants can readily be exemplified by simple examination of a woody plant cross-section. From the cambial zone to the pith, the different cells extending inward are at different stages of cell wall maturation and, hence, degrees of lignification. This means that to investigate this process in actively lignifying intact plants, only an average effect of these different maturation stages (i.e., lignification) can be studied. In order to avoid such limitations, we considered it instructive to develop a relatively homogenous suspension culture system which underwent lignification; it would be useful in studying initial stages of lignification, lignin structure *in situ*, monolignol transport, etc., under carefully defined conditions.

This goal was achieved with *Pinus taeda* cell suspension cultures (unpublished results, this laboratory) as follows: two different cell lines were maintained on Brown and Lawrence medium, and differed only in choice of the growth regulator (2,4-dichlorophenoxyacetic acid, 2,4-D or α-naphthaleneacetic acid, NAA). In the 2,4-D line, the cells had a very low lignin content (~1.6%), as evidenced by the acetyl bromide test,[134] and both PAL and CAD activities were low (<0.033 and 288 pkat mg^{-1} protein, respectively). This low lignin content was confirmed by solid-state ^{13}C nuclear magnetic resonance (NMR) spectroscopic analysis of the *P. taeda* cell walls, which mainly contained cellulose and noncellulosic polysaccharides (data not shown). Further, [U-^{14}C]phenylalanine was not efficiently incorporated into the walls of the 2,4-D-grown *P. taeda* cell line, thus suggesting that lignification had not occurred to any extent. Next, when [1-^{13}C]phenylalanine was administered to this cell line over 14 days, the resulting cell wall tissue showed only a C-13 enhanced resonance at 172.4 ppm, consistent with either unmetabolized Phe or its incorporation into proteinaceous material. No resonances corresponding to lignin were detected. Electron microscopy of ultrathin sections revealed that only primary walls were formed under these cell culture conditions. By contrast, the NAA line was lignified (~9.9%), and both PAL and CAD activities were readily detected (229 and 2800 pkat/mg protein, respectively). [U-^{14}C] Phenylalanine was now efficiently incorporated (~4.5%) into the cell walls of *P. taeda* cultures grown in the presence of NAA. Next, [1-^{13}C], [2-^{13}C], and [3-^{13}C]phenylalanines were individually administered to the NAA-grown *P. taeda* cell lines; subsequent analysis of each cell wall tissue by solid-state ^{13}C NMR spectroscopy revealed signals consistent with that expected from lignified tissue.[1,135-138] The spectra are shown in Figure 9a-c, and assignments correspond to the substructures shown in Scheme 2.

Thus, following administration of [1-^{13}C]Phe, the cell wall tissue had a large C-13-enriched resonance at 63.4 ppm, corresponding to 2-*O*-aryl (B), 2-C'1 (E), phenolic linked coniferyl alcohol (A), and phenylcoumaran (D) substructures; the shoulder at 72.7 ppm was attributed to a pinoresinol-like

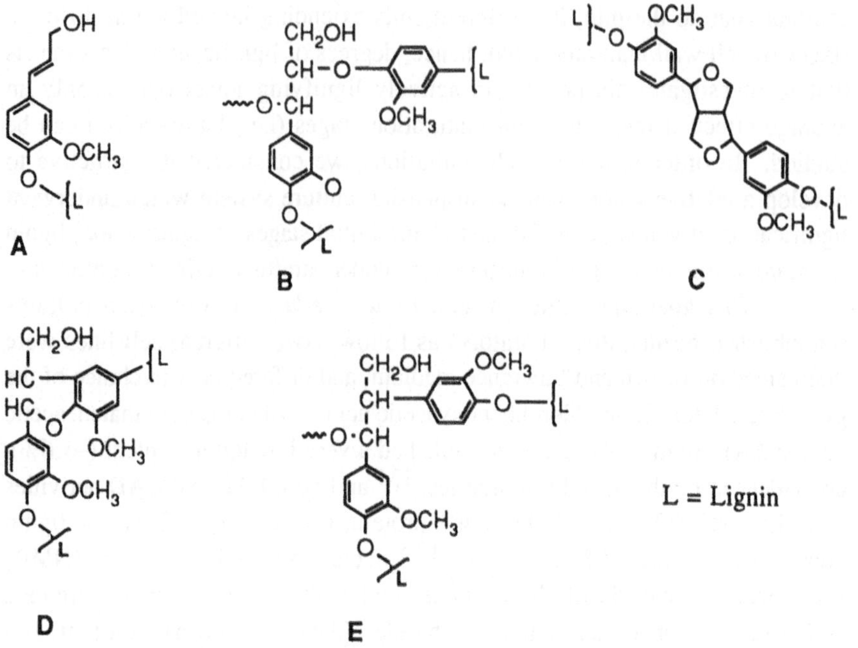

A

B

C

D

E

L = Lignin

Scheme 2. Proposed dominant lignin substructures.

moiety (C) in the lignin polymer (Scheme 2). [The small signals at 173.2 ppm corresponded to incorporation of $[1-^{13}C]$Phe into proteinaceous material.] In the case of $[2-^{13}C]$Phe metabolism, three sets of signals were observed at 55.4, 85.8, and 127.9 ppm; the signal at 55.4 ppm corresponded to pinoresinol (C), phenylcoumaran (D), and 2-C'1 (E) substructures, while that at 85.8 and 127.9 ppm were due to 2-O-aryl (B) and phenolic-linked coniferyl alcohol (A) substructures, respectively. Finally, metabolism of $[3-^{13}C]$Phe afforded resonances at 74.4, 86.4, and 130.7 ppm. These correspond to 2-O-aryl (B) and 2-C' (E) substructures (74.4 ppm), with pinoresinol (C) and phenylcoumaran (D) substructures at 86.4 ppm; the resonance at 130.7 ppm was coincident with that for phenol-linked coniferyl alcohol moieties (A). Electron microscopy revealed the NAA-grown cell walls were ~0.2 μm, consistent with the thickness of a cell wall at the S_1 stage of maturation.

In summary, we have developed an excellent system to initiate the early stages of lignification in cell suspension cultures, and which permits the elucidation of specific bonding environments of lignin in (a) a gymnosperm, and (b) cell culture. This model system should be an excellent means to probe lignification at the "molecular" level.

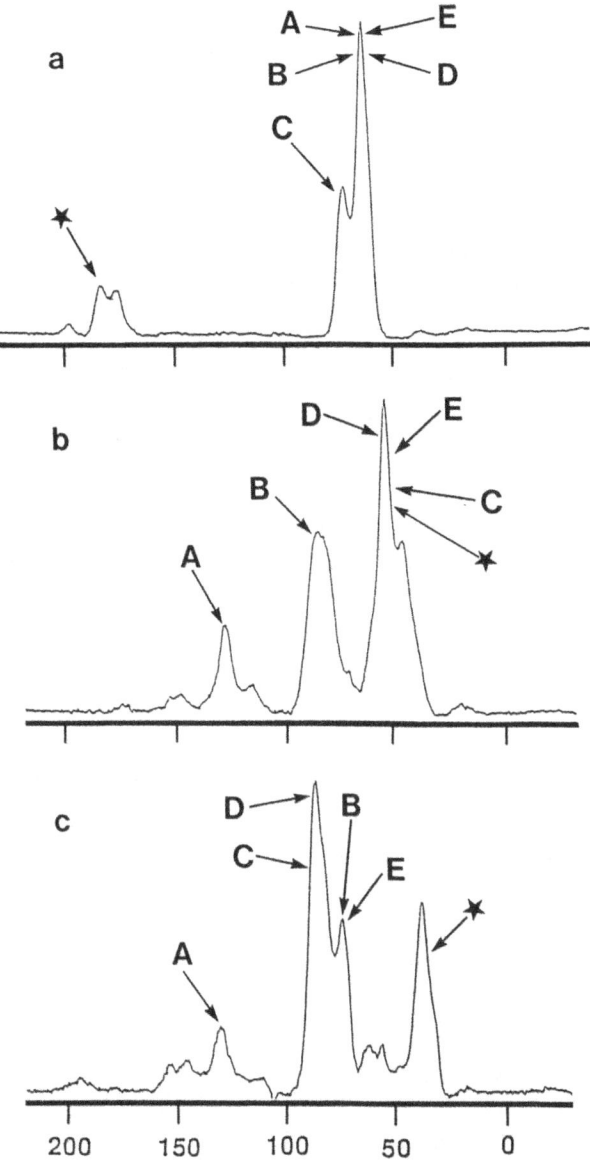

Fig. 9. Solid state ^{13}C-NMR difference spectra of *P.taeda* cell suspension culture cell walls following administration and metabolism of : (a), [1-^{13}C]Phe; (b), [2-^{13}C]Phe and (c), [3-^{13}C]Phe. [In each case, natural abundance resonances were subtracted.] * = tentatively assigned to proteinaceous materiel.

SUBERIZATION

Suberization is a poorly understood process, particularly as far as its chemical identity is concerned. This section addresses our current knowledge of suberization.

Suberin is deposited in a number of tissues as layers alternating with waxy components[139] (Fig. 10). It is found principally in periderm tissues of both nonwoody[140] (e.g., beet, sweet potato etc.) and woody (e.g., oak,[141] Douglas fir[142]) plants, where it can reach concentrations as high as 40-50% of the suberized tissue. Suberized lamellae[139] have also been observed in the epidermis and hypodermis of roots, e.g., corn, the endodermis (Casparian band), bundle sheaths of grasses, idioblast sheaths between secretory organs (e.g., of glands, trichomes, etc.), pigment strands of grains and connections between vascular tissue and seed coats. Suberin is normally visualized by histochemical staining with Sudan Black B.[143-145]

Remarkably, there is no method to isolate suberin in either a pure form or in its native state. Current preparations can best be described as suberin-enriched. An example will suffice to illustrate this point: "suberin" from wound-healing potato (*Solanum tuberosum* L.) periderm contains about 50% cell wall carbohydrates as impurities.[146,147]

In terms of structure, most of our limited knowledge about suberin comes from chemical degradation, isolation and analysis of components released from suberin-enriched tissue. Initially, long-chain di- or multifunctional fatty acids were identified (Fig. 11).[141,146,148,149] Typically, these acids range in length from C_{16} to > C_{28}, with C_{16}, C_{18}, C_{22} and C_{24} being the most prevalent. Functionalities are normally introduced at either end of the molecule (e.g., ω-hydroxy acids or dicarboxylic acids) or within the chain itself (e.g., C=C, CHOH, epoxides, etc.). These findings initially led to the conclusion that suberin was an ester-linked lipid-like material (i.e., a polyester).[150]

Next came the slow recognition that suberized tissue was also rich in phenolic material, with the isolation of sinapic acid from saponified suberin samples, which was presumed linked to lipid material.[151] The subsequent isolation of phellochryseine as its methyl ester (Fig. 12) from cork confirmed that covalent linkages existed between phenolic acids (e.g., ferulic acid) and hydroxyfatty acids (e.g., ω-hydroxydocosanoic acid).[152] Since then, several examples of feruloyl fatty acid esters from Douglas fir (*Pseudotsuga menziesii*) bark,[153-155] potato[154] and peat soil[156] have been reported. Indirect evidence for covalent attachment of hydroxycinnamic acids to suberin came from a study by Riley and Kolattukudy in 1975,[157] where they "depolymerized" potato and

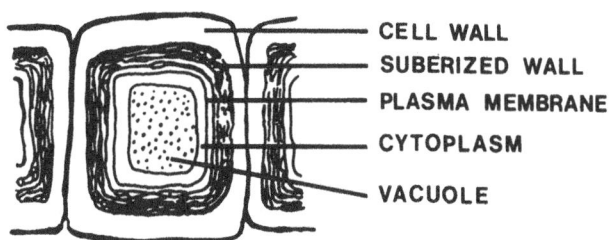

Fig. 10. Schematic representation of the suberized cell wall, redrawn from reference 139.

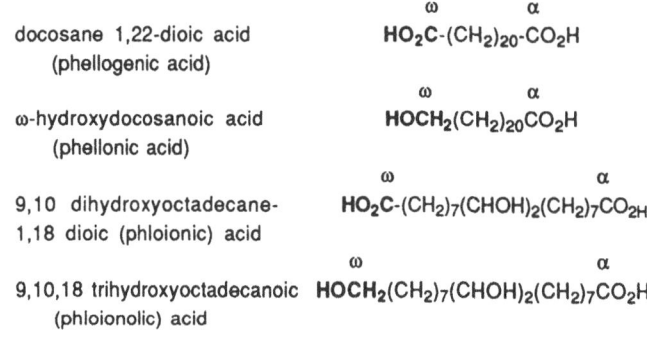

Fig. 11. Representative fatty acids from suberin-enriched tissues.

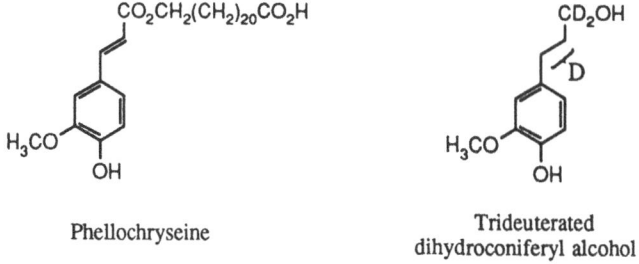

Phellochryseine

Trideuterated
dihydroconiferyl alcohol

Fig. 12. Phellochryseine and trideuterated dihydroconiferyl alcohol.

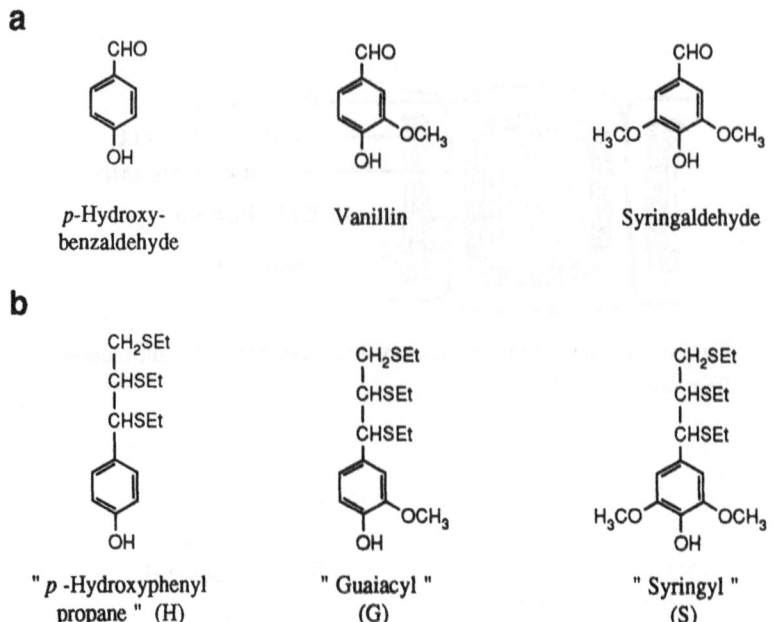

Fig. 13. (a) Nitrobenzene oxidation, and (b) thioacidolysis products from lignins and suberin polymers (see text for explanation).

beet suberin-enriched tissue with deuterated lithium aluminum hydride (LiAlD₄). Following this reductive treatment, trideuterated dihydroconiferyl alcohol (Fig.12) was isolated; importantly, this product could only result by reduction of a feruloyl ester. Interestingly, no p-coumaryl ester (as its dihydrocoumaryl alcohol derivative) was detected. This indicated that, in potato and beet tissues, the phenolic component of suberin was derived only from ferulic acid. However, it should be cautioned that recovery of the phenolic material was very low, suggesting that other monomers/more stable bonding environments were involved.

To further confirm and extend these findings, Cottle and Kolattukudy [158] subjected wound-healing potato periderm tissue to nitrobenzene oxidation. [This technique is widely used to estimate the ratio of different aromatic constituents in lignin, though at best it only degrades ~20-30% of labile linkages within the polymer to afford p-hydroxybenzaldehyde, vanillin and syringaldehyde (see Fig. 13). These aldehydes are derived from p-hydroxyphenyl (H), guaiacyl (G) and syringyl (S) moieties in lignin which, in turn, originate from coumaryl, coniferyl and sinapyl alcohols, respectively. The limitations of this analytical method have been critically reviewed.[1]

Cottle and Kolattukudy found that both p-hydroxybenzaldehyde and vanillin were released from suberized potato periderm tissue, together with traces of syringaldehyde. This indicated that p-coumaryl alcohol or p-coumaric acid may also be part of the suberin polymer. This suggestion was further substantiated by separate administration of radiolabeled [U-^{14}C]phenylalanine and [U-^{14}C]cinnamic acid, obtained by deamination of [U-^{14}C]Phe, to suberizing wound-healing potato tissue. Following nitrobenzene oxidation as before, it was demonstrated that both p-hydroxybenzaldehyde and vanillin were radiolabeled, again suggesting that p-hydroxyphenyl (H) and guaiacyl (G) moieties were present in the suberin polymer.

Based on these results, a tentative model for suberin was proposed where both aromatic and aliphatic components were covalently linked (Fig. 14). As can be seen, the proposed aromatic domains are "lignin-like", being built up from p-coumaryl, caffeoyl and coniferyl alcohols, and ferulic acid. Aliphatic domains were envisaged to consist largely of esterified long-chain (C_{18} to C_{28}) fatty acids or alcohols bonded to each other and to the aromatic domain. The aliphatic domain is thought to be completely saponifiable, i.e., it can be readily cleaved by alkali, whereas almost all envisaged linkages between the phenylpropanoid monomers have alkylaryl ether bonds (designated → in Fig. 14). It must be emphasized, however, that no evidence to substantiate these phenylpropanoid bonding environments has ever been made. Nor is there any convincing scientific evidence to prove the existence of the proposed aliphatic-aromatic linkages in suberin itself. This should not be viewed as a criticism but, instead, an attestation to the difficulties encountered in the analysis of this complex biopolymer at that time.

Monties' group in France[159] recently developed a thioacidolysis procedure to characterize lignins according to their p-hydroxyphenyl (H), guaiacyl (G) and syringyl (S) character. During this thioacidolytic (BF$_3$/C$_2$H$_5$SH) treatment, alkylaryl ether linkages in lignins are cleaved, ultimately affording simple diastereomeric mixtures of the 1,2,3-trithioethane phenylpropanoids shown in Figure 13. Like the nitrobenzene oxidation procedure, recovery of lignin fragments is normally low (~10-30% of original lignin), again due to the presence of more stable interunit linkages (e.g., C-C rather than C-O-C) which resist cleavage.

When thioacidolysis was applied to the analysis of suberized potato periderm tissue, only guaiacyl (G) and syringyl (S) derivatives were released; the corresponding p-hydroxyphenylpropane moieties (H) were not formed. This

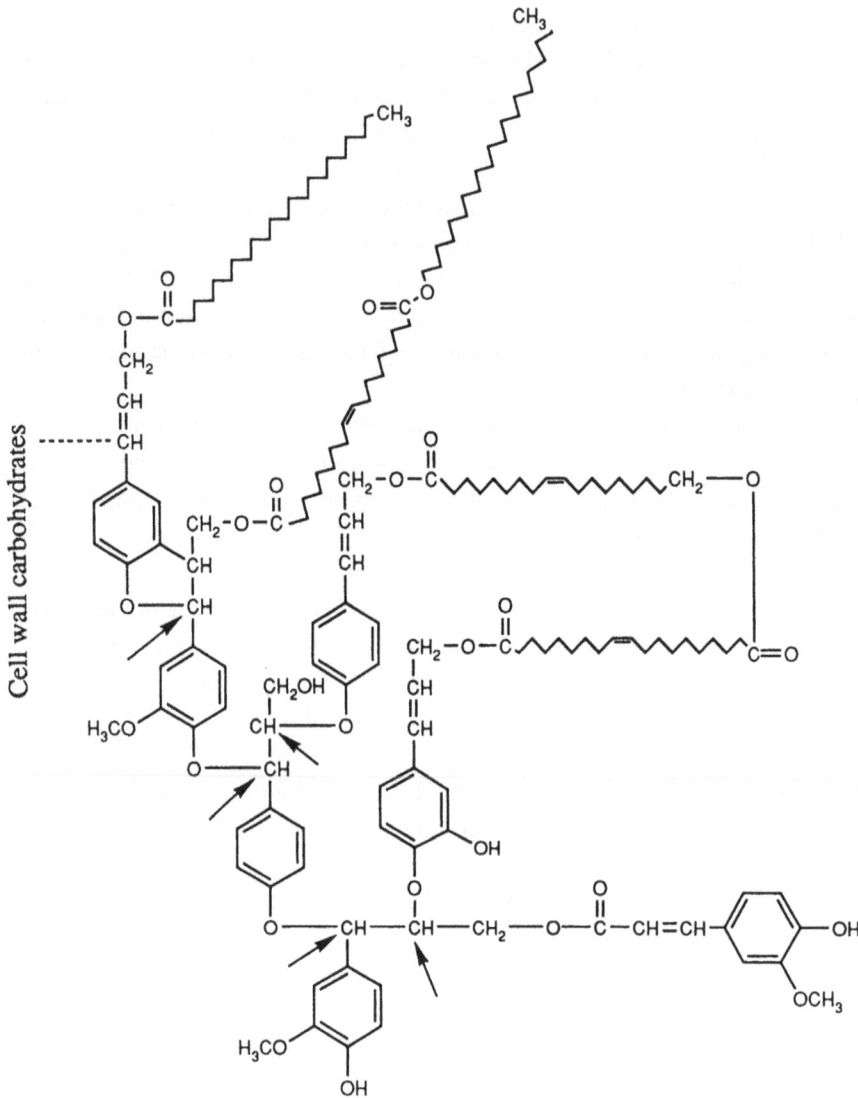

Fig. 14. Tentative suberin structure (redrawn from reference 139). Alkaryl ether bonds are designated with arrows →.

finding is in direct contradiction to that obtained by alkaline nitrobenzene oxidation treatment. Importantly, it proves that there is no *p*-coumaryl alcohol linked via alkylaryl ether bonds as proposed in Figure 14. To explain this discrepancy, the authors attempted to account for the formation of p-hydroxybenzaldehyde by nitrobenzene oxidation, as perhaps resulting from tyramine, tyrosine or p-coumaric acid moieties in the tissue.

As far as tyramine and tyrosine are concerned, there is some partial justification to this notion, since tyrosine can contribute to p-hydroxybenzaldehyde generation, and some pathogens produce a tyrosine-containing hydroxyproline rich protein.[160,161] However, this cannot explain the finding that radiolabelled p-hydroxybenzaldehyde[158] is released (by nitrobenzene oxidation) from suberin-enriched potato periderm tissue previously administered [U-^{14}C]-cinnamic acid. Nor can its formation be due to degradation of p-coumaryl esters, since none had been detected in an earlier study.[157]

Taking into account the limited information obtained from thioacidolysis, alkaline nitrobenzene oxidation and $LiAlH_4$ reduction, it is clear that the monomer composition of suberin and inter-unit bonding patterns are essentially unknown.

Carbon-13 nuclear magnetic resonance (^{13}C-NMR) spectroscopy has recently been used to study suberin structure in potato-tuber wound-healing tissue[147,162] which had been pretreated with cellulases, pectinases and solvent extraction to remove contaminants. Even with this treatment, the resulting solid-state ^{13}C-NMR spectrum mainly contained carbohydrate resonances (~50%). Several weaker signals attributed to suberin were noted, i.e., low-field carboxylic acid, olefinic and aromatic resonances, and high-field methoxyl and methylene signals. However, no definitive information about phenylpropanoid monomer composition or inter-unit linkages was obtained. One significant finding[147] was made, namely that aromatic and other unsaturated linkages outnumbered the bulk methylene groups 2:1. This places an additional constraint on the proposed suberin model (see Fig. 14), which has an aromatic:bulk methylene ratio of 1:4. Indeed, the 2:1 aromatic:methylene ratio observed in wounded potato periderm tissue strongly indicates the material is more similar to a typical lignin than to the proposed suberin model!

In summary, we have no proof of aliphatic-aromatic linkages in suberin, nor do we know how the phenolic moieties are linked together or even the identity of those involved. Further, the aromatic:methylene ratio in the suberized tissue differs by a factor of 8, when the theoretical and experimentally determined figures are compared. Taken together, the body of scientific evidence does not support the proposed structure for suberin, and emphasizes the need for more definitive information to be gained regarding its structure and formation.

CONCLUSION

Although all of the enzymes in the pathway to the hydroxycinnamic acids and monolignols have been detected, the mechanisms of regulation and control of metabolic flux into the different classes of phenylpropanoids remain obscure. In this context, much more definitive and compelling evidence needs to be obtained to support the notion that phenylalanine ammonia-lyase isoenzymes play a key regulatory role: at this point, the evidence in support of this contention is not conclusive.

Moreover, key regulatory points in phenylpropanoid metabolism must be unambiguously identified, since these will ultimately provide opportunities (at the molecular level) to manipulate metabolic flux into lignins, suberins, lignans, neolignans and hydroxycinnamic acids, etc.

It is interesting that the pathways to the monolignol glucosides, lignans and neolignans, lignins and perhaps suberin, all share the same common precursor (i.e., coniferyl alcohol). Given the questions regarding E- and Z-monolignol (glucoside) synthesis and function, lignan and neolignan biogenesis, and monolignol transport, it is clear that the enzymology and tissue specificity of these processes must now be deciphered to explain such phenomena. An excellent start to answering such questions has already been made, with the identification of the subcellular location of CAD and polyphenol oxidases(s), the detection of substrate specific UDPG:(E and Z) coniferyl alcohol glucosyltransferases and lignan-specific enzymes affording (-)-secoisolariciresinol and (-)-matairesinol.

Development of a relatively homogenous cell culture system using $P.$ *taeda* now permits investigation of the early stages of lignification, i.e., from its initiation to ~10% deposition. This process is very well-controlled in cell culture, as evidenced by secondary wall thickening, and C-13 specific labeling of lignin *in situ*. Using molecular biological approaches such cultures will provide an excellent means to examine lignification, and the early stages of cell wall development.

Finally, a critical examination of the literature of suberin formation revealed inconsistencies in the proposed model for its aromatic component; more detailed studies, such as employed in our investigations on lignin structure *in situ*, should clarify this question.

ACKNOWLEDGEMENTS

The authors thank the United States Department of Energy (DEFG06-91ER20022), the National Aeronautics and Space Administration (NAG100086) and the United States Department of Agriculture (91371036638) for financial assistance.

REFERENCES

1.   LEWIS, N.G., YAMAMOTO, E. 1990. Lignins: Occurrence, biosynthesis and biodegradation. Annu. Rev. Plant Physiol. Plant Mol. Biol. 41:455-496.

2.   KOUKOL, J., CONN, E.E. 1961. The metabolism of aromatic compounds in higher plants: IV. Purification and properties of the phenylalanine deaminase of *Hordeum vulgare*. J. Biol. Chem. 236:2692-2698.

3.   HANSON, K.R., HAVIR, E.A. 1979. An introduction to the enzymology of phenylpropanoid biosynthesis. Rec. Adv. Phytochem. 12:91-137, New York, Plenum Press.

4.   HANSON, K.R., HAVIR, E.A. 1981. Phenylalanine ammonia-lyase. In: The Biochemistry of Plants. Vol. 7, Secondary Plants Products (E.E. Cohn, ed.) Academic Press, New York:, pp. 577-625.

5.   JONES, D.H. 1984. Phenylalanine ammonia-lyase: Regulation of its induction, and its role in plant development. Phytochemistry 23:1349-1359.

6.   HAHLBROCK, K., SCHEEL, D. 1989. Physiology and molecular biology of phenylpropanoid metabolism. Annu. Rev. Plant Physiol. Plant Mol. Biol. 40:347-369.

7.   TANAKA, Y., URITANI, I. 1977. Purification and properties of phenylalanine ammonia-lyase in cut-injured sweet potato. J. Biochem. 81:963-970.

8.   TANAKA, Y., MATSUOKA, M., YAMAMOTO, N., OHASHI, Y., KANO-MURAKAMI, Y., OZEKI, Y. 1989. Structure and characterization of a cDNA clone for phenylalanine ammonia-lyase from cut-injured roots of sweet potato. Plant Physiol. 90:1403-1407.

9.   HAVIR, E.A., HANSON, K.R. 1968. L-Phenylalanine ammonia-lyase. I. Purification and molecular size of the enzyme from potato tubers. Biochemistry 7:1896-1903.

10. HAVIR, E.A. 1981. Phenylalanine ammonia-lyase: Purification and characterization from soybean cell suspension cultures. Arch. Biochem. Biophys. 211:556-563.

11. GUPTA, S., ACTON, G.J. 1979. Purification to homogeneity and some properties of L-phenylalanine ammonia-lyase of irradiated mustard (*Sinapis alba* L.) cotyledons. Biochim. Biophys. Acta 570:187-197.

12. HAVIR, E.A., REID, P.D., MARSH, H.V., JR. 1971. L-Phenylalanine ammonia-lyase (Maize). Evidence for a common catalytic site for L-phenylalanine and L-tyrosine. Plant Physiol. 48:130-136.

13. CAMM, E.L., TOWERS, G.H.N. 1973. Phenylalanine ammonia-lyase. Phytochemistry 12:961-973.

14. CHEN, R.-Y., CHANG, T.-C., LIU, M.-S. 1988. Phenylalanine ammonia-lyase of bamboo shoots. Agric. Biol. Chem. 52:2137-2142.

15. JORRIN, J., LOPEZ-VALBUENA, R., TENA, M. 1988. Purification and properties of phenylalanine ammonia-lyase from sunflower (*Helianthus annuus.* L.) hypocotyls. Biochim. Biophys. Acta 964:73-82.

16. GIVEN, N.K., VENIS, M.A., GRIERSON, D. 1988. Purification and properties of phenylalanine ammonia-lyase from strawberry fruit and its synthesis during ripening. J. Plant Physiol. 133:31-37.

17. WHETTEN, R.W., SEDEROFF, R.R. 1992. Phenylalanine ammonia-lyase from loblolly pine. Plant Physiol. 98:380-386.

18. JORRIN, J., DIXON, R.A. 1990. Stress responses in alfalfa (*Medicago sativa* L.). II. Purification, characterization, and induction of phenylalanine ammonia-lyase isoforms from elicitor-treated cell suspension cultures. Plant Physiol. 92:447-455.

19. LOPEZ-VALBUENA, R., OBRERO, R., JORRIN, J., TENA, M. 1991. Isozyme multiplicity in phenylalanine ammonia-lyase from *Vicia faba* leaves. Plant Physiol. Biochem. 29:159-164.

20. BOLWELL, G.P., BELL, J.N., CRAMER, C.L., SCHUCH, W., LAMB, C.J., DIXON, R.A. 1985. Phenylalanine ammonia-lyase from *Phaseolus vulgaris*: Characterization and differential induction of multiple forms from elicitor-treated cell suspension cultures. Eur. J. Biochem. 149:411-419.

21. CRAMER, C.L., EDWARDS, K., DRON, M., LIANG, X., DILDINE, S.L., BOLWELL, G.P., DIXON, R.A., LAMB, C.J., SCHUCH, W. 1989. Phenylalanine ammonia-lyase gene organization and structure. Plant Mol. Biol. 12:367-383.

22. SCHULZ, W., EIBEN, H.-G., HAHLBROCK, K. 1989. Expression in

Escherichia coli of catalytically active phenylalanine ammonia-lyase from parsley. FEBS Lett. 258:335-338.

23. LOIS, R., DIETRICH, A., HAHLBROCK, K., SCHULZ, W. 1989. Phenylalanine ammonia-lyase gene from parsley: Structure, regulation and identification of elicitor and light responsive *cis*-acting elements. EMBO J. 8:1641-1648.

24. MINAMI, E., OZEKI, Y., MATSUOKA, M., KOIZUKA, N., TANAKA, Y. 1989. Structure and some characterization of the gene for phenylalanine ammonia-lyase from rice plants. Eur. J. Biochem. 185:19-25.

25. OHL, S., HEDRICK, S.A., CHORY, J., LAMB, C.J. 1990. Functional properties of a phenylalanine ammonia-lyase promoter from *Arabidopsis*. Plant Cell 2:837-848.

26. LIANG, X., DRON, M., CRAMER, C.L., DIXON, R.A., LAMB, C.J. 1989. Differential regulation of phenylalanine ammonia-lyase genes during plant development and by environmental cues. J. Biol. Chem. 264:14486-14492.

27. LIANG, X., DRON, M., SCHMID, J., DIXON, R.A., LAMB, C.J. 1989. Developmental and environmental regulation of a phenylalanine ammonia-lyase-β-glucuronidase gene fusion in transgenic tobacco plants. Proc. Natl. Acad. Sci. USA 86:9284-9288.

28. KUHN, D.N., CHAPPELL, J., HAHLBROCK, K. 1983. Identification and use of cDNAs of phenylalanine ammonia-lyase and 4-coumarate: CoA ligase mRNAs in studies of the induction of phytoalexin biosynthetic enzymes in cultured parsley cells. In: Structure and Function of Plant Genes. (O. Ciferri, L. Dure, eds.) NATO ASI Series, Chapman and Hall, London, pp. 329-336.

29. KUHN, D.N., CHAPPELL, J., BOUDET, A., HAHLBROCK, K. 1984. Induction of phenylalanine ammonia-lyase and 4-coumarate: CoA ligase mRNAs in cultured plant cells by UV light or fungal elicitor. Proc. Natl. Acad. Sci. USA 81:1102-1106.

30. SHAW, N.M., BOLWELL, G.P., SMITH, C. 1990. Wound-induced phenylalanine ammonia-lyase in potato (*Solanum tuberosum*) tuber discs. Significance of glycosylation and immunolocalization of enzyme subunits. Biochem. J. 267:163-170.

31. GRAND, C. 1984. Ferulic acid 5-hydroxylase: A new cytochrome P-450-dependent enzyme from higher plant microsomes involved in lignin synthesis. FEBS Lett. 169:7-11.

32. SAIMMAIME, I., COULOMB, C., COULOMB, P.-J., ROGGERO, J.-
     P. 1990. Mise en évidence d'une activité cinnamate 4-hydroxylase
     dans des feuilles de *Capsicum annuum*. Plant Physiol. Biochem.
     28:323-331.
33. KOJIMA, M., TAKEUCHI, W. 1989. Detection and characterization of
     p-coumaric acid hydroxylase in mung bean (*Vigna mungo*) seedlings.
     J. Biochem. 105:265-270.
34. GROSS, G.G. 1985. Biosynthesis and metabolism of phenolic acids and
     monolignols. In: Biosynthesis and Biodegradation of Wood Compo-
     nents. (T. Higuchi, ed.) Academic Press, Orlando, pp. 229-271.
35. KAMSTEEG, J., VAN BREDERODE, J., VERSCHUREN, P.M., VAN
     NIGTEVECHT, G. 1981. Identification, properties and genetic
     control of *p*-coumaroyl coenzyme A, 3-hydroxylase isolated from
     petals of *Silene dioica*. Z. Pflanzenphysiol. 102:435-442.
36. BONIWELL, J.M., BUTT, V.S. 1986. Flavin nucleotide-dependent 3-
     hydroxylation of 4-hydroxyphenylpropanoid carboxylic acids by
     particulate preparations from potato tubers. Z. Naturforsch 41c:56-60.
37. DUKE, S.O., VAUGHN, K.C. 1982. Lack of involvement of polyphenol
     oxidase in ortho-hydroxylation of phenolic compounds in mung bean
     seedlings. Physiol. Plant. 54:381-385.
38. SHERMAN, T. D., VAUGHN, K. C., DUKE, S.O. 1991. A limited
     survey of the phylogenetic distribution of polyphenol oxidase.
     Phytochemistry 30:2499-2506.
39. SATO, M. 1967. Metabolism of phenolic substances by the chloroplasts.
     III. Phenolase as an enzyme concerning the formation of esculetin.
     Phytochemistry 6:1363-1373.
40. OHASHI, H., YAMAMOTO, E., LEWIS, N.G., TOWERS, G.H.N.
     1987. 5-Hydroxyferulic acid in *Zea mays* and *Hordeum vulgare* cell
     walls. Phytochemistry 26:1915-1916.
41. KURODA, H. 1983. Comparative studies on *O*-methyltransferases
     involved in lignin biosynthesis. Wood Research 69:91-135.
42. KURODA, H., SHIMADA, M., HIGUCHI, T. 1975. Purification and
     properties of *O*-methyltransferase involved in the biosynthesis of
     gymnosperm lignin. Phytochemistry 14:1759-1763.
43. HERMANN, C., LEGRAND, M., GEOFFROY, P., FRITIG, B. 1987.
     Enzymatic synthesis of lignin: Purification to homogeneity of the
     three *O*-methyltransferases of tobacco and production of specific
     antibodies. Arch. Biochem. Biophys. 253:367-376.
44. SCHMITT, D., PAKUSCH, A. E., MATERN, U. 1991. Molecular

cloning, induction and taxonomic distribution of caffeoyl-CoA-3-*O*-methyltransferase, an enzyme involved in disease resistance. J. Biol. Chem. 266:17416-17423.

45. KURODA, H., SHIMADA, M., HIGUCHI, T. 1981. Characterization of a lignin-specific *O*-methyltransferase in aspen wood. Phytochemistry 20:2635-2639.

46. BUGOS, R.C., CHIANG, V.L.C., CAMPBELL, W.H. 1991. Seasonal expression of a lignin specific *O*-methyltransferase cloned from aspen developing xylem. Plant Physiol. 96(1s):84.

47. GOWRI, G., BUGOS, R.C., CAMPBELL, W.H., MAXWELL, C.A., DIXON, R.A. 1991. Stress responses in alfalfa (*Medicago sativa* L.). X. Molecular cloning and expression of *S*-adenosyl-L-Methionine: Caffeic acid 3-*O*-methyltransferase, a key enzyme of lignin biosynthesis. Plant Physiol. 97:7-14.

48. KNOBLOCH, K.-H., HAHLBROCK, K. 1975. Isoenzymes of *p*-coumarate:CoA ligase from cell suspension culture of *Glycine max*. Eur. J. Biochem. 52:311-320.

49. WALLIS, P.J., RHODES, M.J.C. 1977. Multiple forms of hydroxy-cinnamate:CoA ligase in etiolated pea seedlings. Phytochemistry 16:1891-1894.

50. RANJEVA, R., BOUDET, A.M., FAGGION, R. 1976. Phenolic metabolism in petunia tissues. IV. Properties of *p*-coumarate: coenzyme A ligase isoenzymes. Biochimie 58:1255-1262.

51. GRAND, C., BOUDET, A., BOUDET, A.M. 1983. Isoenzymes of hydroxycinnamate:CoA ligase from poplar stems. Properties and tissue distribution. Planta 158:225-229.

52. LOZOYA, E., HOFFMANN, H., DOUGLAS, C., SCHULZ, W., SCHEEL, D., HAHLBROCK, K. 1988. Primary structures and catalytic properties of isoenzymes encoded by the two 4-coumarate:CoA ligase genes in parsley. Eur. J. Biochem. 176:661-667.

53. DOUGLAS, C., HOFFMANN, H., SCHULZ, W., HAHLBROCK, K. 1987. Structure and elicitor or U.V.-light stimulated expression of two 4-coumarate:CoA ligase genes in parsley. EMBO J. 6:1189-1195.

54. WENGENMAYER, H., EBEL, J., GRISEBACH, H. 1976. Enzymic synthesis of lignin precursors. Purification and properties of a cinnamoyl-CoA:NADPH reductase from cell suspension cultures of soybean (*Glycine max*). Eur. J. Biochem. 65:529-536.

55. LÜDERITZ, T., GRISEBACH, H. 1981. Enzymic synthesis of lignin precursors. Comparison of cinnamoyl-CoA reductase and cinnamyl alcohol: NADP⁺ dehydrogenase from spruce (*Picea abies* L.) and soybean (*Glycine max* L.). Eur. J. Biochem. 119:115-124.

56. SARNI, F., GRAND, C., BOUDET, A.M. 1984. Purification and properties of cinnamoyl-CoA reductase and cinnamyl alcohol dehydrogenase from poplar stems (*Populus X euramericana*). Eur. J. Biochem. 139:259-265.

57. MANSELL, R.L., GROSS, G.G., STÖCKIGT, J., FRANKE, H., ZENK, M.H. 1974. Purification and properties of cinnamyl alcohol dehydrogenase from higher plants involved in lignin biosynthesis. Phytochemistry 13:2427-2435.

58. WYRAMBIK, D., GRISEBACH, H. 1975. Purification and properties of isoenzymes of cinnamyl-alcohol dehydrogenase from soybean-cell-suspension cultures. Eur. J. Biochem. 59:9-15.

59. KUTSUKI, H., SHIMADA, M., HIGUCHI, T. 1982. Regulatory role of cinnamyl alcohol dehydrogenase in the formation of guaiacyl and syringyl lignins. Phytochemistry 21:19-23.

60. GRAND, C., SARNI, F., LAMB, C.J. 1987. Rapid induction by fungal elicitor of the synthesis of cinnamyl-alcohol dehydrogenase, a specific enzyme of lignin synthesis. Eur. J. Biochem. 169:73-77.

61. WALTER, M.H., GRIMA-PETTENATI, J., GRAND, C., BOUDET, A.M., LAMB, C.J. 1988. Cinnamyl-alcohol dehydrogenase, a molecular marker specific for lignin synthesis: cDNA cloning and mRNA induction by fungal elicitor. Proc. Natl. Acad. Sci. USA 85:5546-5550.

62. WALTER, M.H., GRIMA-PETTENATI, J., GRAND, C., BOUDET, A.M., LAMB, C.J. 1990. Extensive sequence similarity of the CAD4 (cinnamyl-alcohol dehydrogenase) to a maize malic enzyme. Plant Mol. Biol. 15:525-526.

63. VAN DOORSSELAERE, J., VILLARROEL, R., VAN MONTAGU, M., INZE, D. 1991. Nucleotide sequence of a cDNA encoding malic enzyme from poplar. Plant Physiol. 96:1385-1386.

64. O'MALLEY, D. M, PORTER, S., SEDEROFF, R.R. 1992. Purification characterization and cloning of cinnamyl alcohol dehydrogenase in loblolly pine (*Pinus taeda* L.). Plant Physiol. 98(in press)

65. HARMATHA, J., LÜBKE, H., RYBARIK, I., MÄHDALIK, M. 1977. *Cis*-coniferyl alcohol and its glucoside from the bark of beech (*Fagus silvatica* L.). Collect. Czech. Chem. Commun. 43:774-780.

66. MORELLI, E., REJ, R.N., LEWIS, N.G., JUST, G., TOWERS, G.H.N. 1986. *Cis*-monolignols in *Fagus grandifolia* and their possible involvement in lignification. Phytochemistry 25:1701-1705.

67. LEWIS, N.G., INCIONG, MA.E.J., OHASHI, H., TOWERS, G.H.N., YAMAMOTO, E. 1988. Exclusive accumulation of Z-isomers of monolignols and their glucosides in bark of *Fagus grandifolia*. Phytochemistry 27:2119-2121.

68. LEWIS, N.G., DUBELSTEN, P., EBERHARDT, T.L., YAMAMOTO, E., TOWERS, G.H.N. 1987. The *E/Z* isomerization step in the biosynthesis of Z-coniferyl alcohol in *Fagus grandifolia*. Phytochemistry 26:2729-2734.

69. LEWIS, N.G., INCIONG, MA, E.J., DHARA, K.P., YAMAMOTO, E. 1989. High-performance liquid chromatographic separation of *E*- and Z-monolignols and their glucosides. J. Chromatogr. 479:345-352.

70. YAMAMOTO, E., INCIONG, MA.E.J., DAVIN, L.B., LEWIS, N.G. 1990. Formation of *cis*-coniferin in cell-free extracts of *Fagus grandifolia* Ehrh bark. Plant Physiol. 94:209-213.

71. WHITING, D.A. 1985. Lignans and neolignans. Nat. Prod. Rep. 2:191-212.

72. MACRAE, W.D., TOWERS, G.H.N. 1984. Biological activities of lignans. Phytochemistry 23:1207-1220.

73. WHITING, D.A. 1987. Lignans, neolignans and related compounds. Nat. Prod. Rep. 4:499-525.

74. WHITING, D.A. 1990. Lignans, neolignans and related compounds. Nat. Prod. Rep. 7:349-364.

75. EL-FERALY, F.S., CHEATHAM, S.F., BREEDLOVE, R.L. 1983. Antimicrobial neolignans of *Sassafras randaiense* roots. J. Nat. Prod. 46:493-498.

76. NAKATANI, N., IKEDA, K., KIKUZAKI, H., KIDO, M., YAMAGUCHI, Y. 1988. Diaryldimethylbutane lignans from *Myristica argentea* and their antimicrobial action against *Streptococcus mutans*. Phytochemistry 27:3127-3129.

77. ARNONE, A., DI MODUGNO, V., NASINI, G., VENTURINI, I. 1988. Isolation and structure of new active neolignans and norneolignans from *Ratanhia*. Gaz. Chim. Ital. 118:675-682.

78. TAKASUGI, M., KATUI, N. 1986. A biphenyl phytoalexin from *Cercidiphyllum japonicum*. Phytochemistry 25:2751-2752.

79. BARATA, L.E.S., BAKER, P.M., GOTTLIEB, O.R., RUVEDA, E.A.

1978. Neolignans of *Virola surinamensis*. Phytochemistry 17:783-786.

80. SHIZURI, Y., NAKAMURA, K., YAMAMURA, S., OHBA, S., YAMASHITA, H., SAITO, Y. 1986. Total syntheses of isodihydrofutoquinol A, futoquinol, and isofutoquinol A and B. Tetrahedron Lett. 27:727-730.

81. HARMATHA, J., NAWROT, J. 1984. Comparison of the feeding deterrent activity of some sesquiterpene lactones and a lignan lactone towards selected insect storage pests. Biochem. Syst. Ecol. 12:95-98.

82. TATEMATSU, H., KUROKAWA, M., NIWA, M., HIRATA, Y. 1984. Piscicidal constituents of *Stellera chamaejasme* L. II. Chem. Pharm. Bull. 32:1612-1613.

83. MUNAKATA, K., MARUMO, S., OHTA, K. 1965. Justicidin A and B, the fish-killing components of *Justicia hayatai* var. Decumbens. Tetrahedron Lett. 47:4167-4170.

84. BINNS, A.N., CHEN, R.H., WOOD, H.N., LYNN, D.G. 1987. Cell division promoting activity of naturally occurring dehydrodiconiferyl glucosides: Do cell wall components control cell division? Proc. Natl. Acad. Sci. USA 84:980-984.

85. SAITO, H., YOSHIKAWA, H., NISHIMURA, Y., KONDO, S., TAKEUCHI, T., UMEZAWA, H. 1986. Studies on lignan lactone antitumor agents. I. Synthesis of aminoglycosidic lignan variants related to podophyllotoxin. Chem. Pharm. Bull. 34:3733-3740.

86. SAITO, H., NISHIMURA, Y., KONDO, S., TAKEUCHI, T., UMEZAWA, H. 1988. Studies on lignan lactone antitumor agents. IV. Synthesis of glycosidic lignan variants related to α–peltatin. Bull. Chem. Soc. Jpn. 61:1259-1263.

87. SAITO, H., NISHIMURA, Y., KONDO, S., KOMURO, K., TAKEUCHI, T. 1988. Studies on lignan lactone antitumor agents. V. 1-*O*-(aminoethyl) ether of 4'-*O*-demethyl-1-epipodophyllotoxin. Bull. Chem. Soc. Jpn. 61:2493-2497.

88. FUKAMIYA, N., LEE, K.-H. 1986. Antitumor agents, 81. Justicidin-A and diphyllin, two cytotoxic principles from *Justicia procumbens*. J. Nat. Prod. 49:348-350.

89. BADAWI, M.M., HANDA, S.S., KINGHORN, A.D., CORDELL, G.A., FARNSWORTH, N.R. 1983. Plant anticancer agents XXVII: Antileukemic and cytotoxic constituents of *Dirca occidentalis* (Thymelaeaceae). J. Pharm. Sci. 72:1285-1287.

90. FANG, X., NANAYAKKARA, N.P.D., PHOEBE, C.H., JR.,

PEZZUTO, J.M., KINGHORN, A.D., FARNSWORTH, N.R. 1985. Plant anticancer agents XXXVII. Constituents of *Amanoa oblongifolia*. Planta Medica 51:346-347.

91. DUH, C.-Y., PHOEBE, C.H., JR., PEZZUTO, J.M., KINGHORN, A.D., FARNSWORTH, N.R. 1986. Plant anticancer agents XLII. Cytotoxic constituents from *Wikstroemia elliptica*. J. Nat. Prod. 49:706-709.

92. LE QUESNE, P.W., LARRAHONDO, J.E., RAFFAUF, R.F. 1980. Antitumor plants X. Constituents of *Nectandra rigida*. J. Nat. Prod. 43:353-359.

93. TOMIOKA, K., ISHIGURO, T., MIZUGUCHI, H., KOMESHIMA, N., KOGA, K., TSUKAGOSHI, S., TSURUO, T., TASHIRO, T., TANIDA, S., KISHI, T. 1991. Absolute structure-cytotoxic activity relationships of steganacin congeners and analogues. J. Med. Chem. 34:54-57.

94. BEDOWS, E., HATFIELD, G.M. 1982. An investigation of the antiviral activity of *Podophyllum peltatum*. J. Nat. Prod. 45:725-729.

95. NIKAIDO, T., OHMOTO, T., KINOSHITA, T., SANKAWA, U., NISHIBE, S., HISADA, S. 1981. Inhibition of cyclic AMP phosphodiesterase by lignans. Chem. Pharm. Bull. 29:3586-3592.

96. BERNARD, C.-B., ARNASON, J.T., PHILOGENE, B.J.R., LAM, J., WADDELL, T. 1989. Effect of lignans and other secondary metabolites of the Asteraceae on the mono-oxygenase activity of the European corn borer. Phytochemistry 28:1373-1377.

97. NISHIBE, S., KINOSHITA, H., TAKEDA, H., OKANO, G. 1990. Phenolic compounds from stem bark of *Acanthaponax senticosus* and their pharmalogical effect in chronic swimming stressed rats. Chem. Pharm. Bull. 38:1763-1765.

98. TAKEDA, S., ARAI, I., KASE, Y., OHKURA, Y., HASEGAWA, M., SEKIGUCHI, Y., SUDO, K., ABURADA, M., HOSOYA, E. 1987. Pharmacological studies on antihepatotoxic action of (+)-(6S,7S,R-Biar)-5,6,7,8-tetrahydro-1,2,3,12-tetramethoxy-6,7-dimethyl-10,11-methylenedioxy-6-dibenzo[a,c]cyclooctenol (TJN-101), a lignan component of *Schisandra* fruits. Influences of resolvents on the efficacy of TJN-101 in the experimental acute hepatic injuries. Yakugaku Zasshi 107:517-524.

99. IKEYA, Y., TAGUCHI, H., MITSUHASHI, H., TAKEDA, H., KASE, Y., ABURABA, M. 1988. A lignan from *Schizandra chinensis*. Phytochemistry 27:569-573.

100. AXELSON, M., SJÖVALL, J., GUSTAFSSON, B.E., SETCHELL, K.D.R. 1982. Origin of lignans in mammals and identification of a precursor from plants. Nature 298:659-660.

101. BANNWART, C., ADLERCREUTZ, H., WÄHÄLÄ, K., BRUNOW, G., HASE, T. 1989. Detection and identification of the plant lignans, lariciresinol, isolariciresinol, and secoisolariciresinol in human urine. Clin. Chim. Acta 180:293-302.

102. ADLERCREUTZ, H., HOCKERSTADT, K., BANNWART, C., HAMAILEN, E., FOTSIS, T., BLOIGU, S. 1988. Association between dietary fiber, urinary excretion of lignans and isoflavonic phytoestrogens, and plasma non-protein bound sex hormones in relation to breast cancer. Prog. Cancer Res. Ther. 35 (Horm. Cancer 3):409-412.

103. ADLERCREUTZ, H. 1984. Does fiber-rich food containing animal lignan precursors protect against both colon and breast cancer? An extension of the "Fiber hypothesis". Gastroenterology 86:761-764.

104. SCHRÖDER, H.C., MERZ, H., STEFFEN, R., MÜLLER, W.E.G., SARIN, P.S., TRUMM, S., SCHULZ, J., EICH, E. 1990. Differential *in vitro* anti-HIV activity of natural lignans. Z. Naturforsch. 45c:1215-1221.

105. RAHMAN, M.M.A., DEWICK, P.M., JACKSON, D.E., LUCAS, J.A. 1990. Biosynthesis of lignans in *Forsythia intermedia*. Phytochemistry 29:1841-1846.

106. SETCHELL, K.D.R., ADLERCREUTZ, H. 1988. Mammalian lignans and phytoestrogens. Recent studies on their formation, metabolism and biological role in health and disease. In: Gut Flora in Toxicology and Cancer (I. R. Rowland, ed.) Academic Press, London, pp. 315-345.

107. RAO, C.B.S. 1978. Chemistry of lignans. Andrha University Press, Waltair, India, pp. 377.

108. KITAGAWA, S., NISHIBE, S., BENECKE, R., THIEME, H. 1988. Phenolic compounds from *Forsythia* leaves II. Chem. Pharm. Bull. 36:3667-3670.

109. UMEZAWA, T., DAVIN, L.B., YAMAMOTO, E., KINGSTON, D.G.I., LEWIS, N.G. 1990. Lignan biosynthesis in *Forsythia* species. J. Chem. Soc., Chem. Comm. 1405-1408.

110. UMEZAWA, T., DAVIN, L.B., LEWIS, N.G. 1991. Formation of the lignans (-)-secoisolariciresinol and (-)-matairesinol with *Forsythia intermedia* cell-free extracts. J. Biol. Chem. 266:10210-10217.

111. FREUDENBERG, K. 1965. Lignin: Its constitution and formation from *p*-hydroxycinnamyl alcohols. Science 148:595-600.

112. YAMAGUCHI, H., NAKATSUBO, F., KATSURA, Y., MURAKAMI, K. 1990. Characterization of (+)- and (-)-syringaresinol di-β-D-glucosides. Holzforschung 44:381-385.

113. LEWIS, N.G., DAVIN, L.B. 1992. Stereoselectivity in polyphenol biosynthesis. In: Plant Polyphenols: Biogenesis, Biochemical Properties and Significance. (R.W. Hemingway, P.E. Laks, eds.) Plenum Press (in press).

114. UMEZAWA, T., DAVIN, L.B., LEWIS, N.G. 1990. Formation of the lignan, (-)-secoisolariciresinol, by cell-free extracts of *Forsythia intermedia*. Biochem. Biophys. Res. Comm. 171:1008-1014.

115. YAMAMOTO, E., BOKELMAN, G.H., LEWIS, N.G. 1989. Phenylpropanoid metabolism in cell walls. In: Plant Cell Wall Polymers, Biogenesis and Biodegradation. (N.G. Lewis, M.G. Paice, eds.) ACS Symp. Ser. 399:68-88.

116. MARCINOWSKI, S., FALK, H., HAMMER, D.K., HOYER, B., GRISEBACH, H. 1979. Appearance and localization of a β-glucosidase hydrolyzing coniferin in spruce (*Picea abies*) seedlings. Planta 144:161-165.

117. BURMEISTER, G., HÖSEL, W. 1981. Immunohistochemical localization of β-glucosidases in lignin and isoflavone metabolism in *Cicer arietinum* L. seedlings. Planta 152:578-586.

118. FREUDENBERG, K. 1968. The constitution and biosynthesis of lignin. In : Constitution and Biosynthesis of Lignin: Molecular Biology, Biochemistry and Biophysics. Vol.2. (K. Freudenberg, A.C. Neish, eds.) Springer Verlag., Berlin, pp 47-122.

119. MARCINOWSKI, S., GRISEBACH, H. 1978. Enzymology of lignification. Cell wall-bound β-glucosidases for coniferin from spruce (*Picea abies*) seedlings. Eur. J. Biochem. 87:37-44.

120. GRISEBACH, H. 1981. Lignins. In: The Biochemistry of Plants. Vol. 7. Secondary Plant Products. (E.E. Conn, ed.) Academic Press, New York, pp. 457-478.

121. HÖSEL, W. 1981. Glycosylation and glycosidases. In: The Biochemistry of Plants. Vol. 7. Secondary Plant Products. (E.E. Conn, ed.) Academic Press, New York, pp. 725-753.

122. MARCINOWSKI, S., GRISEBACH, H. 1977. Turnover of coniferin in pine seedlings. Phytochemistry 16:1665-1667

123. HÖSEL, W., SURHOLT, E., BORGMANN, E. 1978. Characterization

of β-glucosidase isoenzymes possibly involved in lignification from chick pea (*Cicer arietinum* L.) cell suspension cultures. Eur. J. Biochem. 84:487-492.

124. HÖSEL, W., FIEDLER-PREISS, A., BORGMANN, E. 1982. Relationship of coniferin β-glucosidase to lignification in various plant cell suspension cultures. Plant Cell Tissue Organ Cult. 1:137-148.

125. LAGRIMINI, L.M., BURKHART, W., MOYER, M., ROTHSTEIN, S. 1987. Molecular cloning of complementary DNA encoding the lignin-forming peroxidase from tobacco: Molecular analysis and tissue-specific expression. Proc. Natl. Acad. Sci. USA 84:7542-7546.

126. MÄDER, M., NESSEL, A., BOPP, M. 1977. Über die physiologische Bedeutung der Peroxidase Isoenzymgruppen des Tabaks anhand einiger biochemischer Eigenschaften II. Z. Pflanzenphysiol. 82:247-260

127. LAGRIMINI, L.M., ROTHSTEIN, S. 1987. Tissue specificity of tobacco peroxidase isozymes and their induction by wounding and tobacco mosaic virus infection. Plant Physiol. 84:438-442.

128. LAGRIMINI, L.M., BRADFORD, S., ROTHSTEIN, S. 1990. Peroxidase-induced wilting in transgenic tobacco plants. Plant Cell 2:7-18.

129. ROTHSTEIN, S.J., LAGRIMINI, L.M. 1989. Silencing gene expression in plants. In: Oxford Surveys of Plant Molecular and Cell Biology (B.J. Miflin, H.F. Miflin, eds.) Oxford University Press 6:221-246.

130. LAGRIMINI, L.M. Altered phenotypes in plants transformed with chimeric tobacco peroxidase genes. In: Molecular and Physiological Aspects of Plant Peroxidases. II. (C. Penel, T. Gaspar, J. Lobarzewski, eds.) (in press).

131. HIGUCHI, T., ITO, Y., SHIMADA, M., KAWAMURA, J. 1967. Chemical properties of milled wood lignin of grasses. Phytochemistry 6:1551-1556.

132. SCALBERT, A., MONTIES, B., LALLEMAND, J.-Y., GUITTET, E., ROLANDO, C. 1985. Ether linkage between phenolic acids and lignin fractions from wheat straw. Phytochemistry 24:1359-1362.

133. TERASHIMA, N., FUKUSHIMA, K. 1989. Biogenesis and structure of macromolecular lignin in the cell wall of tree xylems as studied by microautoradiography. In: Plant Cell Wall Polymers, Biogenesis and Biodegradation. (N.G. Lewis, M.G. Paice, eds.) ACS Symp. Ser. 399:160-168.

134. MORRISSON, I.M. 1972. Improvements of the acetyl bromide technique

to determine lignin and digestibility and its application to legumes. J. Sci. Food Agric. 23:1463-1469.

135. LEWIS, N.G., YAMAMOTO, E., WOOTEN, J.B., JUST, G., OHASHI, H., TOWERS, G.H.N. 1987. Monitoring biosynthesis of wheat cell-wall phenylpropanoids in situ. Science 237:1344-1346.

136. LEWIS, N.G., RAZAL, R.A., DHARA, K.P., YAMAMOTO, E., BOKELMAN, G.H., WOOTEN, J.B. 1988 Incorporation of [2-^{13}C] ferulic acid, a lignin precursor, into *Leucaena leucocephala* and its analysis by solid-state ^{13}C-NMR. J. Chem. Soc. Chem. Commun. 1626-1628.

137. LEWIS, N.G. 1988. Lignin biosynthesis, biodegradation and utilization. Bull. Liaison Groupe Polyphenols 14:398-410.

138. LEWIS, N.G., RAZAL, R.A., YAMAMOTO, E., BOKELMAN, G.H., WOOTEN, J.B. 1989. ^{13}C-Specific labelling of lignin in intact plants. In: Plant Cell Wall Polymers: Biogenesis and Biodegradation. (N.G. Lewis, M.G. Paice, eds.) ACS Symp. Ser. 399:68-88.

139. KOLATTUKUDY, P.E. 1984. Biochemistry and function of cutin and suberin. Can. J. Bot. 62:2918-2933.

140. KOLATTUKUDY, P.E., KRONMAN, K., POULOSE, A.J. 1975. Determination of structure and composition of suberin from the roots of carrot, parsnip, rutabaga, turnip, red beet and sweet potato by combined gas-liquid chromatography and mass spectrometry. Plant Physiol. 55:567-573.

141. HOLLOWAY, P.J. 1972. The composition of suberin from the corks of *Quercus suber* L. and *Betula pendula* Roth. Chem. Phys. Lipids 9:158-170.

142. LITVAY, J.D., KRAHMER, R.L. 1977. Wall-layering in Douglas fir cork cells. Wood Sci. 9:167-173.

143. TIPPETT, J.T., O'BRIEN, T.P. 1976. The structure of eucalypt roots. Aust. J. Bot. 24:619-632.

144. SCOTT, M.G., PETERSON, R.L. 1979. The root endoderma in *Ranunculus acris*. II. Histochemistry of the endodermis and the synthesis of phenolic compounds in roots. Can. J. Bot. 57:1063-1077.

145. PEARCE, R.B., RUTHERFORD, J. 1981. A wound-associated suberized barrier to the spread of decay in the sapwood of oak (*Quercus robur* l.). Physiol. Plant Pathol. 19:359-369.

146. KOLATTUKUDY, P.E., ESPELIE, K.E. 1985. Biosynthesis of cutin, suberin and associated waxes. In: Biosynthesis and Biodegradation of

Wood Components. (T. Higuchi, ed.) Academic Press, Orlando pp. 161-207.

147. GARBOW, J.R., FERRANTELLO, L.M., STARK, R.E. 1989. ^{13}C Nuclear magnetic resonance study of suberized potato cell wall. Plant Physiol. 90:783-787.

148. JENSEN, W., FREMER, K.E., SIERILDÄ, P., WARTIOVAARA, V. 1963. The chemistry of bark. In: The Chemistry of Wood. (B.L. Browning, ed.) Intersci. Publ. New York, London , pp. 587-666.

149. HOLLOWAY, P.J. 1972. The composition of suberin from the corks of *Quercus suber* L. and *Betula pendula* Roth. Chem. Phys. Lipids 9:158-170.

150. RIBAS, I. 1942. Ion 2, No. 6, 25.

151. HERGERT, H.L. 1958. Chemical composition of cork from White fir bark. Forest. Prod. J., 335-339.

152. GUILLEMONAT, A., TRAYNARD, J.-C. 1962. Sur la constitution chimique du liége. IV Mémoire: Structure de la phellochryséine, Mémoires Présentés à la Société Chimique 142-144.

153. SWAN, E.P. 1968. Alkaline ethanolysis of extractive-free western red cedar bark. TAPPI 51:301-304.

154. ADAMOVICS, J.A., JOHNSON, G., STERMITZ, F.R. 1977. Ferulates from cork layers of *Solanum tuberosum* and *Pseudotsuga menziesii*. Phytochemistry 16:1089-1090.

155. LAVER, M.L., FANG, H.L. 1989. Ferulic acid esters from bark of *Pseudotsuga menziesii*. J. Agric. Food Chem. 37(1):114-116.

156. RIES-KAUTT, M., KINTZINGER, J.P., ALBRECHT, P. 1988. Omega-feruloyloxy acids, a novel class of polar lipids in peat soil. Naturwissenschaften 75:305-307.

157. RILEY, R.G., KOLATTUKUDY, P.E. 1975. Evidence for covalently attached *p*-coumaric acid and ferulic acid in cutins and suberins. Plant Physiol. 56:650-654.

158. COTTLE, W., KOLATTUKUDY, P.E. 1982. Biosynthesis, deposition and partial characterization of potato suberin phenolics. Plant Physiol. 69:393-399.

159. BORG-OLIVIER, O., MONTIES, B. 1989. Caractérisation des lignines, acides phénoliques et tyramine dans les tissus subérisés du périderme naturel et du périderme de blessure de tubercule de pomme de terre. C.R. Acad. Sci. Paris Ser. III 308:141-147.

160. ESQUERRÉ-TUGAYÉ, M.-T., LAMPORT, D.T.A. 1979. Cell-surfaces in plant-microorganism interactions. I. A structural investigation of

cell-wall hydroxyproline-rich glycoproteins which accumulate in fungus infected plants. Plant Physiol. 64:314-319.

161. HAMMERSCHMIDT, R. 1984. Rapid deposition of lignin in potato tuber tissue as a response to fungi non-pathogenic on potato. Physiol. Plant Pathol. 24:33-42.

162. STARK, R.E., ZLOTNIK-MAZORI, T., FERRANTELLO, L.M., GARBOW, J.R. 1989. Molecular structure and dynamics of intact plant polyesters. Solid-state NMR studies. In: Plant Cell Wall Polymers: Biogenesis and Biodegradation. (N.G. Lewis, M.G. Paice, eds.) ACS Symp. Ser. 399:214-229.

Chapter Twelve

THE PHYTOCHEMICAL SOCIETY OF NORTH AMERICA: ITS FIRST THIRTY YEARS, 1961-1991

Stewart A. Brown

Department of Chemistry
Trent University
Peterborough, Ontario, Canada K9J 7B8

The Phytochemical Society of North America (PSNA) evolved from a more specialized organization, the Plant Phenolics Group of North America (PPGNA), which was organized in 1961. Thus, although it is not technically true that the PSNA is celebrating its 30th anniversary, we can regard 1991 as the

Phenolic Metabolism in Plants, Edited by H.A. Stafford
and R.K. Ibrahim, Plenum Press, New York, 1992

30th year of the present society together with its precursor group, many of whose members are still active in the PSNA.

THE FORMATIVE YEARS

Conception

The idea of forming a group with common interests in plant phenolics may well have crossed the minds of a number of people as far back as the mid-1950s, but the first actual steps in that direction were taken as the direct result of conversations just before and during the 9th International Botanical Congress (IBC) held in Montreal in 1959. At that time I was on the staff of the former Prairie Regional Laboratory (PRL) of the National Research Council of Canada in Saskatoon. There the late Arthur Neish and his associates were building the strong phytochemical research program that was the foundation of the present Plant Biotechnology Institute when it supplanted the old PRL in the eighties. The IBC was a big scientific event in Canada that year, and, like phytochemists from other laboratories, our group was well represented. Just previously I had had a memorable visit from Eric Conn, my first meeting with him, and I recall that we flew together from Saskatoon to Montreal for the Congress.

One evening during the meetings I was strolling along one of Montreal's downtown streets in conversation with Neil Towers, then of McGill University, and Vic Runeckles, then with the Imperial Tobacco Company of Canada, both in that city. Vic brought up the idea of organizing a group of those with research interests in plant phenolics, quite a number of whom were then at the Congress, along the lines of the Plant Phenolics Group in Britain. Not long before, Neil had attended one of their meetings, and was, in his own words, so exhilarated on his return that he and Vic had discussed the possibility of forming a similar group here. I was equally keen on the idea, and it was agreed that it should be seriously explored with some of the prominent people in the field. The feeling was that we should aim at an international group, and not one confined to our own country, since an organization encompassing the United States as well as Canada would be potentially much more viable. If such a group were to be envisaged at the outset, we agreed that a prominent phytochemist from the United States should participate in the initial planning. As I felt that an informal committee of three was large enough, it was decided that the other two should explore the idea, co-opting a U.S. representative. The late Ted Geissman of the University of California, Los Angeles, was approached and he consented to be involved.

The following year saw the entry into the picture of Gestur Johnson of Colorado State University, at Fort Collins. In the late fifties that university was operating what were called special programs, directed by chemist Merle Payne, through which funds were sought for seminars and conferences. Gestur suggested to her that, in view of the interest in organizing a plant phenolics group, support might be solicited for a symposium in that field. To get things under way, a conference telephone call was set up with Geissman and Runeckles, who agreed with the idea of a symposium as an excellent way of bringing the proposed organization into existence, and of providing an opportunity for personal contact between many scientists with common interests in plant phenolics. A proposal was submitted to the National Science Foundation for a two-day symposium, of which a half-day was planned for the organizational meeting of the Plant Phenolics Group of North America. The NSF, while responding favorably, lacked sufficient funds in 1960, but when the proposal was resubmitted the next year it was approved. Gestur Johnson and Ted Geissman as co-directors, assisted by the other members of the planning committee, then organized the symposium for 1961, on August 31 and September 1.

Parturition

Two or three days beforehand, an enthusiastic group of six of us phytochemists from the PRL set out in two cars to drive from Saskatoon to Fort Collins, a safari which, I well remember, took us across the indifferent roads of Saskatchewan, the Dakotas and Wyoming (only the barest start on the Interstates then), and even more indifferent roadside eating facilities! Ernst von Rudloff, then beginning his chemosystematic studies on terpenoids, came along with a postdoc who was then working with him on polyphenols, as did two of our colleagues who later moved to other fields of research: Ted Underhill and John Watkin. Takayoshi Higuchi, now long established in Japan as a leading authority on lignin biochemistry, then at Gifu University, who was a visiting scientist in my group that year, accompanied us. Gestur Johnson later recalled that, despite smoother journeys, the arrival in Fort Collins of some of the others was not problem-free: when he took Geissman, Runeckles and Towers to their campus dormitory they found it locked. This problem was soon overcome, but next day there was no hot water. Gestur was sure that this made an unforgettable impression of the organization meeting on those who had to take a cold shower!

The symposium was entitled Biochemistry of Plant Phenolic Substances, and the speakers were E.M. Bickoff, Albert Booth, Stewart Brown,

Robert Horowitz, Nabusuki Kawano, G.A. Swan and Ikuzu Uritani. Notably, two of those participating were from Japan, initiating at the very outset the practice of inviting symposium speakers from outside North America. No written record survives of those who attended this first meeting, but according to Gestur's recollection the attendance approximated 60—certainly a good start! The last afternoon of the gathering saw the organizational meeting under the chairmanship of Vic Runeckles. One of the first matters considered was whether the fledgeling society should espouse complete independence at the outset or attempt to establish itself under the wing of some existing, cognate organization. The group discussed this question at some length, but in a show of hands finally decided that the PPGNA should organize as an independent society. Time has proved the soundness of this decision. Membership fees were discussed and a figure of $5 per year was accepted. It is at first amusing but then sobering to note that the chairman mentioned the possibility that this amount might seem a little excessive! It was agreed that the new executive would draft a consititution for adoption at the next meeting. Other matters discussed at this initial business session included how future meetings were to be organized, their frequency (annually, it was decided), and publication of the proceedings. Interestingly, the idea of a quarterly newsletter, which has since proved so successful, arose as early as this meeting. The final item on the agenda was the election of the first officers of the new society. As chairman of the nominating committee, which also included Neil Towers, and the late Margaret Seikel and Ted Geissman, I presented the slate, which was unanimously adopted. Simon Wender of the University of Oklahoma was elected the Group's first president, with Leonard Jurd of the USDA in Albany, California, as Vice-President and President-Elect, and Vic Runeckles Secretary-Treasurer.

Metamorphosis

During the PPGNA's first few years of existence the view developed widely that the Group, as initially constituted, was too restrictive in scope. Many thought that, by its exclusive concentration on phenolic compounds, it was neglecting other cognate areas of plant chemistry and biochemistry which were not only of some interest to existing members, but were the research fields of many other phytochemists who could be advantageously brought into a society casting a broader net. More interaction with such scientists could, it was argued, be of much mutual benefit, providing to our Group a depth, and making possible certain economies of scale, which the PPGNA could scarcely hope to attain alone, and offering the potential for some stimulating cross-fertilization.

These thoughts led to a formal proposal at the 1965 meeting in Albany, California, that the PPGNA become a phytochemical society. Tom Mabry of the University of Texas at Austin offered to organize the 1966 meeting on condition that the symposium cover an expanded range of topics by international speakers. Although the idea of a change of name and scope was by no means universally accepted as a desirable step, it was agreed to put the question to the membership in a mail ballot, and a favorable vote resulted. Consequently the 1966 Austin symposium was organized to cover not only phenolics, but included as well the biochemistry and chemical systematics of acetylenes, pseudoguaianolides, di- and triterpenes, sulphur compounds, betalains and alkaloids. At this meeting the Phytochemical Society of North America was formally established, supplanting the PPGNA, with appropriate changes to the constitution. In 1967, on January 1, the Phytochemical Society of North America came into legal existence, with Tom Mabry as its first president.

ADMINISTRATIVE STRUCTURE

Since the original formation of the PPGNA, the elected officers have included a President, and a Vice-President who serves as President-Elect and assumes the presidency at the annual business meeting following his election. From 1961 to 1965, in the PPGNA, the offices of Secretary and Treasurer were combined, and were held throughout the period by Vic Runeckles. In 1966, when the PSNA was formed, the combined office was divided *de facto* into Secretary and Treasurer, although the constitution still makes provision for the two offices to be combined. Officers are elected annually. Until recent years election was by vote of those attending the annual meeting, but concern about disenfranchising those not attending these meetings has led since 1986 to election by mail ballot in advance of the meeting. That same year a constitutional amendment was enacted restricting the Society's president to one term of office. A list of all officers up to the present is given in Appendix 1.

Management of the Society is vested in an Executive Committee consisting of these officers plus the immediate Past-President and the Editor-in-Chief of the symposium publications. In 1984 the Executive Committee struck an Advisory Committee to give advice to the Executive Committee on such matters as annual meetings and membership, and consisting of five members with staggered terms of office, one being replaced each year.

THE MEMBERSHIP

Phytochemical research has been, historically, a somewhat neglected area, lacking the more glamorous associations of medical, mammalian, and even some microbial investigation, and having comparatively few connections with secondary industry. Paul Stumpf and Eric Conn, in the preface to their series, *The Biochemistry of Plants*, have commented on past difficulties in getting some of the more established journals and their referees, not to mention many granting agencies, to take plant biochemical research seriously. Thus it is not too surprising that phytochemistry is not one of the more populous disciplines, and that the PSNA has not grown into a group comparable in size to many others in biology and chemistry. There was almost uninterrupted membership growth from inception in 1961 into the mid-seventies, but numbers peaked in the early eighties, followed by a significant decline after 1982. This decline has recently been reversed, and the membership is now the highest in the Society's history, but the total after thirty years has still reached only 425. Meeting attendance has seldom exceeded 100, a circumstance which is, however, by no means devoid of advantages. There certainly remains an untapped pool of potential members, although some are bound to remain inaccessible owing to the PSNA's conscious decision not to encourage inclusion of a few specific fields of phytochemistry, notably photosynthesis, among its areas of concentration. Nevertheless, the prospect of any quantum leap in membership, or even of any rapid expansion in numbers, now appears slight, and the PSNA, while remaining of great importance in its field,is therefore likely to continue as a small, intimate, closely knit society.

Geographically the membership has always been, of course, predominantly North American. Although most of the members have predictably been U.S. phytochemists, it is worthy of note that Canadians are represented out of proportion to their total population, comprising 17% of the membership in 1990, compared to 10% of the whole population north of the Rio Grande. To a considerable degree this has been true throughout the Society's history, and perhaps accounts in part for the fact that a disproportionate number of meetings have been held in Canada. Almost from the outset there has been a significant number of members classed as foreign, mostly European, and varying since 1979, for example, between 10 and 13% of the total membership.

Student members have been an important constituency of the Society for many years. A surge in student membership in the late seventies more than tripled their numbers in the three years up to 1981. Although students made up 12% of the total membership in 1990, their importance has been not only from the standpoint of numbers, as they have long made notable contributions to the Society, especially through contributed papers at the annual meetings. This last fact received practical recognition in 1981 with financial support offered by the Society toward student travel to meetings. Largely on the initiative of Cornelius Steelink, the Society's president at the time, it was decided in 1982 to award travel fellowships to graduate students and recent PhDs for the 1983 meeting in Tucson, based on submitted manuscripts. The experiment proved so successful that these travel awards have since become a permanent feature of the annual meetings, and a number of promising young phytochemists have been recipients. In addition, monetary awards are now given to the best oral talks and posters presented by young investigators.

In the late seventies the PSNA instituted a policy of recognizing outstanding contributions by its members to phytochemistry and to the Society through the award of Life Memberships. Ten members have been honored to date: Stewart Brown, Eric Conn, Gestur Johnson, Leonard Jurd, Frank Loewus, Vic Runeckles, Helen Stafford, Neil Towers, Tien Tso and Simon Wender. As one of the recipients, I can abundantly testify that such recognition by our phytochemist colleagues is highly valued by those so honored.

MEETINGS

Since its formation in 1961 the PPGNA, and later the PSNA, have always held a combined scientific and business meeting annually. Two principles of note have generally been observed: holding the meetings at universities or government laboratories, between late spring and early autumn, and choosing sites to give the widest possible geographic distribution. By adhering to the first principle, our society has kept the costs of attending the meetings minimal, and, since they have usually been held in smaller communities, with attendance generally of the order of 100, a more intimate atmosphere, reminiscent of the Gordon Conferences, has been fostered. By the second, every effort has been made to encourage attendance by members in widely scattered locations. Of the 30 meetings to date, 22 have been held in the United States and six in Canada, ranging from the east to the west coasts, one, in 1971, in Mexico, and one (a joint meeting with the Phytochemical Society of Europe in 1977) in Belgium. Appendix 2 lists the locales of all meetings to date, along with the symposium topics for each year. Mention of the Gordon Conference atmosphere brings to mind the meeting held in August of 1969 at the School of Fine Arts in Banff, Alberta, still considered by many of those who attended as our most successful meeting overall. Its organizers, Vic Runeckles and John Watkin, adopted the Gordon Conference format of scheduling meetings in the mornings and evenings, leaving the afternoons for enjoyment of the magnificent scenery of the Rockies and the many other local attractions. Strangely, despite its success, this format has never been repeated, even though it would have been ideal in several of the later locales.

A central feature of each annual meeting has been a symposium. These symposia, as Appendix 2 shows, have covered a wide range of subjects of phytochemical interest. In the early years they were, of course, confined to phenolics-related topics, but after the formation of the PSNA very diverse fields, of both fundamental and more applied interest, have been treated. The speakers have naturally been drawn primarily from North America, but, as mentioned above, the tradition of importing some speakers was established at the very beginning, and has been maintained throughout. The meetings have often been greatly enriched by participants from overseas, especially those associated with our sister European society.

However, the diversity has been not only geographical, but has been extended in two ways to embrace cognate disciplines. First, scientists from such areas as entomology and ecology have been invited as symposium speakers in a number of meetings organized by our society alone. Second, the PSNA has linked with several other organizations to hold joint meetings. In addition to meeting with the Phytochemical Society of Europe at Ghent in 1977 (with a second such meeting planned for 1992 in Miami), the PSNA has met jointly with the American Society of Pharmacognosy (Stillwater, Oklahoma, 1978), the American Society of Plant Physiologists (Pullman, Washington, 1980) and the International Society of Chemical Ecology (Quebec, 1990). Although more effort was needed to organize such joint meetings, and such efforts have not always borne fruit, the continuing movement in this direction indicates that they are considered beneficial to our society, and the inherent cross-fertilization can be regarded only as advantageous to all participants. There is certainly further scope for co-operative endeavors. The possibility of contacts with Latin American phytochemists, for example, was suggested as early as 1973, and for a time complimentary copies of the Newsletter were sent to prominent members of that community to encourage continued contact. When the Latin American Society of Phytochemistry was established in late 1983, this possibility resurfaced, but their initial organizational difficulties and, probably, the language barrier have thus far thwarted any further action.

PUBLICATIONS

The PSNA Newsletter

Of very great importance in holding the Society together has been the publication of a Newsletter, initiated in about 1970. We have always been handicapped by the sheer geographical immensity of our area of operation. In western Europe, with its greater population density than in most of North America, scientific meetings have always been held much more frequently than is possible here, where only one meeting of an organization annually is often feasible. This has certainly been a restriction on the PSNA, and to keep the widely dispersed membership in mutual contact, a good society publication has been crucial. The PSNA has been fortunate throughout in having dedicated members willing to devote time to editing its Newsletter and keep all its members in touch with plans, progress and new developments of interest. The Society owes a great debt to (in chronological order) Helen Stafford, James Wallace, Constance Nozzolillo, James Saunders, George Wagner and Helen Habermann for their untiring efforts in this regard. Publication, especially in the earlier years, was somewhat irregular, and issues were sometimes skimpy, but the important thing was that they did appear, and kept members informed and in touch. From mimeographed sheets in the seventies to the very professional looking laser-printed product now produced by Helen Habermann, with its two-column format, proportional fonts and half-tone illustrations, the Newsletter has always been invaluable to the membership.

In scope, as well as size and frequency of publication, the Newsletter has expanded over the years. It has always been essential in the organization of annual meetings, with advance information about the dates, sites, the symposia, the circulation of registration and abstract forms, and, since 1981, the abstracts themselves. Minutes of the annual business meetings are regularly published, and much information has also appeared about related meetings of interest to members. Increasingly, in more recent years, the Newletter has been used as a medium for advertising, free to members, of positions open or by those seeking positions. Items of personal interest, such as various honors to members, and casual photographs taken at meetings (mostly by the PSNA's unofficial photographic historians, Connie Nozzolillo and Bruce Stowe, have livened its pages on numerous occasions.

The Symposium Publications

Certainly of no lesser importance than publication of a Newsletter, and in some ways greater, has been the practice, initiated at the time the PPGNA was formed, of publishing papers based on the symposium presentations from 1961 on. The beginning was modest; the first four volumes appeared in ring binders as reproductions of typed manuscripts, produced by the Imperial Tobacco Company of Canada through the efforts of Vic Runeckles. However, papers of the fifth symposium, held in 1965, appeared in a hard-covered, commercially produced volume under the title *Phenolic Compounds and Metabolic Regulation.* After the formation of the PSNA in Austin in 1966, all subsequent symposia comprise a series entitled *Recent Advances in Phytochemistry.* The first four bound volumes of the series were published under a general title only, but each later volume bore specific titles based on that of the symposium. Again, the Society has been fortunate in having capable and dedicated people to edit the symposium publications, and competent publishers of high standing to see them into print; the editors and publishers of each volume are listed in Appendix 3. In 1981, an Editor-in-Chief was appointed. The symposium volumes have enjoyed satisfactory sales over the years, and royalties from them have been the single greatest source of income for the society. Until 1969 copies were provided free to members, but as this practice was not popular with the publishers, they have since been offered instead at a substantial discount.

Recent Advances in Phytochemistry has been a valuable undertaking also from the standpoint of the dissemination of phytochemical information to the wider scientific community, since most copies have been sold outside the PSNA membership. Together with its counterpart published by the Phytochemical Society of Europe and its British predecessor, this series has been a most useful reference in phytochemical research during the past quarter-century, and promises to continue so.

The PSNA Directory

Another extremely valuable publication of the PSNA, initiated in 1976, has been a membership directory, distributed to members biennially. This organ has evolved over the intervening years, but the most recent edition is a veritable mine of information about the Society and its membership. It contains an alphabetical listing of all the members in an easily read format, together with complete addresses, telephone numbers, and, where applicable, contact numbers for Fax, Bitnet and Internet (edu). Thus all the needed data for contacting the vast majority of North American and some foreign phytochemists are literally at one's fingertips. But the Directory contains much more than this. Also given are a complete list of the officers since inception of the PPGNA in 1961, a listing of Life Members, all symposium titles and meeting locations, symposium volumes available, and the current version of the Constitution and Bylaws. Following the alphabetical directory is a geographical directory and a research interest directory, and, finally, a grouping of members according to research areas. Truly an encyclopedia in miniature!

Journal Affiliations

Unlike the Phytochemical Society of Europe, which has affiliated with *Phytochemistry*, the PSNA has chosen to avoid journal affiliations. On several occasions this question has arisen, both with respect to existing journals and a proposed new publication, but interest among members has never been high enough for the Society to recommend proceeding in this direction. The view has evidently prevailed that various suitable vehicles exist for publication of phytochemical papers, and that members' interests are better served without any pressure, however subtle, to channel their published work into any single journal.

FUTURE DIRECTIONS

Although a historian should, strictly speaking, stop writing when the present is reached, if not before, it nevertheless does not seem inappropriate to speculate briefly, in conclusion, on how the PSNA might build on its past as it faces the future. Like the discipline on which it is based, our Society, too, must adapt and evolve, as indeed it has done in its first 30 years. In the area of structure there seems little advantage to further major expansion of scope such as was undertaken in the mid-sixties. Such an expansion, even if feasible, could lead to problems, such as greater overlap and competition with cognate societies, which might very well outweigh the advantages to be gained. The PSNA has, in general, functioned well in its present size and format, and has maintained a community of interest together with what some would describe as a comradeship, while retaining an invaluable scientific function for many of its members. Continuing efforts to attract students with phytochemical interests into its membership and integrate them into meetings and other activities will be of major importance in retaining an important place in the scientific community.

It is notable that several recent presidents of the PSNA were students well within the memories of the Society's founding members.

Scientifically, of course, the Society will and should evolve in step with developments in phytochemistry and related areas. Some of the topics covered in recent symposia, not to mention contributed papers, had not even been conceived in 1961, and certainly many of the topics of the 2021 symposium, could we peep into the future, would be virtually unrecognizable to us now. However, in some cases a renewed look at topics of former interest might be worth considering. In the sixties, for example, there was some concentration of attention on the industrial aspects of phytochemistry which has, unfortunately, largely disappeared. It is questionable whether this decline has been parallelled by any diminution in the subject itself, and in fact, to the extent it is related to agriculture, it has arguably increased. Perhaps there should be a new emphasis in this area in coming decades.

Although plants have always interacted with other organisms in this planet's biosphere, the nature of these interactions is still very incompletely understood, and the recent interest in ecological phytochemistry is certain to grow as its importance is increasingly recognized and the mechanisms of the plant's interactions with its environment become better understood. At the micro level, we are beginning to get a better picture of phytochemical reactions as they relate to the intracellular environment, and the Society's attention to this area, which now goes back at least ten years, will continue and expand. We can confidently expect the development of ever more sophisticated techniques to probe phytochemical pathways in their cellular settings. But quite apart from this aspect, there still remains an enormous potential for studies of metabolic reactions despite the large volume of work with tracer and enzyme techniques in recent decades, which has really just dented the surface of this enormous field.

Above all, the stage seems set for developments in plant molecular genetics and its applications in biotechnology which are scarcely imaginable at this time. To what extent emphasis should be shifted in this direction is a question the Society will have to address, but there is no doubt that some shift will be both necessary and desirable. It is germane in this context to quote the view of one of our founders, Neil Towers, expressed to me in a private communication, with which I concur. "Molecular biology must be integrated into phytochemistry and plant biochemistry but we must never lose sight of the fact that our major interests revolve around (a) what plants are made of chemically, (b) how they are assembled biochemically, how they are put together and how made. We do not want to become another society of molecular biologists, or plant physiologists, or ethnopharmacologists or chemical ecologists or spectroscopists, although these areas will be represented...."

In the next few years we shall no doubt be forming a clearer picture of what will be the prime areas of concentration by the PSNA in the medium term. But whatever these may be, the Society must continue, as in the past, its efforts to focus on the "cutting edge" of our discipline as we approach and enter the new millenium.

ACKNOWLEDGEMENTS

I am especially grateful to the three surviving PSNA members involved in the initial organization of the Society: Gestur Johnson, Vic Runeckles and Neil Towers, for providing much helpful information on areas where my own sources or memory were deficient. Several other long-time members were good enough to comment on drafts of the manuscript, and I am indebted in this context to Eric Conn, Frank Loewus, Tom Mabry, Connie Nozzolillo, Helen Stafford, Tien Tso and Ernst von Rudloff. I also wish to thank the PSNA's current secretary and archivist, Helen Habermann, for providing a set of the Newsletters, invaluable records originally contributed by Eric Conn.

APPENDIX 1 Officers of the Society, 1961-1990

Plant Phenolics Group of North America

	President*	Secretary/Treasurer
1961	Simon Wender	Victor Runeckles
1962	Leonard Jurd	Victor Runeckles
1963	Stewart Brown	Victor Runeckles
1964	Margaret Seikel	Victor Runeckles
1965	Bernard Finkle	Victor Runeckles

Phytochemical Society of North America

	President*	Secretary	Treasurer
1966	Thomas Mabry	A. Merritt	Howard Wright

1967	Victor Runeckles	Sarah Clevenger	Howard Wright
1968	Bruce Bohm	Cornelius Steelink	Howard Wright
1969	Peyton Teague	Cornelius Steelink	Howard Wright
1970	Tien Tso	Helen Stafford	Jerry McClure
1971	Eric Conn	James Wallace	Jerry McClure
1972	Kenneth Hanson	James Wallace	Jerry McClure
1973	Neil Towers	James Wallace	Richard Mansell
1974	Heinz Floss	James Wallace	Richard Mansell
1975	Frank Loewus	Constance Nozzolillo	Richard Mansell
1976	Jerry McClure	Constance Nozzolillo	Richard Mansell
1977	Helen Stafford	Constance Nozzolillo	Richard Mansell
1978	George Waller	Constance Nozzolillo	John Romeo
1979	Leroy Creasy	Constance Nozzolillo	John Romeo
1980	Cornelius Steelink	James Saunders	John Romeo
1981	Constance Nozzolillo	James Saunders	John Romeo
1982	Geza Hrazdina	James Saunders	John Romeo
1983	Ragai Ibrahim	James Saunders	Jonathan Poulton
1984	Richard Mansell	George Wagner	Jonathan Poulton
1985	David Loomis	George Wagner	Jonathan Poulton
1986	Neil Towers	George Wagner	Jonathan Poulton
1987	John Romeo	Helen Habermann	Jonathan Poulton
1988	David Seigler	Helen Habermann	Kelsey Downum
1989	Jonathan Poulton	Helen Habermann	Kelsey Downum
1990	Brian Ellis	Helen Habermann	Kelsey Downum

* Each President except the first PPGNA president served during the preceding year as Vice-President and President-Elect.

APPENDIX 2   Meeting Sites and Symposium Topics

1961	Fort Collins, CO	Biochemistry of plant phenolic substances
1962	Corvallis, OR	Plant phenolics and their industrial significance
1963	Toronto, ON	Aspects of plant phenolic chemistry
1964	Norwood, MA	Phenolics in normal and diseased fruits and vegetables
1965	Albany, CA	Phenolic compounds and metabolic regulation
1966	Austin, TX	Recent advances in phytochemistry
1967	Madison, WI	Phytochemical techniques
1968	Tucson, AZ	Phytochemistry and the plant environment

1969	Banff, AB	Enzymology and biochemistry of phenolics
1970	Beltsville, MD	Structural aspects of phytochemistry
1971	Monterrey, Mexico	Terpenoid chemistry and biochemistry
1972	Syracuse, NY	Chemistry and biochemistry of plant growth regulators
1973	Pacific Grove, CA	Metabolism of secondary plant products including regulation
1974	Cullowhee, NC	Phytochemistry in relation to disease and medicine
1975	Tampa, FL	Biochemical interaction between plants and insects
1976	Vancouver, BC	Structure, biosynthesis and degradation of wood
1977	Ghent, Belgium	Biochemistry of plant phenolics
1978	Stillwater, OK	Topics in the biochemistry of natural products
1979	De Kalb, IL	Resource potential in phytochemistry
1980	Pullman, WA	Phytochemistry of cell recognition and cell surface interactions
1981	Ithaca, NY	Cellular and subcellular localization in plant metabolism
1982	Ottawa, ON	Mobilization of reserves in germination
1983	Tucson, AZ	Phytochemical adaptation to stress
1984	Boston, MA	Chemically mediated interactions between plants and other organisms
1985	Pacific Grove, CA	The shikimate pathway: recent developments
1986	College Park, MD	Phytochemical effects of environmental compounds
1987	Tampa, FL	Opportunities for phytochemistry in plant biotechnology
1988	Iowa City, IA	Plant nitrogen metabolism
1989	Vancouver, BC	Biologically active products of the mevalonic acid pathway
1990	Quebec, PQ	Modern phytochemical methods

APPENDIX 3   The Symposium Publications*

Year*	Volume	Title Editor(s)[#]	Publisher
1962		Biochemistry of Plant Phenolic Substances (No formal editor)	

Year*	Volume	Title Editor(s)#	Publisher
1963		Plant Phenolics and their Industrial Significance V.C. Runeckles	
1964		Aspects of Plant Phenolic Chemistry V.C. Runeckles	
1965		Phenolics in Normal and Diseased Fruits and Vegetables V.C. Runeckles	
1967		Phenolic Compounds and Metabolic Regulation B.J. Finkle, V.C. Runeckles	Appleton-Century-Crofts
1968	1	Recent Advances in Phytochemistry T.J. Mabry, R.E. Alston, V.C. Runeckles	Appleton-Century-Crofts
1969	2	Recent Advances in Phytochemistry M.K. Seikel, V.C. Runeckles	Appleton-Century-Crofts
1970	3	Recent Advances in Phytochemistry C. Steelink, V.C. Runeckles	Appleton-Century-Crofts
1972	4	Recent Advances in Phytochemistry V.C. Runeckles, J.E. Watkin	Appleton-Century-Crofts
1972	5	Structural and Functional Aspects of Phytochemistry V.C. Runeckles, T.C. Tso	Academic
1973	6	Terpenoid Chemistry and Biochemistry V.C. Runeckles	Academic
1974	7	Chemistry and Biochemistry of Plant Growth Regulators V.C. Runeckles	Academic
1974	8	Metabolism of Secondary Plant Products Including Regulation V.C. Runeckles	Academic
1975	9	Phytochemistry in Relation to Disease and Medicine V.C. Runeckles	Plenum
1976	10	Biochemical Interaction Between Plants and Insects J.W. Wallace, R.L. Mansell	Plenum
1977	11	Structure, Biosynthesis and Degradation of Wood V.C. Runeckles, F.A. Loewus	Plenum
1977	12	Biochemistry of Plant Phenolics T. Swain, J.B. Harborne, C.F. Van Sumere	Plenum
1979	13	Topics in the Biochemistry of Natural Products T. Swain, G.R. Waller	Plenum
1981	14	Resource Potential in Phytochemistry T. Swain, R. Kleiman	Plenum
1981	15	Phytochemistry of Cell Recognition and Cell Surface Interactions F.A. Loewus, C.A. Ryan	Plenum

Year*	Volume	Title Editor(s)#	Publisher
1982	16	Cellular and Subcellular Localization in Plant Metabolism L.L. Creasy, G. Hrazdina	Plenum
1983	17	Mobilization of Reserves in Germination C. Nozzolillo, P.J. Lea, F.A. Loewus	Plenum
1984	18	Phytochemical Adaptation to Stress B. Timmermann, C. Steelink, F.A. Loewus	Plenum
1985	19	Chemically Mediated Interactions Between Plants and Other Organisms G.A. Cooper-Driver, T. Swain, E.E. Conn	Plenum
1986	20	The Shikimic Acid Pathway E.E. Conn	Plenum
1987	21	Phytochemical Effects of Environmental Compounds J.A. Saunders, L. Kosak-Channing, E.E. Conn	Plenum
1988	22	Opportunities for Phytochemistry in Plant Biotechnology E.E. Conn	Plenum
1989	23	Plant Nitrogen Metabolism J.E. Poulton, J.T. Romeo, E.E. Conn	Plenum
1990	24	Biochemistry of the Mevalonic Acid Pathway to Terpenoids G.H.N. Towers, H.A. Stafford	Plenum
1991	25	Modern Biochemical Methods N.H. Fischer, M.B. Isman, H.A. Stafford	Plenum

* The year is that in which the volume was published (1-2 years after the symposium). The bound volumes published from 1968 onward comprise the series **RECENT ADVANCES IN PHYTOCHEMISTRY** and are numbered accordingly. The first four volumes in ring binders, and the first hard-cover volume before the inception of this series, were unnumbered.

#General Editors: V.C. Runeckles was the main editor for most volumes between 1962-1976. Editors varied subsequently until 1981. Editors-in-Chief were instituted in 1981: F.A. Loewus, 1981-1983; E.E. Conn, 1984-1988; H.A. Stafford, 1989-.

The manufacturer's authorised representative in the EU is Springer
Nature Customer Service Centre GmbH, Europaplatz 3, 69115 Heidelberg,
Germany. If you have any concerns regarding our products, please
contact ProductSafety@springernature.com

Printed and bound by CPI Group (UK) Ltd, Croydon, CR0 4YY
23/04/2026
02095629-0001